Nutritional Signaling Pathway Activities in Obesity and Diabetes

Food Chemistry, Function and Analysis

Series editors:
Gary Williamson, *Monash University, Australia*
Alejandro G. Marangoni, *University of Guelph, Canada*
Graham A. Bonwick, *AgriFoodX Limited, UK*
Catherine S. Birch, *AgriFoodX Limited, UK*

Titles in the series:
1: Food Biosensors
2: Sensing Techniques for Food Safety and Quality Control
3: Edible Oil Structuring: Concepts, Methods and Applications
4: Food Irradiation Technologies: Concepts, Applications and Outcomes
5: Non-extractable Polyphenols and Carotenoids: Importance in Human Nutrition and Health
6: Cereal Grain-based Functional Foods: Carbohydrate and Phytochemical Components
7: Steviol Glycosides: Cultivation, Processing, Analysis and Applications in Food
8: Legumes: Nutritional Quality, Processing and Potential Health Benefits
9: Tomato Chemistry, Industrial Processing and Product Development
10: Food Contact Materials Analysis: Mass Spectrometry Techniques
11: Vitamin E: Chemistry and Nutritional Benefits
12: Anthocyanins from Natural Sources: Exploiting Targeted Delivery for Improved Health
13: Carotenoid Esters in Foods: Physical, Chemical and Biological Properties
14: Eggs as Functional Foods and Nutraceuticals for Human Health
15: Rapid Antibody-based Technologies in Food Analysis
16: DNA Techniques to Verify Food Authenticity: Applications in Food Fraud
17: Advanced Gas Chromatography in Food Analysis
18: Handbook of Food Structure Development
19: Mitigating Contamination from Food Processing
20: Biogenic Amines in Food: Analysis, Occurrence and Toxicity
21: Nutrition and Cancer Prevention: From Molecular Mechanisms to Dietary Recommendations
22: Health Claims and Food Labelling
23: Nutraceuticals and Human Health: The Food-to-supplement Paradigm
24: Nutritional Signaling Pathway Activities in Obesity and Diabetes

How to obtain future titles on publication:
A standing order plan is available for this series. A standing order will bring delivery of each new volume immediately on publication.

For further information please contact:
Book Sales Department, Royal Society of Chemistry, Thomas Graham House, Science Park, Milton Road, Cambridge, CB4 0WF, UK
Telephone: +44 (0)1223 420066, Fax: +44 (0)1223 420247,
Email: booksales@rsc.org
Visit our website at www.rsc.org/books

Nutritional Signaling Pathway Activities in Obesity and Diabetes

Edited by

Zhiyong Cheng
The University of Florida, USA
Email: z.cheng@ufl.edu

ROYAL SOCIETY
OF **CHEMISTRY**

Food Chemistry, Function and Analysis No. 24

Print ISBN: 978-1-78801-557-8
PDF ISBN: 978-1-83916-060-8
EPUB ISBN: 978-1-83916-061-5
Print ISSN: 2398-0656
Electronic ISSN: 2398-0664

A catalogue record for this book is available from the British Library

The Royal Society of Chemistry is a charity, registered in England and Wales, Number 207890, and a company incorporated in England by Royal Charter (Registered No. RC000524), registered office: Burlington House, Piccadilly, London W1J 0BA, UK, Telephone: +44 (0) 20 7437 8656.

Visit our website at www.rsc.org/books

Printed in the United Kingdom by CPI Group (UK) Ltd, Croydon, CR0 4YY, UK

Preface

The epidemic of obesity and diabetes is growing worldwide. In addition to providing energy, nutrients from our diets can elicit signaling transduction to induce genetic, epigenetic, and posttranslational changes that affect cell function and metabolic homeostasis. As such, advancing the understanding of nutritional signaling pathway and activity modulators may reveal new options to prevent or treat obesity and diabetes. This book serves as a timely forum that presents the cutting-edge research and critical knowledge about nutritional signaling pathways leading to metabolic dysregulation in obesity and diabetes.

This book consists of eleven chapters that can be grouped into three sections: (1) the effects of nutrition (and maternal nutrition) on metabolism and clinical trials of dietary intervention (Chapters 1–3); (2) the molecular mechanisms by which nutrients regulate cellular functions and events such as inflammation responses, nuclear receptors, mitochondrial health, and gut microbiota (Chapters 4–7); and (3) the allometric scaling and omics approaches (*e.g.*, metabolomics and proteomics) in translational and precision nutrition research (Chapters 8–11). Each chapter discusses a perspective of intervention or future application of the relevant pathways or approaches, enhancing the transformation of new knowledge.

This book illustrates the leading edge of basic science and translational research in nutrition and metabolism. At molecular and cellular level, it covers metabolomics, proteomics, nutrigenomics, nuclear receptors and transcription factors, inflammatory pathways, autophagy, mitochondrial quality control, and gut microbiota. From the clinical and translational research perspective, it covers clinical trials, precision nutrition, maternal nutrition and transgenerational health, and allometric scaling of dietary bioactives

Nutritional Signaling Pathway Activities in Obesity and Diabetes
Food Chemistry, Function and Analysis No. 24
Edited by Zhiyong Cheng

Published by the Royal Society of Chemistry, www.rsc.org

in translational metabolic research. These unique features may benefit students, nutritionists, dietitians, biomedical researchers, and healthcare professionals who study the nutritional aspects, cellular and molecular biology, as well as pathophysiology of obesity and diabetes.

I am grateful to my family for their love and support during editing this book. Special thanks go to the authors of the chapters for their outstanding contributions to and passion for nutrition and metabolic research. I also wish to express my gratitude to the Royal Society of Chemistry, especially to Nicki Dennis, Katie Morrey, Janet Freshwater, and Sylvia Pegg, for their assistance with the book planning and editing. Thank you all for making this book possible.

Zhiyong Cheng
University of Florida

Contents

Nutritional Signaling Pathway Activities in Obesity and Diabetes
Food Chemistry, Function and Analysis No. 24
Edited by Zhiyong Cheng

Published by the Royal Society of Chemistry, www.rsc.org

CHAPTER 1

Diabetes and Obesity: An Overview of Nutritional Effects

SONIA DE ALWIS AND MICHAEL A. VIA*

Division of Endocrinology, Diabetes, and Bone Disease, Mount Sinai Beth Israel Medical Center, Icahn School of Medicine at Mount Sinai, New York NY 10003, USA
*E-mail: michael.via@mountsinai.org

1.1 Introduction

Type 2 diabetes (T2D) has evolved into a leading public health concern both for its high prevalence and devastating complications. Current global estimates of 422 million individuals affected by T2D represent a fourfold increase since 1980, with a continued rapid rise.[1] The International Diabetes Federation (IDF) similarly quotes 451 million individuals affected aged 18–99 years in 2017 (8.4%), with a projected prevalence of 693 million or 9.9% by 2045.[2] Interestingly, the rise in incidence comes at a time when accumulated evidence clearly demonstrates that T2D is a preventable disease.[3] Prior to the 1900s, T2D and obesity were exceedingly rare, affecting two individuals per 100 000 population and 1 in 30 persons, respectively.[4] The modern-day lifestyle has largely contributed to the overwhelming disease burden, in part due to the increasing availability of processed foods and a sedentary lifestyle facilitated by technology. Obesity, itself, confers a fourfold increase in risk of T2D, and represents the single greatest modifiable risk factor for development of the disease.[5] In contrast to T2D, Type 1 diabetes (T1D) results from the loss

Nutritional Signaling Pathway Activities in Obesity and Diabetes
Food Chemistry, Function and Analysis No. 24
Edited by Zhiyong Cheng

Published by the Royal Society of Chemistry, www.rsc.org

of pancreatic β-cell production of insulin most often through autoimmune mechanisms.[4] Though patients with T1D also benefit from dietary modification,[5] strategies for T1D are beyond the scope of this chapter and book, which focus on insulin resistance syndromes. Given the extensive disease morbidity, mortality, and global healthcare expenditure, an intensive lifestyle intervention that includes a nutrient-rich dietary pattern, among other modifications, is vital for prevention and management of these chronic metabolic conditions.[6,7]

1.2 Mechanisms of Insulin Resistance and Metabolic Syndrome

The presence of insulin resistance may be demonstrated in patients long before the development of T2D and is believed to serve as the main underlying driver of the disease process.[8] Obesity predisposes to insulin resistance through a number of complex mechanisms, in part related to increased adipocyte size and quantity and to adipocyte dysfunction.[9] Adipose tissue, traditionally viewed as simple fat storage, is now more aptly recognized as an active endocrine organ that secretes adipokines, including adiponectin, leptin and resistin, which regulate multiple metabolic pathways involved in glucose homeostasis.[10–12] The excess caloric intake associated with obesity overwhelms adipocyte mechanisms, resulting in dysregulated adipokine secretion, and ultimately a pro-inflammatory state.[13] Enlarged adipocytes further encourage macrophage recruitment to the tissue, prompting chronic low-grade inflammation.[14] Once established, insulin resistance behaves as a highly disruptive state to regulation of metabolism, exerting widespread effects that include raised glucagon activity during meals that results in inappropriate prandial hepatic glucose release, impaired glucagon-like peptide 1 (GLP-1) release, leptin resistance, reduced ghrelin and adiponectin activity, systemic endoplasmic reticulum stress, systemic inflammation, and pancreatic β-cell dysfunction which culminate clinically as T2D.[15] Similar pathophysiologic mechanisms are observed in other conditions of insulin resistance, including the metabolic syndrome (MetS), hypertriglyceridemia, non-alcoholic hepatosteatosis, and polycystic ovarian syndrome (PCOS).

MetS represents a composite of clinical risk factors that predict insulin resistance, and cumulatively contribute to high risk for atherosclerotic cardiovascular disease (ASCVD).[15] Though published classifications vary slightly, the National Cholesterol Education Program Adult Treatment Panel III defines MetS by the presence of three of the following five parameters: (1) increased abdominal girth (waist circumference >102 cm in men and >88 cm in women); (2) hypertension (systolic blood pressure >130 mmHg and diastolic blood pressure >85 mmHg); (3) elevated circulating triglycerides (>150 mg dL^{-1}); (4) reduced circulating high-density lipoprotein (HDL) levels

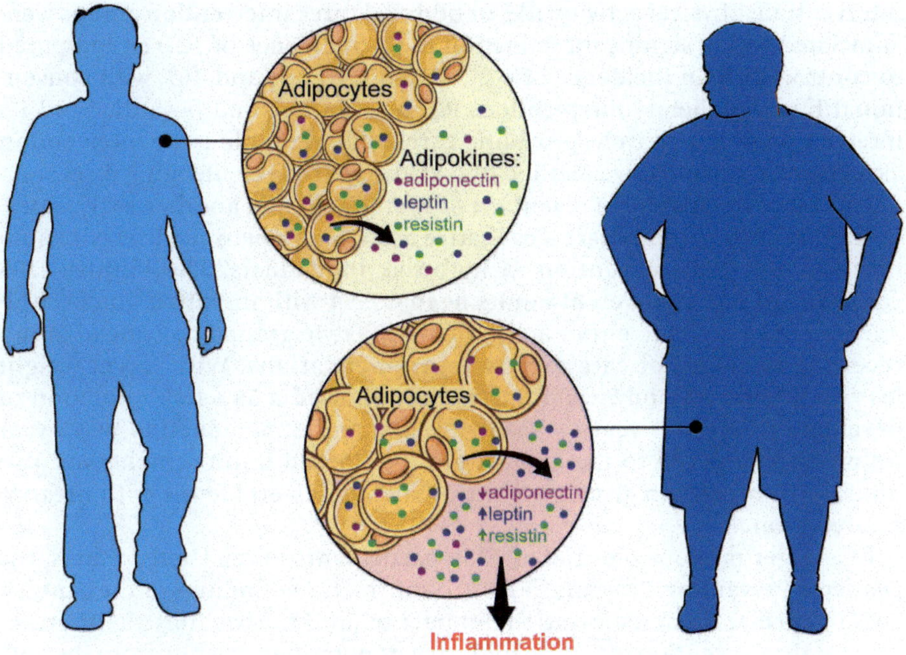

Figure 1.1 Adipocyte function in health and disease.[13] Reproduced with permission © Mount Sinai Health System.

(<40 mg dL^{-1} in men and <50 mg dL^{-1} in women); and (5) elevations in serum glucose concentrations (fasting serum glucose >100 mg dL^{-1}).[16] Using these criteria, MetS was observed in 33% of adults in the US population in 2012, and demonstrates a rising prevalence worldwide.[17] Due to overlapping pathophysiology, hepatosteatosis, seen in approximately 30% of US adults, and PCOS, observed in 7–10% of reproductive-age women, are also highly prevalent in modern populations.[18,19]

Dietary strategies leading to significant and sustained weight loss can successfully hamper the underlying pathophysiologic mechanisms and mitigate both the development and progression of syndromes of insulin resistance, including T2D and MetS, as well as hepatosteatosis and PCOS by extension (Figure 1.1).

1.3 Prevention Trials

Several landmark trials support the hypothesis that lifestyle interventions leading to weight loss can substantially alter metabolic factors that predispose to T2D and MetS. The first, in 1997, the Da Qing Trial[20] in China, involved 577 individuals with impaired glucose tolerance randomly assigned to either conventional lifestyle or one of three treatment groups consisting of dietary intervention, physical activity (PA), or both. All three intervention groups were associated with a significant reduction in the incidence of T2D as compared to control, with an incidence of 67.7%, 43.8%, 41.1%, and 46% with conventional lifestyle, dietary intervention, PA intervention, and both diet and PA interventions, respectively, providing early evidence for lifestyle intervention as a low-cost and effective method of preventing these challenging diseases.

The Finnish Diabetes Prevention Study[21] (DPS) was another early, randomized trial to demonstrate a causative association between lifestyle modification and T2D prevention in high-risk individuals. The Finnish DPS consisted of 522 overweight adults aged 40–64 with impaired glucose tolerance randomized to either an intensive lifestyle group or a control group receiving standard of care dietary recommendations. With target weight loss of at least 5% body weight, the study showed a 58% risk reduction of developing T2D at 3 years, 43% at 7 years, and 32% reduction at 13-year follow-up compared to the control group. Marked improvement was also noted with respect to glycemic control and cholesterol levels with lifestyle intervention.

In a larger randomized trial of 3234 participants in the United States, the Diabetes Prevention Program[22] (DPP) confirmed the findings of the Finnish DPS by further demonstrating the impact of an intensive lifestyle intervention for the prevention of T2D in high-risk populations. Results from the DPP showed a 58% reduction in the incidence of T2D with an intensive lifestyle program as compared with placebo. This reduction was superior to pharmacologic intervention with twice-daily metformin, which conferred a 31% reduced risk. Those who achieved a target weight loss of at least 7% body weight coupled with 150 minutes of weekly moderate-intensity exercise successfully delayed the progression to T2D by an average of 3 years. The effect was particularly pronounced for those aged 60 or older, with a 71% reduction in T2D onset in this subset of the lifestyle intervention group. Moreover, the ongoing Diabetes Prevention Program Outcomes Study (DPPOS) suggests these preventive effects are sustained with a 27% reduction in the progression to T2D for up to at least 15 years.[23]

The Look AHEAD trial (Action for Health in Diabetes) similarly studied the impact of lifestyle intervention and weight loss among patients with T2D with regard to risk of diabetes complications, including cardiovascular outcomes and mortality.[24] This randomized controlled trial (RCT) involved 5145 overweight or obese participants with known T2D who were assigned to either an intensive lifestyle protocol or standard care. Though the study failed to show benefit with respect to its primary outcome of cardiovascular event reduction, it was notable for a host of important secondary outcomes. Specifically, Look AHEAD showed that a target 5-kg weight reduction improved glycemic control and was associated with a 30% reduced risk of kidney dysfunction, a particularly detrimental microvascular complication of T2D.[25] Additionally, scores

Table 1.1 Summary of clinical evidence.[a] Adapted from Via and Mechanick (2016),[7] with permission from Elsevier, copyright 2016.

Trial	Study design	*n*	Duration	Outcomes
Reiser *et al.*, 1979[108]	Crossover	19	6 weeks	Increased fructose intake induces worsened insulin resistance
Pan *et al.*, 1997 (Da Qing trial)[20]	Randomized controlled trial	110 660	6 years	Diet and/or exercise intervention decreases the incidence of diabetes in high-risk individuals with IGT
Lindstrom *et al.*, 2003 (Finnish DPS)[21]	Randomized controlled trial	522	3.2 years	Intensive lifestyle intervention reduced incident T2D by 58% in obese individuals with IGT
DPP Research Group, 2002[22]	Randomized controlled trial	1079	2.8 years	Intensive lifestyle intervention was superior to metformin in the prevention of T2D
Bosch *et al.*, 2012 (ORIGIN)[109]	Randomized controlled trial	12 536	6.2 years	No observable effect of 1 000 mg daily n-3 PUFA supplementation on glucose homeostasis
Tanaka *et al.*, 2013[31]	Cohort	978	8 years	Increased fruit intake is associated with reduced risk of diabetic retinopathy
Look AHEAD Research Group, 2014, 2016[25-29]	Randomized controlled trial	5145	9.6 years	Intensive lifestyle intervention had no observable effect on cardiovascular outcomes in T2D but reduced incidence of retinopathy and nephropathy, improved erectile function, and improved quality of life
Estruch *et al.*, 2013 (PREDIMED)[45]	Randomized controlled trial	7447	4.8 years	Subjects assigned to the Mediterranean diet had a 30% reduction in cardiovascular events and a 30% reduction in incidence of T2D compared with a low-fat diet
Gepner *et al.*, 2015 (CASCADE)[50]	Randomized controlled trial	224	2 years	In addition to the Mediterranean diet, red wine consumption reduced fasting glucose and insulin levels in T2D
Fritzen *et al.*, 2015[60]	Randomized controlled trial	64	26 weeks	Subjects on the NND had reduced fasting glucose, reduced triglycerides, and greater weight loss than controls

[a] IGT, impaired glucose tolerance; DPS, Diabetes Prevention Study; T2D, Type 2 diabetes; DPP, Diabetes Prevention Program; ORIGIN, Outcome Reduction with Initial Glargine Intervention; PUFA, polyunsaturated fatty acids; AHEAD, Action for Health in Diabetes; PREDIMED, Prevención con Dieta Mediterránea; CASCADE, Cardiovascular Diabetes and Ethanol; NND, New Nordic Diet.

of erectile function, depression, and quality of life were overall improved, with 11% lower rate of hospitalization, and 6% reduced medication requirements among the intensive lifestyle group.[26–28] Moreover, a post-hoc analysis showed study participants who successfully lost weight had an approximate 15% reduction in cardiovascular outcomes for every 7% of weight lost.[29] Accordingly, the Look AHEAD trial provides large-scale evidence of the effect of an intensive lifestyle program on clinically relevant and patient-important outcomes in patients with T2D.

Observational studies have further clarified the role of dietary modification in prevention of the microvascular complications of T2D, including blindness, kidney failure, and lower-limb amputation. The Nurses' Health Study demonstrated that a diet high in fruit, vegetables, nuts, and lean meat (Dietary Approaches to Stop Hypertension (DASH)-style pattern) was associated with a 40% reduced risk of microalbuminuria compared with a Western diet.[30] Similarly, a cohort study of 978 participants in Japan demonstrated a 52% reduction in diabetic retinopathy associated with high fruit intake following 8 years, independent of glycemic control.[31] A high-fiber diet also was associated with reduced incidence of both retinopathy and microalbuminuria in a cohort study of 1272 subjects in India.[32] These studies highlight the substantial impact of a nutrient-rich diet on both the onset of T2D as well as its debilitating complications (Table 1.1).

1.4 Adherence

The dietary strategies utilized among these studies primarily focused on a low-fat, high-fiber, calorie-restricted diet in addition to regular moderate-intensity exercise and frequent contact with individual lifestyle coaches. This is in line with the prevailing recommendation of a low-fat diet in past years. However, optimal macronutrient dietary content has been challenged more recently. Though several dietary strategies have been proposed, the dietary pattern individuals choose to adopt is less vital to their outcome than adherence level. Most importantly, dietary strategies should be practical and feasible to encourage long-term adherence. As diet remains a highly personalized choice, influenced by culture, religion, and socioeconomic status, among other factors, the unifying concept is that long-term adherence is ultimately what predicts successful weight loss. In a randomized trial comparing four popular diets in 160 overweight or obese individuals, each diet achieved comparable levels of weight loss and cardiovascular risk factor reduction after 1 year, but amount of weight loss correlated more strongly with adherence level rather than type of diet.[33] All diets performed poorly with respect to self-reported adherence, highlighting the importance of choosing a dietary pattern most in keeping with one's lifestyle and preferences. The large RCTs to date included a level of supervision not feasible in the general population. Adherence often suffers under real-world circumstances, and is a notable barrier to achieving the level of benefit seen in clinical trials.

1.5 Dietary Components Affecting T2D

1.5.1 Fructose

Carbohydrates take on a variety of forms and pervade the modern diet as starch, glycogen, other polysaccharides, or simple sugars (monosaccharides and disaccharides). Fructose has become increasingly prevalent in the modern diet owing to the industrial availability and use of sucrose and high-fructose corn syrup (HFCS). HFCS was introduced in the 1960s as an inexpensive, liquid sweetener, which was liberally applied to commercial products for added flavor.[4] Sugar-containing beverages in particular gained much attention for their high HFCS content, wide availability, and weighty contribution to both childhood and adult obesity. Sucrose, the main industrial alternative to HFCS, has almost the same fructose content, differing by a mere 5%.

Fructose-containing sweetened beverages have been shown to increase postprandial triglycerides, and reduce both leptin levels and insulin secretion.[34] Perhaps unsurprisingly, intake of soft drinks and fruit drinks has been independently associated with the development of the MetS and T2D.[35] Moreover, those with a genetic predisposition to obesity are more prone to develop this disease with high intake of sugar-sweetened beverages.[36] In countries with higher availability of HFCS, an approximate 20% increase risk of T2D is observed in comparison to countries with lower availability.[37]

Mechanistically, fructose fails to trigger both the insulin and leptin responses typical in carbohydrate metabolism.[38] In the short term, intake of fructose seems to have temporizing blood glucose effects without a sharp serum glucose rise, and has therefore been utilized as a common agent for diabetic supplements. However, the benefits are short-lived and the associations with chronically high fructose intake and worsening insulin resistance and T2D become readily apparent over time. Another detriment to fructose is the hunger response and increased caloric intake elicited, due in part to (1) promoting leptin resistance, thereby decreasing satiety and losing regulation of food intake, and (2) stimulating dopamine receptors within the nucleus accumbens, triggering a pleasure response. Over time repeated sugar intake results in a downregulation of dopamine receptors, inducing a withdrawal response when the sugar stimulus subsides.[4] On the cellular level, fructose metabolism is poorly regulated as it enters the glycolysis pathway downstream of an important regulatory and rate-limiting step, resulting in disordered energy metabolism and fatty acid production. Finally, fructose increases *de novo* lipogenesis, reduces insulin sensitivity, promotes dyslipidemia, and drives central adiposity as compared with glucose.[39] By these mechanisms, it is not surprising that a high dietary fructose intake has been strongly implicated in several metabolic conditions, including T2D, MetS, PCOS, obesity, vascular disease, and hepatosteatosis.

In light of current evidence, the American Heart Association recommendation is to limit daily simple sugar intake to no more than 37.5 g, or 9 teaspoons, of sugar for men and 25 g, or 6 teaspoons, for women. Similarly, in 2015, the World Health Organization proposed that sugar comprise no more than 10% of total energy and ideally less than 5%.

Fructose naturally exists in fruits, honey, and agave-derived syrups though in low quantities (4–8 g), and in many cases the presence of fructose is counterbalanced by fiber, which helps to slow absorption, as well as mixed with other antioxidants and vitamins that may offset potential harms.

1.5.2 Fiber

Dietary fiber constitutes another class of carbohydrate that may be further subdivided into polysaccharides and modified polysaccharides, or insoluble and soluble forms. Insoluble fibers are largely indigestible and therefore structurally useful for bulking chyme and stool as well as slowing gastric emptying. Soluble fibers are inherently more viscous and help to line the intestinal tract and delay absorption of dietary carbohydrates and lipids.[40] Further, soluble fiber plays an integral role in altering the flora of the gut microbiome and subsequently in the production of short-chain fatty acids, which both serve as an energy substrate and play a pivotal role in gut immunity.[41]

The gut microbiome is a highly individualized environment vulnerable to type, quality, and source of food.[40] Though the microbiome naturally exists as a symbiosis of bacteria, fungi, and host organisms, it remains susceptible to alterations inflicted by lifestyle. Different dietary sources supply different substrates for microorganism growth, for example, allowing certain bacteria to flourish and others to perish depending on dietary content. Studies comparing the microbiome of children in urban and rural areas identify a relative increase in Firmicutes bacteria and reduction in Bacteroidetes based upon the dietary patterns endemic to those areas. Specifically, children in urban areas were found to have a microbiome dominated by organisms apt to break down animal fat, protein, and sugars, whereas children in more rural areas harbored organisms better equipped to break down dietary fiber and carbohydrates from vegetable sources.[42] Importantly, a diet rich in fat and simple sugars, characteristic of the Western diet, results in a depletion of protective organisms and a disruption of symbiotic mechanisms, which can manifest clinically as obesity, among other diseases.[40] Diets deficient in fiber, the so-called "fiber gap," result in an additional depletion of favorable microbiota which metabolize fiber to short-chain fatty acids.[41,43]

1.5.3 Glycemic Index and Load

Glycemic index is another useful marker by which to classify carbohydrates and signifies the rapidity with which they are absorbed and utilized. High-glycemic foods are typically high in monosaccharides or easily digestible polysaccharides, low in dietary fiber, and have greater potential to evoke harmful periods of serum glucose peaks. High-glycemic foods should be avoided or minimized as able in individuals with T2D, and preferably paired with fat, protein, or other food items which may delay absorption and temper glucose elevations. Perhaps a more comprehensive measure of glucose content is offered by the glycemic load, which takes into account both the glycemic index and portion size, or the amount of simple carbohydrates consumed. Glycemic load may act as a more telling indicator of glucose content of any given food intake under more real-world circumstances.

1.5.4 Dietary Fat

A low-fat diet has long been touted primarily for its potential to reduce low-density lipoprotein (LDL) cholesterol that comprises atherosclerotic plaques and predisposes to cardiovascular disease. Both observational studies and RCTs have elicited mixed results with respect to dietary fat and insulin resistance, obesity, and T2D. More recently, a higher-fat diet has fallen into favor when taking into account the type of fat consumed. Specifically, monounsaturated fatty acids (MUFAs) and polyunsaturated fatty acids (PUFAs) have garnered much attention for their capacity to reduce LDL cholesterol and improve insulin sensitivity, and their benefits are very much apparent in the Mediterranean diet (MedDiet), where healthy MUFAs and PUFAs, *e.g.*, in the form of olive oil and nuts, are abundant.

1.6 Dietary Interventions in T2D

In the midst of widespread obesity, several dietary patterns have been proposed to target excess weight and improve metabolic parameters, including insulin resistance. Though the philosophies of each of the proposed diets are similar at their core, nuances exist that generally differentiate each on the basis of macronutrient ratio. General principles focus on high intake of fruit, vegetables, fiber, and high-quality fats, and liberal intake of antioxidants and vitamins, while minimizing carbohydrates, and in particular fructose.

1.6.1 Mediterranean Diet

The MedDiet has proven one of the most successful dietary strategies to date, both for its weight loss potential and position as the single diet shown to improve cardiovascular outcomes and mortality among overweight and obese individuals, including those at high cardiovascular risk. The MedDiet, first described in the 1960s based on traditional dietary practices common in the Mediterranean region, favors a diet composed primarily of fruits, vegetables, nuts, and whole grain cereals, with an emphasis on fish and moderate red wine intake, and restricted in dairy, red meat, and processed foods. The MedDiet is mainly distinguished by its ample use of olive oil, nuts, red wine, fatty fish, as well as tomatoes, garlic, onions, and spices simmered in olive oil, with these nutrient-rich ingredients likely behaving in synergy to enhance the overall metabolic environment.[44] The PREDIMED randomized, multicenter trial conducted in Spain was a landmark study designed to confirm observational studies that showed promise for the primary prevention of cardiovascular disease with the MedDiet.[45] In the PREDIMED trial, 7447 participants deemed high-risk but without proven cardiovascular disease were randomly assigned to one of three interventions: (1) MedDiet enriched with extra-virgin olive oil; (2) MedDiet enriched with mixed nuts; or (3) a control, low-fat diet. Following 5 years, the two groups assigned a MedDiet supplemented with extra-virgin olive oil or with nuts performed similarly, with a 30% reduction in cardiovascular events in comparison to the control low-fat diet. As coronary disease remains the leading cause of mortality in the United States,[46] results of the PREDIMED trial are highly relevant. With regard to T2D, a post-hoc analysis of the PREDIMED trial demonstrates a 30% reduction in the incidence of T2D, and of the 3614 subjects with known T2D, the MedDiet supplemented with extra-virgin olive oil reduced the incidence of diabetic retinopathy by 40%, though no risk reduction was observed in the incidence of diabetic nephropathy.

Another large, randomized trial compared three common dietary patterns: low-fat diet, low-carbohydrate diet, and the MedDiet among 322 obese subjects. After 2 years, subjects assigned to the low-carbohydrate diet achieved the greatest mean weight loss (4.7 kg), followed closely by the MedDiet (4.4 kg), with the least weight loss observed with a low-fat diet (2.2 kg). Of the 36 patients with T2D included in the trial, the MedDiet proved the most effective for reducing fasting blood glucose levels.[47] Further, a systematic review of eight meta-analyses and five RCTs comparing the MedDiet to a low-fat or control diet concluded a 19–23% reduced risk of incident T2D with the MedDiet, and a reduced glycosylated hemoglobin of 0.3–0.47% in those with established T2D.[48] In this study and others, the MedDiet has the additional potential benefit of ameliorating all components of MetS.[49]

The social ritual of wine with meals is another nutrient-rich custom of the Mediterranean population. The nutritional benefits of wine have long been studied, largely in conjunction with alcohol in general, with cardioprotective effects credited to an increase in HDL cholesterol. More specific to wine is a high polyphenol content, sevenfold greater in red wine than white, which has been associated with improved glycemic control, reduced retinopathy, blood pressure-lowering effects, and reduced markers of inflammation.[44] A study of 224 subjects with T2D on the MedDiet examined the effects of both red and white wine on metabolic parameters. Participants were randomized to receive 150 mL of red wine, white wine, or water daily for a duration of 2 years. Interestingly, the study noted both wine intervention groups to reduce fasting insulin, but with an average reduction of 18 mg dL^{-1} in subjects receiving white wine, and a 2 mg dL^{-1} average increase in HDL cholesterol with red wine intervention.[50]

Resveratrol in particular, a polyphenol present in peanuts, grapes, and red wine, has been extensively studied for its protective role in cancer, cardiovascular disorders, oxidative stress, neurodegenerative disease, glucose homeostasis, and obesity, among others.[51] In animal models, resveratrol consistently improves insulin resistance by activating sirtuin-1 and adenosine monophosphate-activated kinase, and reducing systemic inflammation.[52]

Sirtuins, a class of histone deacetylases, have been implicated in the mechanism of caloric restriction and extended lifespan. Compounds that activate the sirtuin pathway and its metabolic effects, such as resveratrol, are highly favorable.[53,54] Human trials examining oral supplementation with resveratrol in T2D, though limited in number and short in duration, are similarly promising.[55] A study of 62 subjects with T2D randomized to resveratrol 250 mg day^{-1} or standard of care demonstrated improved hemoglobin A1c (HbA_{1c}), systolic blood pressure, and total cholesterol with resveratrol over 3 months, without significant change in HDL or LDL cholesterol.[56] Moreover, a randomized placebo controlled trial of 24 participants with known MetS assigned to either *trans*-resveratrol 500 mg three times daily before meals or placebo for 90 days displayed significant benefit in body weight, body mass index (BMI), fat mass, waist circumference as well as area under the curve (AUC) of insulin, and total insulin secretion with resveratrol intervention, though no differences were found with respect to blood pressure, fasting glucose, postprandial glucose levels, AUC of glucose, or first phase of insulin secretion.[53]

The "French paradox" has been described as a phenomenon of reduced ischemic cardiac disease despite a high intake of saturated fat observed within French culture. The drivers for this are still poorly understood but may include the cardioprotective effects of red wine, among other aspects of French cuisine. Current views suggest red wine may additionally offset a high saturated fat intake by inhibiting platelet aggregation, rather than promoting atherosclerosis, and in this regard may be a particularly beneficial component of the MedDiet.[57]

Olive oil, rich in MUFAs, is an important staple of the MedDiet that contributes to the prevention and management of T2D. A high-MUFA diet, when compared to a high-carbohydrate diet, reduces fasting plasma glucose and favorably impacts triglycerides, weight, systolic blood pressure, and HDL cholesterol.[58] In further support, a meta-analysis of 29 RCTs and four cohort studies including 15 784 patients found the highest intake of olive oil to be associated with a 16% reduced risk of developing T2D as compared with those with the lowest intake.[59] The published clinical trials and studies investigating individual components provide strong evidence for the benefits of the MedDiet. The implementation of the MedDiet is among the best options for the treatment and prevention of metabolic disease, including T2D.

1.6.2 The New Nordic Diet

The New Nordic Diet (NND) models the cuisine typical of the local Scandinavian region and, in many ways, closely resembles the core of the MedDiet, including large quantities of vegetables, fruits, whole grains, and fish. The difference lies in more distinctly Nordic ingredients, including rapeseed oil in place of olive oil, local vegetables and fruit, and a focus on fresh berries, herbs, and wild mushrooms. While large trial data assessing outcomes of NND is limited, a study performed in Denmark that included obese subjects randomized to either the NND or Danish diet, comparable to the Western diet, determined a 6-kg *versus* 2-kg weight loss in the NND and control group, respectively. Moreover, markers of insulin resistance, including fasting serum glucose and insulin levels, as well as serum triglyceride levels, were significantly reduced with NND intervention.[60] Results from the SYSDIET randomized trial assessing the NND in 200 obese subjects with MetS showed the NND diet to improve both lipid profiles and markers of low-grade inflammation at 18–24 weeks as compared with placebo, suggesting a potential role for this strategy in both T2D and MetS.[61]

1.6.3 The Ornish Diet

The Ornish Diet (OD) favors a predominantly plant-based strategy with significant limitation of animal fat and its derivatives, and a subsequent increase in carbohydrates. Specifically, the diet endorses 75% complex carbohydrates, 15% fat, and 10% protein, supplemented with ample fruit and vegetables. The benefit of the OD has largely been observed with respect to LDL cholesterol and atherosclerotic disease. Though data is limited, one randomized control study found a similarly structured whole-food, vegetarian diet, high in complex carbohydrates, to be associated with regression of atherosclerotic plaques and less than half the number of coronary events over 5 years, among patients with known coronary atherosclerotic disease.[62]

Given the high carbohydrate content, the benefit of the OD in T2D is less clear.[63] A brief prospective cohort study assessing the effect of a plant-based, whole-food, low-fat diet on cardiac biomarkers in 131 subjects with established or high-risk coronary artery disease noted improvements in all studied cardiac parameters, including a fasting glucose reduction of 15 mg dL^{-1} and HbA_{1c} reduction of 0.4% in diabetic subjects after 3 months.[64] The Optimal Macronutrient Intake Trial to Prevent Heart Disease (OmniHeart) similarly studied three diets with varied macronutrient ratios on insulin sensitivity in overweight and obese subjects, including a high-carbohydrate diet, comparable to the DASH diet, a diet partially replacing carbohydrates with protein, and a diet partially replacing carbohydrates with unsaturated fats, similar to the MedDiet. The study concluded that the diet with a higher ratio of unsaturated fats increased markers of insulin sensitivity more effectively than either the high-carbohydrate diet or protein diet.[63]

1.6.4 Ketogenic Diet

The very-low-carbohydrate ketogenic diet has reemerged in recent years for its potential to induce profound weight loss and widespread metabolic effects. Prior to the advent of insulin in the 1920s, severe caloric and carbohydrate restriction, the so-called "starvation diet," was the main therapeutic option by which those with diabetes could potentially delay complications of the disease.[65,66] Traditionally developed for refractory epilepsy, the ketogenic diet has since expanded to treat a multitude of disorders, including neurodegenerative disease, traumatic brain injury, cancer, acne, diabetes, and obesity.[67]

The ketogenic diet proposes a stringently low carbohydrate intake, < 20 g day^{-1} or $<10\%$ of daily caloric intake, with calories replaced primarily by fat in an effort to induce a state of nutritional ketosis. Limited in available carbohydrates, fat becomes the predominant fuel source, and is metabolized by the liver into fatty acids and ketone bodies, in the form of acetoacetate and 3-beta-hydroxybutyrate. These compounds enter circulation and are then utilized for systemic energy needs.[67,68]

Current large-scale evidence on the effect of the ketogenic diet on T2D remains limited. One randomized trial of 84 obese subjects with T2D assigned participants to either a low-carbohydrate ketogenic diet or a low-glycemic index diet for a duration of 24 weeks. While both diets led to significant weight loss and improved metabolic markers, the low-carbohydrate ketogenic diet was far superior in reducing HbA_{1c} (−1.5% *vs.* −0.5%) and weight (−11.1 *vs.* −6.9 kg), with notable gains in HDL cholesterol (+5.6 *vs.* 0 mg dL^{-1}). Moreover, the low-carbohydrate ketogenic diet reduced or eliminated medication requirements in almost all participants (95.2%) *versus* 62% observed with a low-glycemic index diet.[69]

A larger study of 363 participants with excess weight enabled participants to select either a low-carbohydrate ketogenic diet or low-calorie diet for a duration of 24 weeks. Here, too, both diets improved the tested parameters, including body weight, BMI, waist circumference, blood glucose level, HbA_{1c}, and lipid profile, but effects on blood glucose level and HbA_{1c} were substantially greater with the low-carbohydrate ketogenic diet, with an additional 22.5 mg dL^{-1} reduction in blood glucose (99 *vs.* 121.5 mg dL^{-1}), and 1.5% reduction in HbA_{1c} (6.25% *vs.* 7.75%).[70]

In a comparison of the efficacy of a low-carbohydrate diet *versus* low-fat diet in the treatment of T2D and obesity, an important study of 132 severely obese subjects (BMI >43 kg m^{-2}) randomized to low-carbohydrate or low-fat diet for 6 months found subjects on the low-carbohydrate diet to have a significantly greater weight loss (−5.3 *vs.* −1.9 kg) and improvement in both triglycerides (−38 *vs.* −7 mg dL^{-1}) and insulin sensitivity than their fat-restricted counterparts. In addition, subjects with T2D who underwent a low-carbohydrate diet reduced fasting blood glucose by 26% as compared with 5% on a low-fat diet.[71]

Given the potential cardiac consequences of a high-fat diet, a randomized trial of 53 obese women investigated a very-low-carbohydrate diet consisting of no more than 20 g carbohydrates per day, or a calorie-restricted, low-fat diet for 6 months with the primary outcome of weight loss and cardiovascular risk factors. The very-low-carbohydrate diet achieved greater weight loss without detriment to blood pressure, lipids, serum glucose, and insulin levels despite high fat intake.[72]

A meta-analysis of 13 RCTs including 1415 patients comparing a very-low-carbohydrate ketone-generating diet with a low-fat diet over at least 12 months demonstrates greater weight loss, greater reduction in triglycerides, lower diastolic blood pressure, greater HDL, though slightly higher LDL with a weighted mean difference of 4.6 mg dL^{-1} among individuals assigned to a ketogenic diet in comparison to those on a low-fat diet.[73] The observed increase in LDL cholesterol was attributed to an increase in saturated fat intake, though this is notably contradicted by additional studies which associate a reduction in LDL cholesterol with the ketogenic diet.[74,75] Interestingly, one study showed an increase in LDL cholesterol observed with a high-fat, low-carbohydrate diet to primarily affect the less atherogenic larger-sized LDL cholesterol, rather than the more harmful small, dense LDL cholesterol.[76]

Thus, the ketogenic diet has proven generally safe, when used judiciously. Caution should be used in patients with renal dysfunction due to a potentially larger protein intake, or those on insulin for whom hypoglycemia remains a high risk. For appropriate candidates, the general recommendation is that the diet be undertaken alongside a healthcare professional with frequent metabolic monitoring (Figure 1.2).

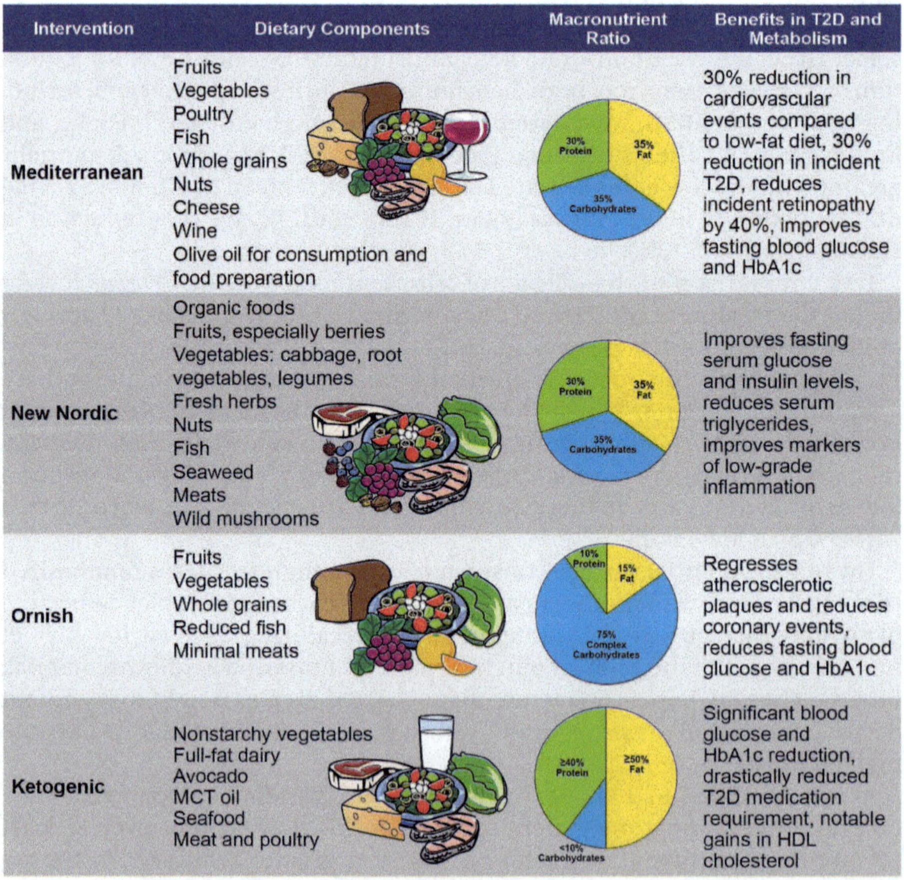

Figure 1.2 Composition of proposed dietary patterns. T2D, Type 2 diabetes; HbA_{1c}, hemoglobin A1c; MCT, medium-chain triglycerides; HDL, high-density lipoprotein.[44,60,62,66] Reproduced with permission © Mount Sinai Health System.

1.7 Micronutrients

Micronutrients, encompassing electrolytes, vitamins, trace elements, and plant alkaloid compounds, are an essential yet often overlooked component of a comprehensive and nourishing diet. Micronutrients are responsible for a wide array of metabolic processes that promote overall health, including glucose metabolism, free radical scavenging, and clearance of advanced glycosylated end products (AGEs).[77] Several micronutrients have garnered interest as a natural adjunct to traditional T2D treatment strategies. The impact of micronutrients may perhaps be most clearly seen in the context of obesity and T2D, where deficiencies are common.

1.7.1 Chromium

Chromium (Cr), present in the diet within yeast, egg yolks, meat, whole grains, and cereals, is a known cofactor of insulin signaling and has been investigated for its potential role in glucose and lipid metabolism.[78] Chromium, bound to the oligopeptide chromodulin, is present intracellularly, where it prolongs insulin action by activating insulin receptor kinases and inhibiting insulin receptor phosphatases. It has further been postulated that chromium plays a role in the upregulation of insulin receptors, and facilitates uptake of glucose at target tissues.[79,80] The overall effect is enhanced insulin sensitivity, and reduced insulin requirements in individuals with T2D.[81]

Chromium was first recognized for its role in carbohydrate metabolism in animal models in the 1950s and similar effects were subsequently demonstrated in humans in the late 1970s. Observations of patients receiving prolonged parenteral nutrition that did not include chromium developed severe insulin resistance and hyperglycemia, which quickly resolved with chromium supplementation. Initial high insulin requirements declined to zero.[82] Chromium is now a standard component of current parenteral nutrition formulations.[83]

Clinically, chromium levels are distinctively lower in patients with T2D as compared with normal subjects, by as much as 20–40% lower in the serum and 40–50% lower in scalp hair.[77] Lower concentrations of chromium in toenails have additionally been associated with a higher risk of developing T2D.[84] Chromium supplementation in the form of chromium picolinate has shown marginal benefit in markers of glycemia and lipids. A randomized trial of 180 subjects with T2D assigned to chromium 100 μg or 500 μg twice daily *versus* placebo showed chromium supplementation to afford minor improvements in glucose parameters following 4 months.[85]

More recently, a randomized trial of 71 patients with poorly controlled T2D (HbA_{1c} >7%) receiving 600 μg of chromium supplementation demonstrated a similarly small but significant improvement in glycemic measures, including fasting and postprandial glucose reductions of 17 mg dL^{-1} and 25 mg dL^{-1}, respectively, and HbA_{1c} reduction of 0.9% as compared with placebo without demonstrable effects on serum lipid concentrations.[86] A pooled analysis of 28 RCTs indicated a significant reduction in fasting blood glucose, HbA_{1c}, triglycerides, and an increase in HDL cholesterol with chromium supplementation.[87] At the same time, other published trials fail to demonstrate benefit of chromium supplementation in patients with T2D,[78] and current knowledge does not yet substantiate its routine use in the management of T2D.

1.7.2 Selenium

Selenium, obtained through the diet as nuts, meats, eggs, and cereal, is another trace element implicated in T2D, though data here, too, remains mixed. Early rodent studies demonstrated reduced insulin resistance when treated with selenate, an inorganic form of selenium.[88] In humans, one RCT of selenium supplementation in obesity revealed a 10% reduction in fasting insulin levels,[89] and in subjects with gestational diabetes, selenium reduced overall fasting plasma glucose by 10 mg dL^{-1} and fasting plasma insulin by 1.98 IU mL^{-1}.[90]

Studies have recognized selenoproteins, particularly glutathione peroxidase (GPx-1), as the key mechanism by which selenium exerts its effects. GPx-1 has a somewhat paradoxical effect; this enzyme scavenges free radicals and prevents oxidative damage, and it catalyzes biochemical redox reactions which result in antioxidant production.[77] On the other hand, rodent models have identified overexpression of GPx-1 to cause hyperglycemia, and GPx-1 knockout mice accordingly have better glycemic control than their wild-type counterparts. Given these dual aspects, selenium may most judiciously be consumed in quantities that ensure satisfactory levels while avoiding oversupplementation.[91]

Importantly, a number of studies have demonstrated either no benefit or, in fact, an increase in incidence of T2D with selenium supplementation.[92] In the late 1990s, selenium gained attention in the Nutritional Prevention of Cancer trial in which supplementation with selenium 200 µg daily was assessed for the primary outcome of non-melanoma skin cancer. In secondary analyses, selenium demonstrated a notable reduction in the incidence of colorectal and prostate cancers, though at the expense of an unexpected increase in incident T2D. Specifically, the study reports 58 new cases of T2D in the selenium intervention group as compared with 39 cases in the placebo group (hazard ratio 1.55 (95% confidence interval, 1.03–2.33)).[93] A systematic review of 13 observational studies and three RCTs indicates conflicting results, with observational studies concluding a positive correlation between serum selenium levels and T2D and RCTs failing to show an association.[91] Current data is insufficient to make recommendations regarding selenium intake or supplementation for benefits in T2D.

1.7.3 Vitamin D

Though classically associated with calcium homeostasis and bone metabolism, vitamin D has more recently been implicated in a number of other conditions, including cardiovascular disease, cancer, autoimmune disease, T2D, and obesity.[94] Vitamin D insufficiency and overt deficiency, defined as serum levels <20 mg dL^{-1} and <30 mg dL^{-1} respectively, are highly prevalent in obesity, affecting up to 80–90% of obese individuals.[95,96] The presence of vitamin D receptors and 1-alpha hydroxylase activity, the enzyme responsible for vitamin D activation, in human pancreatic beta cells provides compelling evidence for a potential mechanism of vitamin D in glucose homeostasis.[7]

Current data is overall insufficient to either advocate or invalidate the role of vitamin D in obesity, MetS, and T2D. Observational studies showing initial promise have since been called into question by randomized clinical trials and meta-analyses. Still, some studies do offer the potential for benefit, if only small. A recent meta-analysis of 24 RCTs assessing the effect of vitamin D supplementation on markers of glycemic control demonstrates statistically significant reductions in fasting plasma glucose (−4.9 mg dL^{-1}), HbA1c (−0.30%), and homeostasis model assessment of insulin resistance (HOMA-IR) (−0.66) with vitamin D supplementation.[97,98] Moreover, supplementation with vitamin D 500 IU daily demonstrated a notable 13% reduction in the risk of T2D when compared with a slightly lower dose of 200 IU daily. Subjects with higher serum 25OH-vitamin D levels, above 25 ng mL^{-1}, were shown to have a 43% risk reduction for developing T2D compared to those with lowest levels below 14 ng mL^{-1}.[94]

More commonly, RCTs that study vitamin D supplementation in T2D fail to demonstrate a consistent and significant benefit.[99] An RCT of 62 vitamin D-deficient participants with T2D treated with high-dose supplementation *versus* placebo found no difference in insulin sensitivity or glucose homeostasis markers as measured by euglycemic clamp method following 6 months.[97] Similarly disappointing, studies assessing vitamin D supplementation in obese individuals have proved largely unsuccessful in demonstrating weight loss, despite the well-established association between obesity and vitamin D deficiency.[99–101] High-dose vitamin D_2 supplementation in subjects with MetS additionally failed to show improvement in HOMA-IR after 8 weeks, suggesting vitamin D may not influence this composite of metabolic risk factors.[102] Future studies may more clearly delineate the role of vitamin D by using optimal and standardized dosing, longer durations, larger sample sizes, consistency in baseline vitamin D status, and more direct measurement rather than surrogate markers of glucose homeostasis.[103]

1.7.4 Cinnamon

Cinnamon is a common household spice that has been postulated to facilitate glycemic control in T2D.[104] The first potential role for cinnamon in glucose metabolism originated in 2003 in a clinical trial of 60 men and women with T2D conducted in Pakistan.[104] In this population, cinnamon, supplied in three separate dosages, was equally effective in reducing mean fasting plasma glucose (18–29%), triglycerides (23–30%), LDL cholesterol (7–27%), and total cholesterol (12–26%) levels over a 40-day span. However, a study in the United States that randomized 57 subjects with T2D to 1 g cinnamon daily or placebo for 3 months failed to show significant differences between the two groups with respect to BMI, HbA_{1c}, cholesterol, triglycerides, or insulin levels in the background of a Western diet.[105] A meta-analysis of 10 RCTs published in 2013 noted variable doses of cinnamon to be associated with reduced fasting plasma glucose, total cholesterol, LDL cholesterol, and triglyceride levels with a statistically significant increase in HDL cholesterol. However, no difference was found with respect to glycosylated hemoglobin, and studies were widely disparate in terms of dosing and follow-up times.[106] The American Diabetes Association currently adopts the position that there is insufficient evidence to support the use of cinnamon or other natural supplements in the treatment of T2D, though a potential benefit should be considered along with a minimal potential risk.

1.8 Plant Alkaloids

Plant alkaloids are a large group of naturally occurring substances with unique biochemical properties. Many exhibit antioxidant effects, and some may have benefits for the treatment and prevention of T2D, though further study is needed (Figure 1.3).

Compound	Dietary Sources	Purported Health Benefits in T2D and Metabolism
Resveratrol	Red grapes, red wine, blueberries, pistachios, hops, and peanuts	Increases insulin sensitivity, promotes endothelial function, reduces LDL cholesterol, anti-inflammatory
Quercetin	Onions, fennel, cilantro, capers, and dill	Reduces fasting glucose levels in streptozotocin treated mice, potentially by inhibiting intestinal α-glucosidase
Epigallocatechin gallate	Grapes, tea, wine, chocolate	Improves insulin secretion and mitochondrial activity, improves glucose tolerance in db/db mice
Bilberry Extract	Anthocyanin-rich component of nutraceutical products	Reduces fasting glucose and HbA_{1c} levels in mouse models of T2D
Naringin and Hesperidin	Citrus fruits	Resolve hyperglycemia in streptozotocin-induced diabetes models and in db/db mice
Berberine	Chinese herbs: Coptidis Rhizome	Improves insulin resistance, lowers blood glucose by targeting pancreatic β cell function
Piperine	Black pepper	Anti-inflammatory, antioxidant, glucose lowering in diabetes
Curcumin	Turmeric	Anticancer activity, reduces LDL cholesterol, increases circulating HDL levels, anti-inflammatory
Rutin	Apples, tea, buckwheat	Metabolized to quercetin, improves insulin resistance and pancreatic β cell function, inhibits formation of AGEs

Figure 1.3 Select plant alkaloids. T2D, Type 2 diabetes; LDL, low-density lipoprotein; HbA_{1c}, hemoglobin A1c; HDL, high-density lipoprotein; AGEs, advanced glycation end products.[7,107] Reproduced with permission © Mount Sinai Health System.

1.9 Future Directions

T2D and obesity are well known for their capacity to induce extensive morbidity, costly medical care, and ultimately, shortened lifespan. Of further urgency, these conditions have reached epidemic proportions globally. As with other chronic conditions, the most effective strategy for combating these diseases is often prevention. An intensive lifestyle intervention affords an opportunity to both prevent T2D and mitigate its potential complications. Moreover, nutrition therapy leading to weight loss targets the underlying process of insulin resistance, and therefore additionally improves parameters of MetS, PCOS, and hepatosteatosis, among other conditions.

For these reasons, dietary strategies are increasingly deserving of scientific attention. Future studies, particularly long-term studies, can provide a better understanding of the role of nutrition, specific dietary patterns, and micronutrients in these common disorders of insulin resistance. In particular, future studies may address more personalized approaches to diet, in which specific dietary patterns are undertaken to target specific metabolic derangements, and investigate efforts to improve adherence. Dietary adherence remains a major barrier to achieving sustained weight loss, and thus a reasonable, palatable diet should be a central focus of management.

References

1. G. Roglic, *Global Report on Diabetes*, World Health Organization, Geneva, Switzerland, 2016.
2. N. H. Cho, J. E. Shaw, S. Karuranga, Y. Huang, J. D. da Rocha Fernandes, A. W. Ohlrogge, *et al.*, *Diabetes Res. Clin. Pract.*, 2018, **138**, 271–281.
3. J. Tuomilehto, P. Schwarz and J. Lindstrom, *Diabetes Care*, 2011, **34**(Suppl 2), S210–S214.
4. R. J. Johnson, L. G. Sanchez-Lozada, P. Andrews and M. A. Lanaspa, *Adv. Nutr.*, 2017, **8**(3), 412–422.
5. J. Salas-Salvado, M. A. Martinez-Gonzalez, M. Bullo and E. Ros, *Nutr. Metab. Cardiovasc. Dis.*, 2011, **21**(Suppl 2), B32–B48.
6. Y. Handelsman, Z. T. Bloomgarden, G. Grunberger, G. Umpierrez, R. S. Zimmerman, T. S. Bailey, *et al.*, *Endocr. Pract.*, 2015, **21** (Suppl 1), 1–87.
7. M. A. Via and J. I. Mechanick, *Med. Clin. North Am.*, 2016, **100**(6), 1285–1302.
8. R. A. DeFronzo, *Med. Clin. North Am.*, 2004, **88**(4), 787–835.
9. J. I. Mechanick, D. L. Hurley and W. T. Garvey, *Endocr. Pract.*, 2017, **23**(3), 372–378.
10. K. Smitka and D. Maresova, *Prague Med Rep.*, 2015, **116**(2), 87–111.
11. C. McGown, A. Birerdinc and Z. M. Younossi, *Clin. Liver Dis.*, 2014, **18**(1), 41–58.
12. M. Potenza, M. A. Via and R. T. Yanagisawa, *Endocr. Pract.*, 2009, **15**(3), 254–262.

13. R. T. Atawia, K. L. Bunch, H. A. Toque, R. B. Caldwell and R. W. Caldwell, *Front. Biosci., Landmark Ed.*, 2019, **24**, 890–934.
14. V. Bourlier and A. Bouloumie, *Diabetes Metab.*, 2009, **35**(4), 251–260.
15. D. W. Lam and D. LeRoith, Metabolic Syndrome, in K. R. Feingold, B. Anawalt, A. Boyce, G. Chrousos, K. Dungan, A. Grossman, *et al.*, *Diabetes Mellitus and Carbohydrate Metabolism*, Endotext, South Dartmouth, MA, 2000.
16. L. S. Sperling, J. I. Mechanick, I. J. Neeland, C. J. Herrick, J. P. Despres, C. E. Ndumele, *et al.*, *J. Am. Coll. Cardiol.*, 2015, **66**(9), 1050–1067.
17. M. Aguilar, T. Bhuket, S. Torres, B. Liu and R. J.Wong, *JAMA, J. Am. Med. Assoc.*, 2015, **313**(19), 1973–1974.
18. M. M. Smits, G. N. Ioannou, E. J. Boyko and K. M. Utzschneider, *J. Gastroenterol. Hepatol.*, 2013, **28**(4), 664–670.
19. S. Franks, *Int. J. Obes.*, 2008, **32**(7), 1035–1041.
20. X. R. Pan, G. W. Li, Y. H. Hu, J. X. Wang, W. Y. Yang, Z. X. An, *et al.*, *Diabetes Care*, 1997, **20**(4), 537–544.
21. J. Lindstrom, A. Louheranta, M. Mannelin, M. Rastas, V. Salminen, J. Eriksson, *et al.*, *Diabetes Care*, 2003, **26**(12), 3230–3236.
22. Diabetes Prevention Program Research Group, *Diabetes Care*, 2002, **25**(12), 2165–2171.
23. Diabetes Prevention Program Research Group, *Lancet Diabetes Endocrinol.*, 2015, **3**(11), 866–875.
24. A. R. G. Look, R. R. Wing, P. Bolin, F. L. Brancati, G. A. Bray, J. M. Clark, *et al.*, *N. Engl. J. Med.*, 2013, **369**(2), 145–154.
25. Look AHEAD Research Group, *Lancet Diabetes Endocrinol.*, 2014, **2**(10), 801–809.
26. R. R. Wing, R. C. Rosen, J. L. Fava, J. Bahnson, F. Brancati, I. N. Gendrano III, *et al.*, *J. Sex Med.*, 2010, **7**(1 Pt 1), 156–165.
27. R. R. Rubin, T. A. Wadden, J. L. Bahnson, G. L. Blackburn, F. L. Brancati, G. A. Bray, *et al.*, *Diabetes Care*, 2014, **37**(6), 1544–1553.
28. J. B. Redmon, A. G. Bertoni, S. Connelly, P. A. Feeney, S. P. Glasser, H. Glick, *et al.*, *Diabetes Care*, 2010, **33**(6), 1153–1158.
29. Look AHEAD Research Group, E. W. Gregg, J. M. Jakicic, G. Blackburn, P. Bloomquist, G. A. Bray, *et al.*, *Lancet Diabetes Endocrinol.*, 2016, **4**(11), 913–921.
30. J. Lin, T. T. Fung, F. B. Hu and G. C. Curhan, *Am. J. Kidney Dis.*, 2011, **57**(2), 245–254.
31. S. Tanaka, Y. Yoshimura, R. Kawasaki, C. Kamada, S. Tanaka, C. Horikawa, *et al.*, *Epidemiology*, 2013, **24**(2), 204–211.
32. S. Ganesan, R. Raman, V. Kulothungan and T. Sharma, *Clin. Exp. Ophthalmol.*, 2012, **40**(3), 288–294.
33. M. L. Dansinger, J. A. Gleason, J. L. Griffith, H. P. Selker and E. J. Schaefer, *JAMA, J. Am. Med. Assoc.*, 2005, **293**(1), 43–53.
34. K. L. Teff, J. Grudziak, R. R. Townsend, T. N. Dunn, R. W. Grant, S. H. Adams, *et al.*, *J. Clin. Endocrinol. Metab.*, 2009, **94**(5), 1562–1569.

35. V. S. Malik, B. M. Popkin, G. A. Bray, J. P. Despres, W. C. Willett and F. B. Hu, *Diabetes Care*, 2010, **33**(11), 2477–2483.
36. Q. Qi, A. Y. Chu, J. H. Kang, M. K. Jensen, G. C. Curhan, L. R. Pasquale, *et al.*, *N. Engl. J. Med.*, 2012, **367**(15), 1387–1396.
37. M. I. Goran, S. J. Ulijaszek and E. E. Ventura, *Glob. Public Health*, 2013, **8**(1), 55–64.
38. S. Luo, J. R. Monterosso, K. Sarpelleh and K. A. Page, *Proc. Natl. Acad. Sci. U. S. A.*, 2015, **112**(20), 6509–6514.
39. K. L. Stanhope, J. M. Schwarz, N. L. Keim, S. C. Griffen, A. A. Bremer, J. L. Graham, *et al.*, *J. Clin. Invest.*, 2009, **119**(5), 1322–1334.
40. K. Makki, E. C. Deehan, J. Walter and F. Backhed, *Cell Host Microbe*, 2018, **23**(6), 705–715.
41. A. Koh, F. De Vadder, P. Kovatcheva-Datchary and F. Backhed, *Cell*, 2016, **165**(6), 1332–1345.
42. C. De Filippo, M. Di Paola, M. Ramazzotti, D. Albanese, G. Pieraccini, E. Banci, *et al.*, *Front. Microbiol.*, 2017, **8**, 1979.
43. M. Han, C. Wang, P. Liu, D. Li, Y. Li and X. Ma, *Protein Pept. Lett.*, 2017, **24**(5), 388–396.
44. J. Salas-Salvado, M. Guasch-Ferre, C. H. Lee, R. Estruch, C. B. Clish and E. Ros, *J. Nutr.*, 2016, **146**, 920S–927S.
45. R. Estruch, E. Ros, J. Salas-Salvado, M. I. Covas, D. Corella, F. Aros, *et al.*, *N. Engl. J. Med.*, 2013, **368**(14), 1279–1290.
46. *The Top 10 Causes of Death*. World Health Organization, Geneva, 2018.
47. I. Shai, D. Schwarzfuchs, Y. Henkin, D. R. Shahar, S. Witkow, I. Greenberg, *et al.*, *N. Engl. J. Med.*, 2008, **359**(3), 229–241.
48. K. Esposito, M. I. Maiorino, G. Bellastella, P. Chiodini, D. Panagiotakos and D. Giugliano, *BMJ Open*, 2015, **5**(8), e008222.
49. C. M. Kastorini, H. J. Milionis, K. Esposito, D. Giugliano, J. A. Goudevenos and D. B. Panagiotakos, *J. Am. Coll. Cardiol.*, 2011, **57**(11), 1299–1313.
50. Y. Gepner, R. Golan, I. Harman-Boehm, Y. Henkin, D. Schwarzfuchs, I. Shelef, *et al.*, *Ann. Intern. Med.*, 2015, **163**(8), 569–579.
51. A. Rauf, M. Imran, H. A. R. Suleria, B. Ahmad, D. G. Peters and M. S. Mubarak, *Food Funct.*, 2017, **8**(12), 4284–4305.
52. T. Szkudelski and K. Szkudelska, *Biochim. Biophys. Acta*, 2015, **1852**(6), 1145–1154.
53. M. Mendez-del Villar, M. Gonzalez-Ortiz, E. Martinez-Abundis, K. G. Perez-Rubio and R. Lizarraga-Valdez, *Metab. Syndr. Relat. Disord.*, 2014, **12**(10), 497–501.
54. J. S. Allard, E. Perez, S. Zou and R. de Cabo, *Mol. Cell Endocrinol.*, 2009, **299**(1), 58–63.
55. E. Ozturk, A. K. K. Arslan, M. B. Yerer and A. Bishayee, *Biomed. Pharmacother.*, 2017, **95**, 230–234.
56. J. K. Bhatt, S. Thomas and M. J. Nanjan, *Nutr. Res.*, 2012, **32**(7), 537–541.
57. S. Renaud and M. de Lorgeril, *Lancet*, 1992, **339**(8808), 1523–1526.
58. F. Qian, A. A. Korat, V. Malik and F. B. Hu, *Diabetes Care*, 2016, **39**(8), 1448–1457.

59. L. Schwingshackl, A. M. Lampousi, M. P. Portillo, D. Romaguera, G. Hoffmann and H. Boeing, *Nutr. Diabetes*, 2017, 7(4), e262.
60. A. M. Fritzen, A. M. Lundsgaard, A. B. Jordy, S. K. Poulsen, S. Stender, H. Pilegaard, *et al.*, *J. Clin. Endocrinol. Metab.*, 2015, **100**(9), 3509–3519.
61. M. Uusitupa, K. Hermansen, M. J. Savolainen, U. Schwab, M. Kolehmainen, L. Brader, *et al.*, *J. Intern. Med.*, 2013, **274**(1), 52–66.
62. D. Ornish, L. W. Scherwitz, J. H. Billings, S. E. Brown, K. L. Gould, T. A. Merritt, *et al.*, *JAMA, J. Am. Med. Assoc.*, 1998, **280**(23), 2001–2017.
63. M. D. Gadgil, L. J. Appel, E. Yeung, C. A. Anderson, F. M. Sacks and E. R. Miller, 3rd., *Diabetes Care*, 2013, **36**(5), 1132–1137.
64. N. Chainani-Wu, G. Weidner, D. M. Purnell, S. Frenda, T. Merritt-Worden, C. Pischke, *et al.*, *Am. J. Cardiol.*, 2011, **108**(4), 498–507.
65. I. Vecchio, C. Tornali, N. L. Bragazzi and M. Martini, *Front. Endocrinol. (Lausanne)*, 2018, **9**, 613.
66. A. Mazur, *Nutr. J.*, 2011, **10**, 23.
67. L. Gupta, D. Khandelwal, S. Kalra, P. Gupta, D. Dutta and S. Aggarwal, *J. Postgrad. Med.*, 2017, **63**(4), 242–251.
68. L. Laffel, *Diabetes Metab. Res. Rev.*, 1999, **15**(6), 412–426.
69. E. C. Westman, W. S. Yancy, Jr., J. C. Mavropoulos, M. Marquart and J. R. McDuffie, *Nutr. Metab. (Lond.)*, 2008, **5**, 36.
70. T. A. Hussain, T. C. Mathew, A. A. Dashti, S. Asfar, N. Al-Zaid and H. M. Dashti, *Nutrition*, 2012, **28**(10), 1016–1021.
71. F. F. Samaha, N. Iqbal, P. Seshadri, K. L. Chicano, D. A. Daily, J. McGrory, *et al.*, *N. Engl. J. Med.*, 2003, **348**(21), 2074–2081.
72. B. J. Brehm, R. J. Seeley, S. R. Daniels and D. A. D'Alessio, *J. Clin. Endocrinol. Metab.*, 2003, **88**(4), 1617–1623.
73. N. B. Bueno, I. S. de Melo, S. L. de Oliveira and T. da Rocha Ataide, *Br. J. Nutr.*, 2013, **110**(7), 1178–1187.
74. H. M. Dashti, N. S. Al-Zaid, T. C. Mathew, M. Al-Mousawi, H. Talib, S. K. Asfar, *et al.*, *Mol. Cell Biochem.*, 2006, **286**(1–2), 1–9.
75. H. M. Dashti, T. C. Mathew, M. Khadada, M. Al-Mousawi, H. Talib, S. K. Asfar, *et al.*, *Mol. Cell Biochem.*, 2007, **302**(1–2), 249–256.
76. R. M. Krauss, P. J. Blanche, R. S. Rawlings, H. S. Fernstrom and P. T. Williams, *Am. J. Clin. Nutr.*, 2006, **83**(5), 1025–1031.
77. M. Via, *ISRN Endocrinol.*, 2012, **2012**, 103472.
78. W. T. Cefalu and F. B. Hu, *Diabetes Care*, 2004, **27**(11), 2741–2751.
79. R. A. Anderson, *Diabetes Metab.*, 2000, **26**(1), 22–27.
80. R. A. Ngala, M. A. Awe and P. Nsiah, *PLoS One*, 2018, **13**(7), e0197977.
81. R. A. Anderson, *Nutr. Res. Rev.*, 2003, **16**(2), 267–275.
82. K. N. Jeejeebhoy, R. C. Chu, E. B. Marliss, G. R. Greenberg and A. Bruce-Robertson, *Am. J. Clin. Nutr.*, 1977, **30**(4), 531–538.
83. H. Freund, S. Atamian and J. E. Fischer, *JAMA, J. Am. Med. Assoc.*, 1979, **241**(5), 496–498.
84. S. Rajpathak, E. B. Rimm, T. Li, J. S. Morris, M. J. Stampfer, W. C. Willett, *et al.*, *Diabetes Care*, 2004, **27**(9), 2211–2216.
85. R. A. Anderson, N. Cheng, N. A. Bryden, M. M. Polansky, N. Cheng, J. Chi, *et al.*, *Diabetes*, 1997, **46**(11), 1786–1791.

86. A. N. Paiva, J. G. Lima, A. C. Medeiros, H. A. Figueiredo, R. L. Andrade, M. A. Ururahy, *et al.*, *J. Trace Elem. Med. Biol.*, 2015, **32**, 66–72.
87. H. Huang, G. Chen, Y. Dong, Y. Zhu and H. Chen, *Mol. Nutr. Food Res.*, 2018, **62**(1), 1700438.
88. J. Zhou, K. Huang and X. G. Lei, *Free Radic. Biol. Med.*, 2013, **65**, 1548–1556.
89. M. Alizadeh, A. Safaeiyan, A. Ostadrahimi, R. Estakhri, S. Daneghian, A. Ghaffari, *et al.*, *Ann. Nutr. Metab.*, 2012, **60**(2), 157–168.
90. Z. Asemi, M. Jamilian, E. Mesdaghinia and A. Esmaillzadeh, *Nutrition*, 2015, **31**(10), 1235–1242.
91. L. N. Kohler, J. Foote, C. P. Kelley, A. Florea, C. Shelly, H. S. Chow, *et al.*, *Nutrients*, 2018, **10**(12), 1924.
92. E. T. Jacobs, P. Lance, L. J. Mandarino, N. A. Ellis, H. S. Chow, J. Foote, *et al.*, *BMJ Open Diabetes Res. Care*, 2019, **7**(1), e000613.
93. S. Stranges, J. R. Marshall, R. Natarajan, R. P. Donahue, M. Trevisan, G. F. Combs, *et al.*, *Ann. Intern. Med.*, 2007, **147**(4), 217–223.
94. J. Mitri, M. D. Muraru and A. G. Pittas, *Eur. J. Clin. Nutr.*, 2011, **65**(9), 1005–1015.
95. O. Kaidar-Person, B. Person, S. Szomstein and R. J. Rosenthal, *Obes. Surg.*, 2008, **18**(7), 870–876.
96. E. Strohmayer, M. A. Via and R. Yanagisawa, *Mt Sinai J. Med.*, 2010, **77**(5), 431–445.
97. H. L. Gulseth, C. Wium, K. Angel, E. F. Eriksen and K. I. Birkeland, *Diabetes Care*, 2017, **40**(7), 872–878.
98. N. Mirhosseini, H. Vatanparast, M. Mazidi and S. M. Kimball, *J. Clin. Endocrinol. Metab.*, 2017, **102**(9), 3097–3110.
99. L. Wamberg, U. Kampmann, H. Stodkilde-Jorgensen, L. Rejnmark, S. B. Pedersen and B. Richelsen, *Eur. J. Intern. Med.*, 2013, **24**(7), 644–649.
100. J. L. Rosenblum, V. M. Castro, C. E. Moore and L. M. Kaplan, *Am. J. Clin. Nutr.*, 2012, **95**(1), 101–108.
101. M. Sneve, Y. Figenschau and R. Jorde, *Eur. J. Endocrinol.*, 2008, **159**(6), 675–684.
102. S. Wongwiwatthananukit, N. Sansanayudh, N. Phetkrajaysang and S. Krittiyanunt, *J. Endocrinol. Invest.*, 2013, **36**(8), 558–563.
103. A. Mousa, N. Naderpoor, M. P. de Courten, H. Teede, N. Kellow, K. Walker, *et al.*, *Am. J. Clin. Nutr.*, 2017, **105**(6), 1372–1381.
104. Z. Yilmaz, F. Piracha, L. Anderson and N. Mazzola, *J. Pharm. Pract.*, 2017, **30**(6), 631–638.
105. S. M. Blevins, M. J. Leyva, J. Brown, J. Wright, R. H. Scofield and C. E. Aston, *Diabetes Care*, 2007, **30**(9), 2236–2237.
106. R. W. Allen, E. Schwartzman, W. L. Baker, C. I. Coleman and O. J. Phung, *Ann. Fam. Med.*, 2013, **11**(5), 452–459.
107. S. Qiu, H. Sun, A. Zhang, H. Xu, G. Yan, Y. Han, *et al.*, *Chinese J. Nat. Med.*, 2014, **12**(6), 401–406.
108. S. Reiser, H. B. Handler, L. B. Gardner, J. G. Hallfrisch, O. E. Michaelis and E. S. Prather, *Am. J. Clin. Nutr.*, 1979, **32**(11), 2206–2216.
109. J. Bosch, H. C. Gerstein, G. R. Dagenais, R. Díaz, L. Dyal, H. Jung, *et al.*, *N. Engl. J. Med.*, 2012, **367**(4), 309–318.

CHAPTER 2

Maternal Nutrition and the Risk of Obesity and Diabetes in the Offspring

C. M. REYNOLDS, P. E. BRIDGE-COMER AND M. H. VICKERS*

Liggins Institute, University of Auckland, Auckland 1142, New Zealand
*Email: m.vickers@auckland.ac.nz

2.1 Introduction

The current epidemic of obesity and related cardiometabolic diseases is often linked to an obesogenic environment and sedentary behaviours. However, increasing evidence from human observational studies and experimental animal models suggests that alterations in the early-life nutritional environment can increase the susceptibility for the development of such disorders in adult life. This developmental programming falls under the "Developmental Origins of Health and Disease" (DOHaD) framework whereby adverse environmental impacts during the early-life period of developmental plasticity can result in adaptations that can have long-term adverse trade-offs in offspring for health outcomes across the lifecourse (Figure 2.1).[1,2] The DOHaD model suggests that predictive adaptations are made in early life in response to environmental cues (*e.g.* malnutrition, resulting in the development of a "thrifty phenotype"[3]) in order to maximise survival in the expected postnatal environment. However, when the interpretations of such prenatal cues

Nutritional Signaling Pathway Activities in Obesity and Diabetes
Food Chemistry, Function and Analysis No. 24
Edited by Zhiyong Cheng

Published by the Royal Society of Chemistry, www.rsc.org

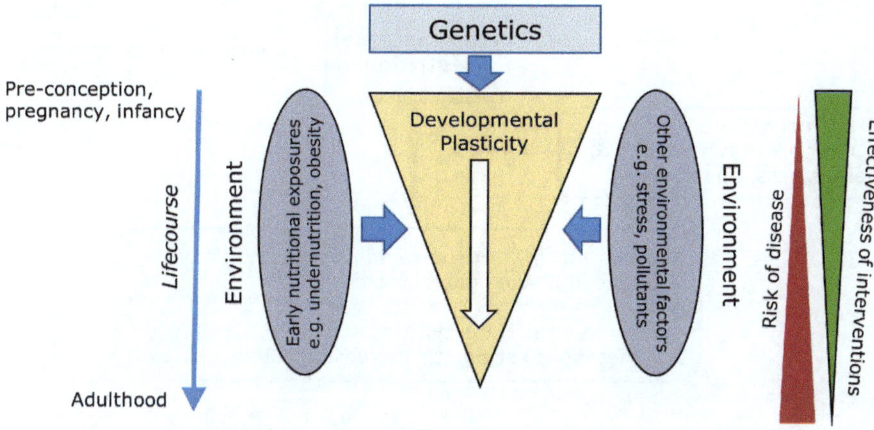

Figure 2.1 Gene–environment interactions across the lifecourse. Disease risk increases throughout the lifecourse due to declining plasticity and diminishing responses to new interventions. The early-life period therefore offers the best avenue for prevention and, although requiring a longer-term investment, is more likely to be efficacious than treating disease once it has already manifest in adulthood.

are inappropriate or there are changes to that immediate environment, this results in a mismatch between the prenatal prediction and postnatal reality. Consequently, these adaptations, termed predictive adaptive responses (PARS),[4,5] can ultimately be disadvantageous as they may lead to an increased susceptibility to develop a range of diseases in later life and/or the inheritance of disease traits that can be transmitted to future generations.[6–8] Of note, the effects of PARS during early development may not manifest as an aberrant phenotype until later in life, particularly if they impact pathways that modulate the responsiveness to later environmental challenges, including exposure to energy-rich diets.[9,10]

2.2 Evidence from Epidemiology and Human Studies

In human populations, most findings that have supported a relationship between nutrition in early life and later risk of Type 2 diabetes mellitus (T2DM) and obesity have been derived from quasi-experimental studies ("natural experiments").[11] Exposure in early life to adverse environmental factors such as suboptimal nutrient supply during critical windows of early development may lead to programming of long-term appetite regulation, energy expenditure, as well as pancreatic β-cell dysfunction and altered adipogenesis in the developing fetus and neonate (Figure 2.2).[12] According to the PARS hypothesis, such metabolic (mal)adaptation may be beneficial in the short term to ensure survival in a suboptimal postnatal environment *via* an enhanced capacity to store fat under conditions of limited food availability, but may increase the susceptibility of offspring to the risk of developing T2DM when food is

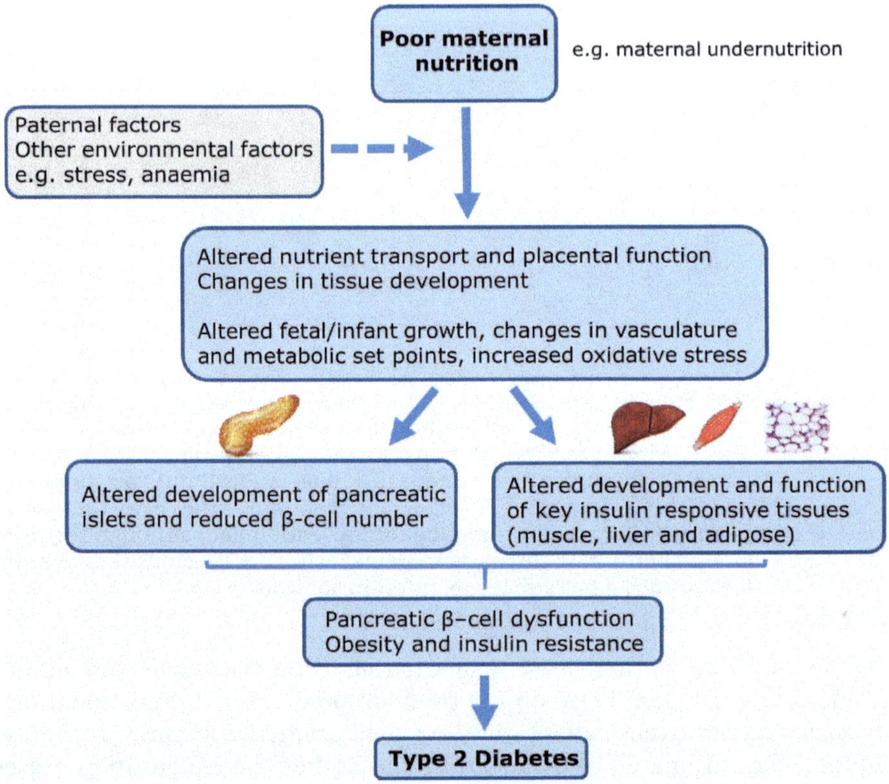

Figure 2.2 Developmental programming of Type 2 diabetes mellitus (T2DM). An adverse early-life environment, including suboptimal maternal nutrition, will result in a range of placental and fetal adaptations that can impact upon pancreatic development and function and responsiveness of insulin-sensitive issues. This can ultimately lead to beta-cell dysfunction, insulin resistance and development of T2DM concomitant with other cardiometabolic disorders, including obesity.

abundant in postnatal life. In particular, in the case of fetal growth restriction, the fetal adaptations to undernutrition are characterised by a reprogramming of mechanisms responsible for energy balance and glucose homeostasis, such as enhanced insulin sensitivity for glucose uptake and utilisation, lowered insulin sensitivity for muscle protein synthesis, increased hepatic glucose production and impaired pancreatic development.[13] In postnatal life, these early adaptations have the potential to promote nutrient uptake and storage, thus leading to insulin resistance and obesity.

In addition to T2DM, being born of low birth weight is a reliable predictor of other outcomes associated with T2DM in later life, including increased adiposity, insulin resistance, atherogenic lipid profiles, hypertension and cardiovascular disease in adulthood.[11] Of note, the relationship between birth weight and later T2DM risk is not linear but is "U-shaped", with high birth weight (> 4 kg) associated with an increased risk of T2DM to the same level as being born at a low birth weight (< 2.5 kg).[14]

The most widely cited study in a human population is that of the Dutch famine of 1944–1945 (the Dutch Hunger Winter), a 5-month period of extreme food shortage, where a number of adverse cardiometabolic outcomes were reported in famine survivors.[15] These outcomes were specific to the timing of famine exposure during pregnancy and in some cases were sex-specific. Prenatal famine exposure, particularly during late gestation, was associated with impaired glucose tolerance and insulin resistance in adulthood.[16] Growth restriction during the period of late gestation may have resulted in a permanent reduction in insulin production capacity as this is a critical window of proliferation for the endocrine pancreas.[17] Obesity prevalence was more common in young men exposed to famine in early gestation.[18] Malnutrition of mothers during the early gestational period was also associated with higher body mass index (BMI) and increased waist circumference in women but not in men at 50 years of age.[19] Of interest, prenatal famine exposure was linked to altered food preferences in later life (a preference for energy-dense fatty foods) and associated with a more atherogenic lipid profile.[20] Thus while exposure of the fetus to an environment of limited substrates at developmental stages leads to dysregulation of metabolism in adulthood, the precise mechanisms responsible appear to vary with the timing and duration of the exposure.

Similar results have been reported for the Great Chinese Famine (1959–1961) where famine exposure during fetal life was associated with an elevated risk for developing hyperglycaemia or diabetes in adulthood.[21] These effects were independent of the severity of the famine exposure, with diabetes risk similar between severely affected and less severely affected areas, although exposure to severe famine during fetal life appeared to be predictive for higher hemoglobin A1c (HbA_{1c}) in later life.[22] Similar to the Dutch famine study, exposure to famine in early development was associated with an increased BMI and increased risk for overweight/obesity in later life, particularly in females.[23,24] However, interpretation of the Chinese famine data has been limited due to uncontrolled age differences between famine and post-famine births that could potentially explain most of the effects that have been commonly attributed to the famine.[25]

In addition to the known implications for increasing complications of pregnancy, including gestational diabetes (GDM) and macrosomia, increasing evidence has highlighted maternal overweight/obesity as a key determinant of the health of offspring during infancy, childhood and later adult life,[26] with observational studies linking maternal overweight/obesity to an increased risk of obesity and T2DM. Although in human studies controlling for potentially confounding variables remains a challenge, evidence to date across a range of cohorts suggests that pre-pregnancy BMI and excessive weight gain during pregnancy are associated with an elevated risk for the development of obesity and insulin resistance in offspring during childhood.[26] In the Helsinki Birth Cohort, maternal BMI was positively associated with BMI in offspring at age 60 years.[27] Similarly, in a study of mother–offspring pairs it was shown that increased pre-pregnancy BMI was associated with an increase in the BMI of offspring at 30 years of age.[28] With up to two-thirds of women of reproductive age (18–44 years) either overweight or obese,[29] and these rates increasing annually, there is an urgent need for studies investigating causality, potential mechanisms and effective strategies for intervention to break the cycle of obesity in women of reproductive age and to reduce consequences for offspring in later life.

2.3 Evidence from Experimental Animal Models

Epidemiological observations formed the foundation for the original "Barker hypothesis", the concept of "fetal origins of disease/developmental programming" and establishment of the DOHaD framework.[2] Animal models have been essential for providing empirical data to support the DOHaD hypothesis and have enabled investigation of the underlying mechanisms involved, and have utility to examine programming effects across the lifecourse and the potential for transgenerational impacts.[30] The rodent has been the most widely used model to date. For undernutrition studies the most commonly used approaches have been global nutrient restriction (30–70% of ad-libitum intake) or low-protein diets (LPD, typically 8–9% vs. 20% protein). Although DOHaD research focused initially on the impact of maternal undernutrition and the link between birth weight and later health outcomes, given the global rise in overweight/obesity, particularly in women of reproductive age, there has been increasing focus on experimental models utilising a maternal obesogenic environment. In general, two main experimental models have been utilised, those using high-fat and/or high-sugar purified diets or those that use a "cafeteria-style diet" that is designed to parallel that of a complex Western-style diet.[31,32] The composition of the experimental diets used differs widely across the range of models, with some debate over the robustness of the different dietary approaches used, *e.g.* the single-fat-source diets *versus* the "cafeteria" or "junk food"-style diets.[33] However, both approaches have been extensively characterised over recent years, with similar phenotypes described, and have provided important mechanistic insights into the development of a range of diseases, particularly as relates to the programming of obesity and T2DM.

2.3.1 Global Nutrient Restriction

Global nutrition restriction (typically restriction to 30–70% of ad-libitum intake) has been used widely as a model to induce programming of T2DM and obesity. In the rat, maternal dietary restriction and normal or restricted postnatal nutrition lead to a reduction in β-cell mass both at birth and throughout early postnatal development[34,35] with increased adiposity, insulin resistance and disorders of appetite regulation and energy balance in adult life.[10] A post-weaning obesogenic environment can exacerbate these effects.[10] Using a rat model of severe maternal undernutrition (30% of ad-libitum through pregnancy),[36] offspring show significant growth restriction but rapid catch-up growth in the pre-weaning period and in adulthood are characterised by increased adiposity, hyperleptinaemia, insulin resistance and hypertension.[10] These offspring also display appetite dysregulation (hyperphagia) and reduced locomotor activity with the phenotype markedly amplified in the setting of a post-weaning high-fat diet (HFD).[10,37,38] Other reports using this model have shown that adult male rat offspring are obesity-prone and display hyperleptinaemia with modifications in leptin and leptin receptor (Ob-R) messenger RNA (mRNA) levels and a dysregulation of the apelinergic system in white adipose tissue (WAT).[39] A similar model using 65% food restriction in the rat showed that offspring of undernourished mothers that were fed a cafeteria diet post-weaning had increased visceral *versus* subcutaneous fat accretion, dislipidaemia and ectopic fat deposition and showed increased signs of inflammation and macrophage infiltration in visceral fat pads compared to control offspring.[40]

Even a moderate dietary restriction has significant lasting effects on offspring. Feeding a 50% restricted maternal diet results in obesity and glucose intolerance in offspring that are independent of the level of postnatal nutrition. Increases in fat mass were associated with adipocyte hypertrophy concomitant with a marked proinflammatory phenotype in adipose tissue which correlated with a decrease in insulin sensitivity.[41] Interestingly, if catch-up growth in the pre-weaning period was prevented by maintenance of the maternal dietary restriction throughout the period of lactation, offspring do not develop the obese or insulin-resistant phenotype in later life.[42] In studies using cross-fostering of litters to isolate the effects of different periods of dietary restriction (*i.e.* pre- and postnatal 50% dietary restriction), prenatal dietary restriction resulted in an approximate 50% reduction in β-cell mass whereas dietary restriction in the postnatal period resulted in decreased body weights, but an increase in both β-cell mass and β-cell fractional area as compared to control animals.[43,44] These observations suggested that prenatal dietary restriction largely determines development of the pancreatic endocrine cells while postnatal dietary restriction primarily impacts development of the exocrine pancreas.[44]

2.3.2 Protein Restriction

Varying the level of maternal protein supply during pregnancy and lactation has been extensively demonstrated to play a key role in pancreatic development with offspring of low-protein (LP) rat dams displaying reduced islet size, neonatal β-cell proliferation and altered islet vasculature in addition to impaired glucose tolerance in adulthood.[45,46] Fetal protein restriction in the rat, followed by catch-up growth postnally, can increase fat mass and the capacity for preadipocytes to proliferate.[47] Similarly, male rat offspring who had demonstrated rapid catch-up growth (*via* litter size reduction) following fetal protein restriction were characterised by an increase in fat mass, hyperglycaemia and adverse lipoid profiles in adulthood. These outcomes were accompanied by significant alterations in the expression of genes that play key roles in lipid metabolism and the differentiation and function of adipose tissue.[48] Altered adipocyte sensitivity to insulin in LP offspring has also been well characterised with alterations in expression of the phosphatidylinositol 3-kinase (PI3K) subunit suggestive of adipocytes of male LP offspring being resistant to the anti-lipolytic effects of insulin.[49] Data to date suggests that a maternal LP diet alters the balance of β-cell replication and apoptosis during fetal and neonatal development concomitant with reduced pancreatic expression of insulin-like growth factor (IGF)-2, which may involve an altered β-cell cycle length, and contribute to the reduction in islet size and impaired insulin release observed in later life.[35] As such, in adulthood, these animals develop diabetic phenotypes that are characterised by β-cell failure that arises due to impaired insulin secretion, glucose intolerance, insufficient expansion of β-cell mass and fasting hyperglycaemia.[50,51]

There is some evidence for a dysregulation in circadian physiology following maternal LP diet exposure that may underpin some of the postnatal effects observed in offspring. In a mouse maternal LP model (6% protein), offspring displayed growth restriction but early rapid catch-up growth and displayed impaired insulin sensitivity and obesity in adulthood. These effects were preceded by abnormal rhythms in daily food intake and related metabolism, increased lipogenesis and a proinflammatory phenotype and therefore suggest that altered circadian physiology may be a causal factor in the metabolic dysregulation associated with growth restriction, and altered appetite behaviour and substrate metabolism may contribute to the obese and insulin-resistant phenotype.[52] Further, as with the global restriction model, it has been shown that, if the LP dietary exposure is continued throughout the period of lactation, it can prevent the development of the metabolic phenotype observed in offspring in adulthood, thus highlighting again the possible adverse consequences of early rapid catch-up growth.[53]

2.3.3 Maternal Obesity

Maternal obesity during the preconception period and pregnancy programs an increased susceptibility to obesity in offspring and leads to what has been termed "an intergenerational cycle of obesity".[54,55] Even mild maternal overnutrition can induce obesity, altered central appetite regulation and glucose intolerance in offspring.[56] A number of models using a maternal obesogenic environment have been developed, primarily in the rat, which show a range of cardiovascular and metabolic disorders in offspring, although the magnitude of phenotypic effects observed can be sex-specific and differ with the timing of the challenge and composition of the experimental diets. It also needs to be noted that, although maternal obesity is commonly associated with large or macrosomic infants, maternal obesity also commonly results in fetal growth restriction and infants born of low birth weight, possibly a result of impaired placental function that has been reported experimentally in the rat.[57]

Maternal obesity in mice results in early-onset hyperphagia and increased adiposity in offspring with food intake significantly increased in the early post-weaning period compared to offspring of the non-obese control group, although fed the same diet.[58] These offspring also developed hypertension, insulin resistance, reduced locomotor activity and endothelial dysfunction. In the rat models, similar observations have been made whereby a maternal obesogenic diet programs for obesity, cardiometabolic and reproductive disorders in offspring in later life independently of the postnatal dietary environment.[59] Similar to the undernutrition models, maternal obesity programs for adipocyte hypertrophy, hyperleptinaemia and increased leptin gene expression in a depot-specific manner.[60] Of note, different windows of maternal exposure to an HFD, *i.e.* pregnancy alone or during both pregnancy and lactation can lead to similar broad phenotypic outcomes of obesity and insulin resistance in offspring, although appears to be mediated *via* different mechanisms.[59]

As with the undernutrition models, developmental programming *via* exposure to HFDs results in a range of diabetogenic phenotypes that vary dependent on the timing and duration of the dietary insult.[61] Exposure to a diet high in saturated fat during the fetal and/or lactation periods has been shown to alter islet cell architecture in neonate and weanling rats.[62] HF programming, through acute hyperglycaemia, initially results in β-cell compensation, although eventually leads to chronic hyperglycaemia that ultimately triggers β-cell exhaustion, dysfunction and death.[61] Further work by this group showed that HFD dietary exposure during late pregnancy enhanced omega-6 fatty acid profiles in the fetal liver concomitant with insulin resistance and hyperglycaemia, thereby resulting in a metabolically compromised phenotype.[63] Maternal programming by HFD can also induce hypothalamic mitochondria fusion, metabolic dysfunction and fat accumulation in male offspring, which is exacerbated by exposure to a post-weaning HFD, potentially arising due to mitochondria calcium overflux.[64] Recent work by Nguyen *et al.* also provided evidence for a mechanistic role of sirtuin (SIRT)1 in maternal HFD-induced metabolic dysfunction in offspring and suggested that SIRT1 was a promising target for developmental reprogramming.[65] One caveat around the term "overnutrition" is that in many instances intake of obesogenic diets may result in specific micronutrient deficiencies and the phenotypic similarities observed across the seemingly disparate models of both maternal under- or overnutrition may reflect in part the consequences of "obesity malnutrition".[66]

2.3.4 Micronutrient Deficiencies

The role of altered maternal micronutrient status is less studied, although work to date implicates micronutrient deficiencies to a range of metabolic conditions. In the rat, maternal iron deficiency leads to several characteristics of the metabolic syndrome in offspring, including raised blood pressure,[67] although glucose tolerance has been reported to be improved in offspring of iron-restricted mothers.[68] In the setting of maternal anaemia, the timing of iron supplementation has been shown to be critical in ameliorating the effects on the developing fetus and postnatal sequalae in offspring.[69] Of note, in humans, iron status in the neonate is impaired in the setting of maternal obesity and excessive gestation weight gain.[70] In humans, vitamin D status has been associated with offspring being born small for gestation age and low maternal vitamin D status increased the risk for pregnancy complications, including pre-eclampsia and GDM.[71,72] Given the established effects of vitamin D and related deficiencies on glucose tolerance and insulin sensitivity, it is not surprising that a deficiency of vitamin D in some populations is associated with parallel increases in T2DM rates.[73] Reduced intake of vitamins during pregnancy can also increase adiposity and altered lipid profiles in rat offspring along with increases in markers of oxidative stress.[74] In rats, maternal magnesium restriction can predispose offspring to increased adiposity, glucose intolerance and impaired insulin sensitivity.[75,76] Similarly, maternal chromium restriction has been shown to lead to increased body weight and adiposity in male and female rat offspring.[77,78] As detailed above, it needs to be noted that a maternal obesogenic environment can also be reflective of micronutrient/secondary deficiencies and may account for some of the phenotypic commonalities observed in offspring of obese mothers *versus* those with specific nutrient deficiencies.[66,79]

2.3.5 Other Models

Although the rodent is the most widely used programming model, there is also extensive evidence for aberrant nutritional programming derived from large animal models of undernutrition and overnutrition, including the sheep and non-human primate.[80] In the sheep, undernutrition during the preconceptional period led to programmed changes in the thermogenic capacity of brown fat and the insulin and fatty acid oxidation signaling pathways in visceral fat which are key determinants for the risk of developing insulin resistance and obesity in later life.[81] Using a similar model, it was also reported that periconceptional undernutrition induced changes in expression of microRNAs that may alter the abundance of the key insulin-signaling markers in skeletal muscle and thus the association between periconceptional undernutrition and insulin resistance in later life.[82] Similarly to the rodent data, maternal undernutrition in the sheep also results in decreased voluntary locomotor activity and this may represent a predisposing factor to obesity observed in this model.[83] In a model of late gestation undernutrition in the sheep, offspring showed metabolic dysfunction with development of obesity in adulthood which also exacerbated the programming of glucose insulin insensitivity in male offspring.[84] In sheep fed at 50% of total energy requirement, lambs showed an increased risk for abdominal adiposity and different physiological adaptations to fasting compared to controls,[85] with effects more pronounced in females than males.[86,87] As with the rodent models, the effects of undernutrition in the sheep models on adiposity and insulin sensitivity differ between tissues and sexes depending upon on the window and timing of nutrient restriction.[88]

In a sheep model of maternal obesity, accelerated fetal pancreatic β-cell but not α-cell development was observed.[89] In this model, overfeeding of ewes induced obesity at conception and throughout the period of gestation. At mid-gestation, fetuses from obese ewes displayed hyperinsulinaemia and hyperglycaemia and had increased body weights compared to offspring of control ewes. They also exhibited significantly increased pancreatic β-cell numbers and pancreatic weights. This work also highlighted differential timing impacts of maternal obesity on pancreatic growth and β-cell numbers in the developing fetus during early and late gestation. Early gestation was characterised by a significant increase in β-cell proliferation, pancreatic growth and insulin secretion while late gestation saw a reduction in pancreatic growth and β-cell numbers, resulting in a reduction in circulating insulin at term.[89,90]

As with the small-animal models, different windows of weight gain and overnutrition during the periconceptional and pregnancy periods can exert different effects on fetal growth and tissue development.[91] In the sheep, overnutrition during late pregnancy results in an increase in the expression of peroxisome proliferator-activated receptor (PPAR)-γ-activated genes in the visceral fat of the fetus and a subsequent increase in subcutaneous fat mass in the lamb postnatally. Maternal overnutrition during the periconceptional period alone, however, resulted in an increase in total body fat mass only in female lambs, with the primary effect seen in visceral fat depots. This work suggested that early programming of later-life obesity may result from "two hits", with the first occurring as a result of maternal periconceptional overnutrition and the second occurring due to increased fetal nutrition during late pregnancy.[91]

In a recently described non-human primate model of maternal protein restriction, a 50% dietary protein reduction throughout gestation resulted in fetal growth restriction, reduced placental perfusion and a high (50%) rate of pregnancy loss[92] but did not report on later offspring outcomes. Early work in a baboon model showed that overnutrition in the pre-weaning period permanently increased fat mass in offspring mediated through adipocyte hypertrophy, although these effects were sex-specific and changes were observed in females only.[93] In macaques, a maternal HFD can also trigger lipotoxicity in the fetal liver and increased the risk for offspring to develop hepatic steatosis.[94] These observations parallel the rodent work whereby a maternal HFD has been shown to lead to developmental priming of fatty liver disease in offspring.[95,96]

2.4 Epigenetic Mechanisms

Epigenetic processes have emerged as key regulatory mechanisms *via* which the memory of early developmental exposures is held, with lasting consequences for various tissues and physiological systems.[97] As such, modifications to epigenetic regulatory processes (*i.e.* DNA methylation, histone modifications and non-coding RNAs) have been proposed as a key mechanism that links altered maternal nutrition to outcomes in offspring. However, most data to date has been derived from experimental models as, at a mechanistic level, there is a paucity of studies in humans that causally link epigenetic change to metabolic disease risk, although there is evidence for inheritance of tissue-specific DNA methylation patterns[98] and data showing that epigenetic gene promoter methylation status at birth is associated with later adiposity in children at 9 years of age.[99] The most studied epigenetic mechanisms identified in the adaptive nutritional programming of obesity and diabetes risk is that of altered DNA methylation and histone modifications, although alterations in miRNAs are also emerging as potential mediators of programming effects with evidence in sheep, for example, that undernutrition induces changes in miRNAs that appear to play a role in the development of insulin resistance in later life.[82] In particular, altered availability of dietary methyl donors and cofactors during critical windows of early development may influence patterns of DNA methylation. Thereby, deficiencies in methyl donors during early periods of development, arising from either excess nutrition or undernutrition, could effectively lead to premature epigenetic ageing (changes in age-related DNA methylation patterns) and confer an increased susceptibility to adult disease in later life.[100]

The role of the one-carbon cycle in the programming of T2DM and obesity has gained increasing attention, with experimental animal data highlighting that altered methyl donor availability in early development may disrupt DNA methylation with lasting adverse consequences for the offspring. It is now clear that an altered early-life nutritional environment can impact upon DNA methylation patterns as one-carbon metabolism is critically dependent upon availability of dietary methyl donors and cofactors, including vitamin B_{12}, choline and folic acid.[101–103] As an example, increased concentrations of vitamin B_{12} during human pregnancy have been associated with a decrease in global DNA methylation in newborns while an increase in newborn serum B_{12} concentrations is associated with a reduction in the methylation of the IGF-binding protein (IGFBP)-3 gene that plays a role in intrauterine growth.[104]

A suboptimal nutritional environment during fetal development in the rat can suppress the transcription of key factors involved in the regulation of β-cell development.[105] An LP diet fed to pregnant females leads to hypomethylation of genes encoding PPAR-γ and the glucocorticoid receptor in the livers of offspring – effects that can be reversed *via* folate supplementation to the mother.[106] Maternal dietary restriction in the rats also results in a significant reduction in the expression of genes encoding key transcription factors regulating embryonic β-cell development, including pancreatic and duodenal homeobox 1 (PDX-1).[107] Reduced PDX-1 activity in the rat is associated with histone modifications[108] and such changes on an epigenetic level are paralleled by a reduction in β-cell formation postnatally and a reduced capability to expand β-cell mass in response to metabolic stress. Perinatal undernutrition can also affect DNA methylation patterns and expression of the leptin gene in later life (*via* hypomethylation of the leptin promoter), thus causing disruption to the adipoinsular axis feedback loop and predisposing to insulin resistance in later life.[109]

In addition to the epigenetic changes observed in models of maternal undernutrition, maternal obesity can lead to epigenetic changes in offspring that can mediate longer-term risk for obesity and T2DM. Maternal obesity has been shown to enhance WAT differentiation and alter genome-scale DNA methylation in male rat offspring.[110] Overnutrition during the early neonatal period (*via* litter size manipulation) can manifest as epigenetic modifications in key genes known to be involved in insulin signaling in skeletal muscle (*e.g.* insulin receptor substrate (IRS)-1 and glucose transport (Glut)-4) and later development of insulin resistance in female offspring.[111] Maternal obesity in the mouse can induce epigenetic modifications that can enhance differentiation of adipocytes, thus potentiating adipose deposition and metabolic dysfunction in adult life.[112] These effects appear to be mediated *via* epigenetic changes in the zinc finger protein Zfp423 promoter, a key transcription factor that commits cells to the adipogenic lineage with reduced DNA methylation of Zpof423, resulting in increased adipogenic differentiation of progenitor cells in adult tissue and thus programming for obesity.

A key feature of many experimental programming models is that of leptin resistance in offspring that parallels increased adiposity and altered appetite and energy expenditure. Maternal obesity and GDM have been associated with alterations in DNA methylation of the leptin promoter in a tissue- and sex-specific manner, including altered placental leptin DNA methylation, with implications for later metabolic dysfunction.[113,114] Work by Quilter *et al.* suggested that similar epigenetic modifications may underlie a broad range of early-life nutritional exposures and increase long-term susceptibility to obesity and related diseases such as T2DM, although the precise mechanisms remain to be defined.[115]

Although a primary focus has been on altered DNA methylation, there is increasing evidence for the role of altered nutrition on histone structure and function as well as non-coding RNAs. Methylation of histones requires methyl donors derived from dietary intake of folate and the presence of choline, methionine and B vitamins (B_6 and B_{12}) along with a range of methyltransferases and represents a key structural alteration that impacts upon gene expression.[116,117] As an example, maternal food restriction in the rat has been shown to alter IGF-1 mRNA expression and histone structure with the effects mediated by the rate of catch-up growth, with modulation of neonatal catch-up growth appearing to confer protection against epigenetic modifications of IGF-1 and thus mitigating risk for later obesity and related disorders.[116] Maternal undernutrition in the rat can also lead to altered miRNA profiles, including the LET-7, family which has been implicated in a range of cardiometabolic disorders.[118] A maternal LP diet alters miRNA expression in offspring and alters glucose homeostasis and insulin secretion *via* a defect in insulin secretion rather than a reduction in β-cell mass.[119] Interestingly, blockage of this subset of miRNAs restores insulin secretion to normal and it is now established that a specific set of miRNAs play key roles in β-cell differentiation and are essential for the regulation of insulin secretion and expansion of β-cell mass as a compensatory response to insulin resistance.[120–122] In a mouse HFD model, male offspring appear to have an increased vulnerability to chromatin remodelling processes but a link to the development of later metabolic disorders has yet to be established.[123] In the setting of maternal chromium restriction, altered regulation of miRNAs appears to be related to disorders of lipid metabolism in offspring *via* persistent effects on the mitogen-activated protein kinase (MAPK) pathway.[124] The experimental data to date mirrors the available human data whereby changes in miRNA profiles in offspring of obese mothers appear to be mediators of early epigenetic programming of metabolic disorders in these children.[125]

2.5 Intervention Strategies

While current preventive strategies for cardiometabolic disease focus on high-risk individuals in mid-life, recent work by Fall and Kumaran suggests that DOHaD concepts offer a "primordial" preventive strategy to reduce disease in future generations by improving fetal and infant development.[126] The effects of aberrant early-life nutritional programming were once considered to represent a permanent change in developmental trajectory, the consequences of which had to be managed across the lifecourse. However, at least in experimental models, programming has been shown to be reversible *via* a range of intervention modalities during critical periods of developmental plasticity.[127] These cover a range of pharmacologic and nutritional interventions, including growth hormone (GH),[128,129] leptin,[130] fatty acids/lipids,[131–133] antioxidants/polyphenols,[134–136] methyl donors[106,137] and insulin sensitisers.[138,139] The adipokine leptin, in particular, has been a focus of attention as it is proposed that deficiencies in leptin during critical early windows of developmental plasticity could hardwire for obesity and metabolic disease in later life.[140,141]

A number of studies across a range of experimental models and species have now reported on the protective effects of neonatal leptin intervention in the setting of maternal undernutrition in reversing later-life effects on obesity and diabetes risk.[130,142–144] Leptin treatment to rats following maternal undernutrition normalised body weight, fat mass, caloric intake, voluntary locomotor activity and fasting concentrations of glucose, leptin and insulin.[130] Similar effects have been observed in a growth-restricted piglet model where treatment with leptin partially reversed the growth-restricted phenotype *via* beneficial effects of leptin on the hypothalamic structures involved in metabolic regulation.[143] The window of leptin treatment in the rodent correlates with the well-known leptin surge during the first 2 weeks of life, with alteration in the maternal nutritional environment shown to alter the timing and duration of the leptin surge, with LP offspring showing a delayed leptin surge[145] and offspring of obese rats showing a prolonged and amplified leptin surge.[146] The effects of leptin treatment are sex-specific and directionally dependent upon the nutritional background of the mother and there was a trend for leptin treatment to induce an adverse metabolic phenotype in male offspring of normal pregnancies.[142] Of note, treatment with a leptin antagonist in offspring of obese mothers partially prevented the development of the programmed phenotype.[147] Translation to the human setting remains unclear as the presence or absence of a leptin surge in the human is not well defined and many of the developmental effects that occur in the early postnatal period in the rodent occur *in utero* in larger model species.

One of the earliest nutritional intervention strategies reported was that of taurine, a "conditionally essential" amino acid that has been shown to have beneficial effects across a range of programming models. including both under- and overnutrition. In the maternal LP model, β-cell mass is reduced at birth, with lasting consequences for insulin action throughout life; supplementing mothers with taurine resulted in normal pancreatic function, including normalised vascularisation of the endocrine pancreas and restoration of normal insulin release from fetal islets.[148,149]

More recent studies in the setting of a maternal obesogenic diet have also shown beneficial effects in rat mothers and offspring, including prevention of hyperinsulinaemia in mothers and beneficial effects on lipid metabolism in neonates. However, as with leptin treatment, these effects were directionally dependent upon maternal nutritional status, with adverse effects of taurine (increased neonatal mortality) observed in offspring of normal pregnancies.[150,151]

Maternal methyl donor supplementation has also been shown to be protective in the setting of maternal LP diets. Maternal choline supplementation normalised fat mass and blood pressure in offspring and folic acid supplementation to LP-fed mothers normalised the hypomethylation status of the glucocorticoid and PPAR-α receptors in offspring, although the effects on postnatal outcomes were not examined. The role of omega-3s has also been shown, with fish oil supplementation to rat mothers fed an HFD during pregnancy preventing the development of impaired insulin sensitivity in male offspring.[131]

Maternal supplementation with conjugated linoleic acid (CLA) has also been shown to confer beneficial effects, with bodyweight, fat deposition, insulin sensitivity, gut inflammatory profiles and adipocyte size all normalised in offspring of obese mothers following maternal CLA supplementation.[132,133,152] There are also a range of reports on the benefits of supplementation with the polyphenol resveratrol in the setting of both maternal undernutrition and maternal obesity.[134] In the maternal LP model, resveratrol supplementation partially prevented oxidative stress in both mother and fetus and ameliorated LP diet-induced increases in circulating insulin and leptin, albeit in a sex-specific manner.[136] In the setting of maternal obesity, treatment of male rat offspring with resveratrol rescued hyperleptinaemia and improved hypothalamic leptin signaling.[135]

Manipulation of the somatotropic axis has been highlighted, with GH and IGF-1 treatment showing reversal effects in the setting of nutrition programming.[129] Pre-weaning GH treatment in the rat following maternal undernutrition normalised most of the adverse programming effects observed in adult offspring, including prevention of adipocyte hypertrophy, insulin resistance in adipose tissue and inflammation.[41] Treatment of adult rat offspring of undernourished mothers with IGF-1 resulted in a complete normalisation of adiposity, appetite regulation, insulin sensitivity and leptin concentrations. Data derived from clinical observations, albeit limited, has suggested that treatment with GH to children with GH deficiency may increase the later risk for development of T2DM, but this may reflect an acceleration of the disorder in predisposed individuals.[153]

Work by Simmons *et al.* has demonstrated that neonatal treatment with the glucagon-like peptide 1 (GLP-1) analogue, Exendin-4 (Ex-4) can prevent the progressive reduction in β-cell mass that is observed in rat offspring following *in utero* growth restriction and normalises the expression of PDX-1, a critical mediator of pancreatic development and islet differentiation.[107,154,155] However, with the exception of small studies of bariatric surgery between pregnancies in women with obesity, there is a paucity of controlled intervention studies in which maternal obesity is reversed and the longer-term consequences for offspring health are investigated. However, weight loss in obese women prior to pregnancy does lead to offspring who have a reduced risk of developing obesity in later life, along with improvements in a range of risk markers for metabolic disease.[156] Similar observations were made in rats, whereby dietary intervention prior to conception was able to reverse metabolic programming in obese rats, including normalisation of leptin, insulin and lipid profiles in male offspring.[157]

2.6 Conclusions

Alteration in early nutrition is a fundamental early exposure that can influence the development and subsequent function of essentially all body systems, including programming of metabolism and energy balance, development of the immune system, neurodevelopment and a range of other physiological processes.[158] Epidemiological observations and experimental animal models have clearly highlighted that risk for developing obesity and T2DM in later life is increased when development during early life has been adversely affected.[127] However, despite a considerable number of studies examining the link between maternal nutrition and later-life risk for obesity and T2DM, how the *in utero* and neonatal environment is (mal)programmed to induce persistent changes in the structure and function of β-cells and adipocytes remains largely undefined at a mechanistic level.

Defining the mechanisms underpinning the programming of aberrant developmental traits is an area requiring urgent attention as it offers significant opportunities for the prevention of disease during early development rather than (largely ineffectual) treatment of disease once manifest in adult life. Such research is particularly relevant to populations that are undergoing nutritional transitions between traditional and Western lifestyles and the potential impact of inappropriate PARs in these populations with a changing nutritional environment. Although initial research in the DOHaD area linking early developmental influences with obesity and diabetes risk in later life focused on the effects of fetal/neonatal undernutrition, increasing evidence suggests that exposure to maternal obesity similarly leads to an increased risk of disease in offspring. Of note, a wide range of nutritional insults in the early-life period (*e.g.* total caloric restriction, protein restricted, obesogenic) produce similar detrimental consequences in adult life, suggestive of a common mechanism underlying the early-life programming of later-life disease.

Mechanistic insights support causal effects of altered maternal nutrition on offspring that are mediated through changes in epigenetic processes, although more work is required as many studies to date remain associative. It is clear from work to date that early methyl donor malnutrition can lead to premature epigenetic ageing, thereby conferring an increased susceptibility to obesity and T2DM in later life.[100] More recent work on miRNAs also suggests that an altered maternal nutritional environment leads to changes that may underlie β-cell dysfunction in offspring.[119] The role of paternal health is also gaining increasing attention. Work by Morris *et al.* showed that a chronic HFD programmed β-cell dysfunction in female rat offspring,[159] while in the human, paternal obesity has been associated with IGF-2 hypomethylation in newborns.[160] Of note, significant and independent associations observed between paternal obesity and methylation status of offspring have highlighted the potential impact of reprogramming of imprint marks during spermatogenesis and the susceptibility of the developing sperm to environmental insults.

Targeted re-programming strategies, including dietary or pharmacologic approaches in experimental animal models, will contribute to new intervention modalities which can be translated to application in humans.[161] That a number of intervention approaches in experimental animal models (leptin, methyl donors, GH, EX-4, taurine, *etc.*) during key early periods of development can induce marked reversal effects in offspring across a wide range of experimental models serves to highlight the importance of

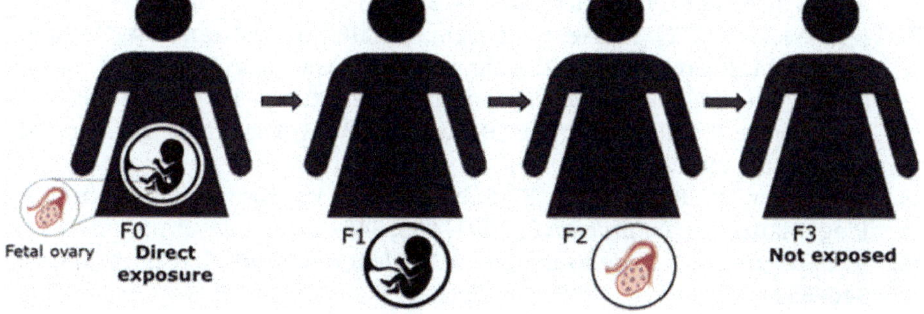

Figure 2.3 The effects of a single environmental exposure can be transmitted across generations. An adverse maternal environment (F0) affects not only the development of the fetus (F1), but can also affect the germ cells which form the F2 generation.

the early-life developmental window as a key target for early preventive strategies.

A limitation of studies to date has been interrogation of sex-specific effects, as many studies have focused on one sex while there is clearly evidence for sexually dimorphic responses to early-life programming stimuli. This will be hopefully addressed *via* a recent National Institutes of Health initiative necessitating the evaluation of outcomes in both sexes[162] and adherence to experimental guidelines such as ARRIVE for balanced experimental design.[163] The consistency of findings across epidemiological studies and the wide range of experimental animal models and species further highlight the need for a better understanding of the mechanisms by which the risk for obesity and T2DM is developmentally programmed *via* early-life exposures. Further, given transgenerational impacts of such programming (Figure 2.3),[6,8] understanding such processes is key to breaking the cycle of disease and development of effective and safe intervention strategies to reduce disease risk for future generations.

References

1. J. L. Tarry-Adkins and S. E. Ozanne, *Proc. Nutr. Soc.*, 2014, 1–13.
2. K. Suzuki, *J. Dev. Origins Health Dis.*, 2018, **9**, 266–269.
3. C. N. Hales and D. J. Barker, *Br. Med. Bull.*, 2001, **60**, 5–20.
4. P. Bateson, P. Gluckman and M. Hanson, *J. Physiol.*, 2014, **592**, 2357–2368.
5. P. D. Gluckman, M. A. Hanson and H. G. Spencer, *Trends Ecol. Evol.*, 2005, **20**, 527–533.
6. C. E. Aiken and S. E. Ozanne, *Hum. Reprod. Update*, 2014, **20**, 63–75.
7. C. E. Aiken, J. L. Tarry-Adkins and S. E. Ozanne, *Mamm. Genome*, 2016, **27**, 430–439.
8. M. H. Vickers, *Ann. Nutr. Metab.*, 2014, **64**, 26–34.
9. M. H. Vickers, *Nutrients*, 2014, **6**, 2165–2178.
10. M. H. Vickers, B. H. Breier, W. S. Cutfield, P. L. Hofman and P. D. Gluckman, *Am. J. Physiol.: Endocrinol. Metab.*, 2000, **279**, E83–E87.
11. A. M. Vaiserman, *Nutrients*, 2017, **9**, 236–252.
12. J. H. Nielsen, T. N. Haase, C. Jaksch, A. Nalla, B. Sostrup, A. A. Nalla, L. Larsen, M. Rasmussen, L. T. Dalgaard, L. W. Gaarn, P. Thams, H. Kofod and N. Billestrup, *Acta Obstet. Gynecol. Scand.*, 2014, **93**, 1109–1122.
13. S. R. Thorn, P. J. Rozance, L. D. Brown and W. W. Hay, Jr., *Semin. Reprod. Med.*, 2011, **29**, 225–236.
14. T. Harder, E. Rodekamp, K. Schellong, J. W. Dudenhausen and A. Plagemann, *Am. J. Epidemiol.*, 2007, **165**, 849–857.
15. R. C. Painter, T. J. Roseboom and O. P. Bleker, *Reprod. Toxicol.*, 2005, **20**, 345–352.
16. A. C. Ravelli, J. H. van der Meulen, R. P. Michels, C. Osmond, D. J. Barker, C. N. Hales and O. P. Bleker, *Lancet*, 1998, **351**, 173–177.
17. Y. Stefan, S. Grasso, A. Perrelet and L. Orci, *Diabetes*, 1983, **32**, 293–301.
18. G. P. Ravelli, Z. A. Stein and M. W. Susser, *N. Engl. J. Med.*, 1976, **295**, 349–353.

19. A. C. Ravelli, J. H. van Der Meulen, C. Osmond, D. J. Barker and O. P. Bleker, *Am. J. Clin. Nutr.*, 1999, **70**, 811–816.
20. F. Lussana, R. C. Painter, M. C. Ocke, H. R. Buller, P. M. Bossuyt and T. J. Roseboom, *Am. J. Clin. Nutr.*, 2008, **88**, 1648–1652.
21. Z. Wang, Z. Zou, Z. Yang, Y. Dong, J. Song, B. Dong, J. Ma and L. Arnold, *BMC Public Health*, 2018, **18**, 1205–1212.
22. N. Wang, J. Cheng, B. Han, Q. Li, Y. Chen, F. Xia, B. Jiang, M. D. Jensen and Y. Lu, *Diabetologia*, 2017, **60**, 262–269.
23. L. Liu, Z. C. Pang, J. P. Sun, B. Xue, S. J. Wang, F. Ning and Q. Qiao, *Nutr. Metab. Cardiovasc. Dis.*, 2017, **27**, 154–160.
24. J. Zhou, L. Zhang, P. Xuan, Y. Fan, L. Yang, C. Hu, Q. Bo, G. Wang, J. Sheng and S. Wang, *PLoS One*, 2018, **13**, e0192212.
25. C. Li and L. H. Lumey, *Int. J. Epidemiol.*, 2017, **46**, 1157–1170.
26. K. M. Godfrey, R. M. Reynolds, S. L. Prescott, M. Nyirenda, V. W. Jaddoe, J. G. Eriksson and B. F. Broekman, *Lancet Diabetes Endocrinol.*, 2017, **5**, 53–64.
27. J. G. Eriksson, S. Sandboge, M. Salonen, E. Kajantie and C. Osmond, *Ann. Med.*, 2015, **47**, 94–99.
28. H. Hochner, Y. Friedlander, R. Calderon-Margalit, V. Meiner, Y. Sagy, M. Avgil-Tsadok, A. Burger, B. Savitsky, D. S. Siscovick and O. Manor, *Circulation*, 2012, **125**, 1381–1389.
29. K. M. Flegal, M. D. Carroll, B. K. Kit and C. L. Ogden, *JAMA, J. Am. Med. Assoc.*, 2012, **307**, 491–497.
30. T. Bianco-Miotto, J. M. Craig, Y. P. Gasser, S. J. van Dijk and S. E. Ozanne, *J. Dev. Origins Health Dis.*, 2017, **8**, 513–519.
31. S. A. Bayol, B. H. Simbi, R. C. Fowkes and N. C. Stickland, *Endocrinology*, 2010, **151**, 1451–1461.
32. Z. Y. Ong and B. S. Muhlhausler, *FASEB J.*, 2011, **25**, 2167–79
33. B. P. Sampey, A. M. Vanhoose, H. M. Winfield, A. J. Freemerman, M. J. Muehlbauer, P. T. Fueger, C. B. Newgard and L. Makowski, *Obesity (Silver Spring)*, 2011, **19**, 1109–1117.
34. O. Dumortier, B. Blondeau, B. Duvillie, B. Reusens, B. Breant and C. Remacle, *Diabetologia*, 2007, **50**, 2495–2503.
35. J. Petrik, B. Reusens, E. Arany, C. Remacle, C. Coelho, J. J. Hoet and D. J. Hill, *Endocrinology*, 1999, **140**, 4861–4873.
36. S. M. Woodall, B. H. Breier, B. M. Johnston and P. D. Gluckman, *J. Endocrinol.*, 1996, **150**, 231–242.
37. M. H. Vickers, S. Reddy, B. A. Ikenasio and B. H. Breier, *J. Endocrinol.*, 2001, **170**, 323–332.
38. M. H. Vickers, B. H. Breier, D. McCarthy and P. D. Gluckman, *Am. J. Physiol.: Regul. Integr. Comp. Physiol.*, 2003, **285**, R271–R273.
39. S. Lecoutre, L. Marousez, A. Drougard, C. Knauf, C. Guinez, D. Eberle, C. Laborie, D. Vieau, J. Lesage and C. Breton, *J. Dev. Origins Health Dis.*, 2017, **8**, 3–7.
40. E. Lizarraga-Mollinedo, E. Fernandez-Millan, M. Garcia-San Frutos, J. de Toro-Martin, T. Fernandez-Agullo, M. Ros, C. Alvarez and F. Escriva, *J. Biol. Chem.*, 2015, **290**, 19353–19366.

41. C. M. Reynolds, M. Li, C. Gray and M. H. Vickers, *Endocrinology*, 2013, **154**, 2676–2686.
42. G. J. Howie, D. M. Sloboda and M. H. Vickers, *Br. J. Nutr.*, 2012, **108**, 298–307.
43. M. Thamotharan, B. C. Shin, D. T. Suddirikku, S. Thamotharan, M. Garg and S. U. Devaskar, *Am. J. Physiol.: Endocrinol. Metab.*, 2005, **288**, E935–E947.
44. A. V. Matveyenko, I. Singh, B. C. Shin, S. Georgia and S. U. Devaskar, *Endocrinology*, 2010, **151**, 5647–5656.
45. A. Snoeck, C. Remacle, B. Reusens and J. J. Hoet, *Biol. Neonate*, 1990, **57**, 107–118.
46. D. S. Fernandez-Twinn and S. E. Ozanne, *Ann. N.Y. Acad. Sci.*, 2010, **1212**, 78–96.
47. V. V. Bol, B. M. Reusens and C. A. Remacle, *Obesity (Silver Spring)*, 2008, **16**, 2760–2763.
48. V. V. Bol, A. I. Delattre, B. Reusens, M. Raes and C. Remacle, *Am. J. Physiol.: Regul. Integr. Comp. Physiol.*, 2009, **297**, R291–R299.
49. S. E. Ozanne, M. W. Dorling, C. L. Wang and B. T. Nave, *Am. J. Physiol.: Endocrinol. Metab.*, 2001, **280**, E534–E539.
50. B. Blondeau, A. Garofano, P. Czernichow and B. Breant, *Endocrinology*, 1999, **140**, 4208–4213.
51. A. Garofano, P. Czernichow and B. Breant, *Diabetologia*, 1999, **42**, 711–718.
52. G. M. Sutton, A. V. Centanni and A. A. Butler, *Endocrinology*, 2010, **151**, 1570–1580.
53. M. Desai, D. Gayle, J. Babu and M. G. Ross, *Am. J. Physiol.: Regul. Integr. Comp. Physiol.*, 2005, **288**, R91–R96.
54. G. J. Howie, D. M. Sloboda, T. Kamal and M. H. Vickers, *J. Physiol.*, 2009, **587**, 905–915.
55. S. A. Segovia, M. H. Vickers and C. M. Reynolds, *J. Dev. Origins Health Dis.*, 2017, **8**, 529–540.
56. S. Rajia, H. Chen and M. J. Morris, *J. Neuroendocrinol*, 2010, **22**, 905–914.
57. P. J. Mark, C. Sisala, K. Connor, R. Patel, J. L. Lewis, M. H. Vickers, B. J. Waddell and D. M. Sloboda, *J. Dev. Origins Health Dis.*, 2011, **2**, 63–70.
58. A. M. Samuelsson, P. A. Matthews, M. Argenton, M. R. Christie, J. M. McConnell, E. H. Jansen, A. H. Piersma, S. E. Ozanne, D. F. Twinn, C. Remacle, A. Rowlerson, L. Poston and P. D. Taylor, *Hypertension*, 2008, **51**, 383–392.
59. G. J. Howie, D. M. Sloboda, C. M. Reynolds and M. H. Vickers, *J. Nutr. Metab.*, 2013, **2013**, 517384.
60. S. Lecoutre, F. Oger, C. Pourpe, L. Butruille, L. Marousez, A. Dickes-Coopman, C. Laborie, C. Guinez, J. Lesage, D. Vieau, C. Junien, D. Eberle, A. Gabory, J. Eeckhoute and C. Breton, *Mol. Metab.*, 2017, **6**, 922–930.
61. M. E. Cerf, *Pediatr. Diabetes*, 2015, **16**, 71–78.
62. M. E. Cerf and J. Louw, *J. Pancreas*, 2014, **15**, 228–236.
63. M. E. Cerf, J. Louw and E. Herrera, *Nutrients*, 2015, **7**, 7231–7241.

64. R. E. Cardenas-Perez, L. Fuentes-Mera, A. L. de la Garza, I. Torre-Villalvazo, L. A. Reyes-Castro, H. Rodriguez-Rocha, A. Garcia-Garcia, J. C. Corona-Castillo, A. R. Tovar, E. Zambrano, R. Ortiz-Lopez, J. Saville, M. Fuller and A. Camacho, *Nutr. Metab. (Lond.)*, 2018, **15**, 38.
65. L. T. Nguyen, H. Chen, A. Zaky, C. Pollock and S. Saad, *J. Physiol.*, 2019, **597**, 467–480.
66. M. Via, *ISRN Endocrinol.*, 2012, **2012**, 103472.
67. L. Gambling, S. Dunford, D. I. Wallace, G. Zuur, N. Solanky, S. K. Srai and H. J. McArdle, *J. Physiol.*, 2003, **552**, 603–610.
68. R. M. Lewis, C. J. Petry, S. E. Ozanne and C. N. Hales, *Metabolism*, 2001, **50**, 562–567.
69. L. Gambling, H. S. Andersen, A. Czopek, R. Wojciak, Z. Krejpcio and H. J. McArdle, *J. Physiol.*, 2004, **561**, 195–203.
70. A. K. Phillips, S. C. Roy, R. Lundberg, T. W. Guilbert, A. P. Auger, S. E. Blohowiak, C. L. Coe and P. J. Kling, *J. Perinatol.*, 2014, **34**, 513–518.
71. A. D. Gernand, H. N. Simhan, S. Caritis and L. M. Bodnar, *Obstet. Gynecol.*, 2014, **123**, 40–48.
72. K. von Websky, A. A. Hasan, C. Reichetzeder, O. Tsuprykov and B. Hocher, *J. Steroid Biochem. Mol. Biol.*, 2018, **180**, 51–64.
73. R. G and A. Gupta, *Nutrients*, 2014, **6**, 729–775.
74. L. Venu, N. Harishankar, T. Prasanna Krishna and M. Raghunath, *Diabetologia*, 2004, **47**, 1493–1501.
75. L. Venu, Y. D. Kishore and M. Raghunath, *J. Nutr.*, 2005, **135**, 1353–1358.
76. L. Venu, I. J. Padmavathi, Y. D. Kishore, N. V. Bhanu, K. R. Rao, P. B. Sainath, M. Ganeshan and M. Raghunath, *Obesity (Silver Spring)*, 2008, **16**, 1270–1276.
77. I. J. Padmavathi, K. R. Rao, L. Venu, M. Ganeshan, K. A. Kumar, N. Rao Ch, N. Harishankar, A. Ismail and M. Raghunath, *Diabetes*, 2010, **59**, 98–104.
78. I. J. Padmavathi, K. R. Rao, L. Venu, A. Ismail and M. Raghunath, *Exp. Biol. Med. (Maywood, N.J.)*, 2010, **235**, 349–355.
79. C. Ngaruiya, A. Hayward, L. Post and H. Mowafi, *Pan African Med. J.*, 2017, **28**, 49.
80. A. Gonzalez-Bulnes and P. Chavatte-Palmer, *Curr. Pharm. Biotechnol.*, 2017, **18**, 541–551.
81. S. Lie, J. L. Morrison, O. Williams-Wyss, S. E. Ozanne, S. Zhang, S. K. Walker, D. O. Kleemann, S. M. MacLaughlin, C. T. Roberts and I. C. McMillen, *Am. J. Physiol.: Endocrinol. Metab.*, 2013, **305**, E931–E941.
82. S. Lie, J. L. Morrison, O. Williams-Wyss, C. M. Suter, D. T. Humphreys, S. E. Ozanne, S. Zhang, S. M. Maclaughlin, D. O. Kleemann, S. K. Walker, C. T. Roberts and I. C. McMillen, *Biol. Reprod.*, 2014, **90**, 5.
83. E. L. Donovan, C. E. Hernandez, L. R. Matthews, M. H. Oliver, A. L. Jaquiery, F. H. Bloomfield and J. E. Harding, *J. Dev. Origins Health Dis.*, 2013, **4**, 296–299.
84. P. Rhodes, J. Craigon, C. Gray, S. M. Rhind, P. T. Loughna and D. S. Gardner, *PLoS One*, 2009, **4**, e7393.

85. P. Khanal, S. V. Husted, A. M. Axel, L. Johnsen, K. L. Pedersen, M. S. Mortensen, A. H. Kongsted and M. O. Nielsen, *Acta Physiol. (Oxf.)*, 2014, **210**, 110–126.
86. P. Khanal, L. Johnsen, A. M. Axel, P. W. Hansen, A. H. Kongsted, N. B. Lyckegaard and M. O. Nielsen, *PLoS One*, 2016, **11**, e0156700.
87. D. S. Gardner, K. Tingey, B. W. Van Bon, S. E. Ozanne, V. Wilson, J. Dandrea, D. H. Keisler, T. Stephenson and M. E. Symonds, *Am. J. Physiol.: Regul. Integr. Comp. Physiol.*, 2005, **289**, R947–R954.
88. K. R. Poore, L. J. Hollis, R. J. Murray, A. Warlow, A. Brewin, L. Fulford, J. K. Cleal, K. A. Lillycrop, G. C. Burdge, M. A. Hanson and L. R. Green, *PLoS One*, 2014, **9**, e90994.
89. S. P. Ford, L. Zhang, M. Zhu, M. M. Miller, D. T. Smith, B. W. Hess, G. E. Moss, P. W. Nathanielsz and M. J. Nijland, *Am. J. Physiol.: Regul. Integr. Comp. Physiol.*, 2009, **297**, R835–R843.
90. L. Zhang, N. M. Long, S. M. Hein, Y. Ma, P. W. Nathanielsz and S. P. Ford, *Domest. Anim. Endocrinol.*, 2011, **40**, 30–39.
91. L. A. George, A. B. Uthlaut, N. M. Long, L. Zhang, Y. Ma, D. T. Smith, P. W. Nathanielsz and S. P. Ford, *Reprod. Biol. Endocrinol.*, 2010, **8**, 75.
92. V. H. J. Roberts, J. O. Lo, K. S. Lewandowski, P. Blundell, K. L. Grove, C. D. Kroenke, E. L. Sullivan, C. T. Roberts, Jr. and A. E. Frias, *Reprod. Sci.*, 2018, **25**, 110–119.
93. D. S. Lewis, H. A. Bertrand, C. A. McMahan, H. C. McGill, Jr., K. D. Carey and E. J. Masoro, *J. Clin. Invest.*, 1986, **78**, 899–905.
94. C. E. McCurdy, J. M. Bishop, S. M. Williams, B. E. Grayson, M. S. Smith, J. E. Friedman and K. L. Grove, *J. Clin. Invest.*, 2009, **119**, 323–335.
95. K. D. Bruce, F. R. Cagampang, M. Argenton, J. Zhang, P. L. Ethirajan, G. C. Burdge, A. C. Bateman, G. F. Clough, L. Poston, M. A. Hanson, J. M. McConnell and C. D. Byrne, *Hepatology*, 2009, **50**, 1796–1808.
96. M. Li, C. M. Reynolds, S. A. Segovia, C. Gray and M. H. Vickers, *Biomed. Res. Int.*, 2015, **2015**, 437107.
97. K. M. Godfrey, P. M. Costello and K. A. Lillycrop, *J. Dev. Origins Health Dis.*, 2015, **6**, 399–406.
98. A. J. Silva and R. White, *Cell*, 1988, **54**, 145–152.
99. K. M. Godfrey, A. Sheppard, P. D. Gluckman, K. A. Lillycrop, G. C. Burdge, C. McLean, J. Rodford, J. L. Slater-Jefferies, E. Garratt, S. R. Crozier, B. S. Emerald, C. R. Gale, H. M. Inskip, C. Cooper and M. A. Hanson, *Diabetes*, 2011, **60**, 1528–1534.
100. R. A. Waterland and R. L. Jirtle, *Nutrition*, 2004, **20**, 63–68.
101. K. Vanhees, I. G. Vonhogen, F. J. van Schooten and R. W. Godschalk, *Cell Mol. Life Sci.*, 2014, **71**, 271–285.
102. C. Gicquel, A. El-Osta and Y. Le Bouc, *Best Pract. Res. Clin. Endocrinol. Metab.*, 2008, **22**, 1–16.
103. N. K. MacLennan, S. J. James, S. Melnyk, A. Piroozi, S. Jernigan, J. L. Hsu, S. M. Janke, T. D. Pham and R. H. Lane, *Physiol. Genomics*, 2004, **18**, 43–50.

104. J. A. McKay, A. Groom, C. Potter, L. J. Coneyworth, D. Ford, J. C. Mathers and C. L. Relton, *PLoS One*, 2012, **7**, e33290.
105. R. A. Simmons, *Rev. Endocr. Metab. Disord.*, 2007, **8**, 95–104.
106. K. A. Lillycrop, E. S. Phillips, A. A. Jackson, M. A. Hanson and G. C. Burdge, *J. Nutr.*, 2005, **135**, 1382–1386.
107. J. H. Park, D. A. Stoffers, R. D. Nicholls and R. A. Simmons, *J. Clin. Invest.*, 2008, **118**, 2316–2324.
108. S. E. Pinney and R. A. Simmons, *Trends Endocrinol. Metab.*, 2010, **21**, 223–229.
109. C. Jousse, L. Parry, S. Lambert-Langlais, A. C. Maurin, J. Averous, A. Bruhat, V. Carraro, J. Tost, P. Letteron, P. Chen, R. Jockers, J. M. Launay, J. Mallet and P. Fafournoux, *FASEB J.*, 2011, **25**, 3271–3278.
110. S. J. Borengasser, Y. Zhong, P. Kang, F. Lindsey, M. J. Ronis, T. M. Badger, H. Gomez-Acevedo and K. Shankar, *Endocrinology*, 2013, **154**, 4113–4125.
111. H. W. Liu, S. Mahmood, M. Srinivasan, D. J. Smiraglia and M. S. Patel, *J. Nutr. Biochem.*, 2013, **24**, 1859–1869.
112. Q. Y. Yang, J. F. Liang, C. J. Rogers, J. X. Zhao, M. J. Zhu and M. Du, *Diabetes*, 2013, **62**, 3727–3735.
113. C. Lesseur, D. A. Armstrong, A. G. Paquette, D. C. Koestler, J. F. Padbury and C. J. Marsit, *Mol. Cell Endocrinol.*, 2013, **381**, 160–167.
114. C. Lesseur, D. A. Armstrong, A. G. Paquette, Z. Li, J. F. Padbury and C. J. Marsit, *Am. J. Obstet. Gynecol.*, 2014, **211**, 654 e651–e659.
115. C. R. Quilter, W. N. Cooper, K. M. Cliffe, B. M. Skinner, P. M. Prentice, L. Nelson, J. Bauer, K. K. Ong, M. Constancia, W. L. Lowe, N. A. Affara and D. B. Dunger, *FASEB J.*, 2014, **28**, 4868–4879.
116. D. N. Tosh, Q. Fu, C. W. Callaway, R. A. McKnight, I. C. McMillen, M. G. Ross, R. H. Lane and M. Desai, *Am. J. Physiol.: Gastrointest. Liver Physiol.*, 2010, **299**, G1023–G1029.
117. A. P. Wolffe, *J. Exp. Zool.*, 1998, **282**, 239–244.
118. C. Gray, M. Li, C. M. Reynolds and M. H. Vickers, *Endocrinology*, 2014, **155**, 4808–4817.
119. E. U. Alejandro, B. Gregg, T. Wallen, D. Kumusoglu, D. Meister, A. Chen, M. J. Merrins, L. S. Satin, M. Liu, P. Arvan and E. Bernal-Mizrachi, *J. Clin. Invest.*, 2014, **124**, 4395–4410.
120. C. Guay and R. Regazzi, *Diabetologia*, 2014, **58**, 456–463.
121. C. Guay, E. Roggli, V. Nesca, C. Jacovetti and R. Regazzi, *Transl. Res.*, 2011, **157**, 253–264.
122. V. Nesca, C. Guay, C. Jacovetti, V. Menoud, M. L. Peyot, D. R. Laybutt, M. Prentki and R. Regazzi, *Diabetologia*, 2013, **56**, 2203–2212.
123. K. A. Glendining and C. L. Jasoni, *Int. J. Mol. Sci.*, 2019, **20**(2), 329.
124. Q. Zhang, X. Xiao, J. Zheng, M. Li, M. Yu, F. Ping, Z. Wang, C. Qi, T. Wang and X. Wang, *Exp. Biol. Med. (Maywood)*, 2017, **242**, 1444–1452.
125. A. Mendez-Mancilla, V. Lima-Rogel, J. C. Toro-Ortiz, F. Escalante-Padron, A. E. Monsivais-Urenda, D. E. Noyola and M. Salgado-Bustamante, *Pediatr Obes.*, 2018, **13**, 168–174.

126. C. H. D. Fall and K. Kumaran, *Philos. Trans. R. Soc., B: Biol. Sci.*, 2019, **374**, 20180123.
127. M. H. Vickers and D. M. Sloboda, *Front. Physiol.*, 2012, **3**, 242.
128. M. Li, C. M. Reynolds, C. Gray and M. H. Vickers, *Endocrinology*, 2015, **156**, 3228–3238.
129. C. M. Reynolds, J. K. Perry and M. H. Vickers, *Int. J. Mol. Sci.*, 2017, **18**(8), 1729.
130. M. H. Vickers, P. D. Gluckman, A. H. Coveny, P. L. Hofman, W. S. Cutfield, A. Gertler, B. H. Breier and M. Harris, *Endocrinology*, 2005, **146**, 4211–4216.
131. B. B. Albert, M. H. Vickers, C. Gray, C. M. Reynolds, S. A. Segovia, J. G. B. Derraik, M. L. Garg, D. Cameron-Smith, P. L. Hofman and W. S. Cutfield, *Sci. Rep.*, 2017, **7**, 5595.
132. C. M. Reynolds, S. A. Segovia, X. D. Zhang, C. Gray and M. H. Vickers, *Biol. Reprod.*, 2015, **92**, 40.
133. C. M. Reynolds, S. A. Segovia, X. D. Zhang, C. Gray and M. H. Vickers, *Physiol. Rep.*, 2015, **3**(10), e12588.
134. Y. L. Tain and C. N. Hsu, *Int. J. Mol. Sci.*, 2018, **19**(9), 2584.
135. J. G. Franco, C. P. Dias-Rocha, T. P. Fernandes, L. Albuquerque Maia, P. C. Lisboa, E. G. Moura, C. C. Pazos-Moura and I. H. Trevenzoli, *Eur. J. Nutr.*, 2016, **55**, 601–610.
136. C. C. Vega, L. A. Reyes-Castro, G. L. Rodriguez-Gonzalez, C. J. Bautista, M. Vazquez-Martinez, F. Larrea, G. A. Chamorro-Cevallos, P. W. Nathanielsz and E. Zambrano, *J. Physiol.*, 2016, **594**, 1483–1499.
137. S. Bai, D. Briggs and M. H. Vickers, *J. Dev. Origins Health Dis.*, 2012, **3**, 1–8.
138. Y. L. Tain, K. L. H. Wu, W. C. Lee, S. Leu and J. Y. H. Chan, *Int. J. Mol. Sci.*, 2018, **19**(4), 1066.
139. M. H. Vickers, P. L. Hofman, P. D. Gluckman, P. E. Lobie and W. S. Cutfield, *Am. J. Physiol.: Endocrinol. Metab.*, 2006, **291**, E1212–E1219.
140. T. L. Horvath and J. C. Bruning, *Nat. Med.*, 2006, **12**, 52–53; discussion 53.
141. S. G. Bouret, S. J. Draper and R. B. Simerly, *Science*, 2004, **304**, 108–110.
142. M. H. Vickers, P. D. Gluckman, A. H. Coveny, P. L. Hofman, W. S. Cutfield, A. Gertler, B. H. Breier and M. Harris, *Endocrinology*, 2008, **149**, 8.
143. L. Attig, J. Djiane, A. Gertler, O. Rampin, T. Larcher, S. Boukthir, P. M. Anton, J. Y. Madec, I. Gourdou and L. Abdennebi-Najar, *Am. J. Physiol.: Endocrinol. Metab.*, 2008, **295**, E1117–E1125.
144. M. H. Vickers and D. M. Sloboda, *Best Pract. Res., Clin. Endocrinol. Metab.*, 2012, **26**, 677–687.
145. C. J. Bautista, L. Boeck, F. Larrea, P. W. Nathanielsz and E. Zambrano, *Pediatr. Res.*, 2008, **63**, 358–363.
146. S. L. Kirk, A. M. Samuelsson, M. Argenton, H. Dhonye, T. Kalamatianos, L. Poston, P. D. Taylor and C. W. Coen, *PLoS One*, 2009, **4**, e5870.
147. J. Beltrand, D. Sloboda, K. Connor, M. Truong and M. Vickers, *J. Nutr. Metab.*, 2012, **2012**, 296935.

148. S. Boujendar, E. Arany, D. Hill, C. Remacle and B. Reusens, *J. Nutr.*, 2003, **133**, 2820–2825.
149. H. Cherif, B. Reusens, M. T. Ahn, J. J. Hoet and C. Remacle, *J. Endocrinol.*, 1998, **159**, 341–348.
150. M. Li, C. M. Reynolds, D. M. Sloboda, C. Gray and M. H. Vickers, *PLoS One*, 2013, **8**, e76961.
151. M. Li, C. M. Reynolds, D. M. Sloboda, C. Gray and M. H. Vickers, *J. Nutr. Biochem.*, 2015, **26**, 267–276.
152. S. A. Segovia, M. H. Vickers, C. Gray, X. D. Zhang and C. M. Reynolds, *Sci. Rep.*, 2017, **7**, 6663.
153. W. S. Cutfield, P. Wilton, H. Bennmarker, K. Albertsson-Wikland, P. Chatelain, M. B. Ranke and D. A. Price, *Lancet*, 2000, **355**, 610–613.
154. S. E. Pinney, L. J. Jaeckle Santos, Y. Han, D. A. Stoffers and R. A. Simmons, *Diabetologia*, 2011, **54**, 2606–2614.
155. D. A. Stoffers, B. M. Desai, D. D. DeLeon and R. A. Simmons, *Diabetes*, 2003, **52**, 734–740.
156. J. Smith, K. Cianflone, S. Biron, F. S. Hould, S. Lebel, S. Marceau, O. Lescelleur, L. Biertho, S. Simard, J. G. Kral and P. Marceau, *J. Clin. Endocrinol. Metab.*, 2009, **94**, 4275–4283.
157. E. Zambrano, P. M. Martinez-Samayoa, G. L. Rodriguez-Gonzalez and P. W. Nathanielsz, *J. Physiol.*, 2010, **588**, 1791–1799.
158. I. Cetin, K. Buhling, C. Demir, A. Kortam, S. L. Prescott, Y. Yamashiro, M. Yarmolinskaya and B. Koletzko, *Ann. Nutr. Metab.*, 2019, **74**, 269–278.
159. S. F. Ng, R. C. Lin, D. R. Laybutt, R. Barres, J. A. Owens and M. J. Morris, *Nature*, 2010, **467**, 963–966.
160. A. Soubry, J. M. Schildkraut, A. Murtha, F. Wang, Z. Huang, A. Bernal, J. Kurtzberg, R. L. Jirtle, S. K. Murphy and C. Hoyo, *BMC Med.*, 2013, **11**, 29.
161. E. Nusken, J. Dotsch, L. T. Weber and K. D. Nusken, *Front. Pediatr.*, 2018, **6**, 36.
162. J. A. Clayton and F. S. Collins, *Nature*, 2014, **509**, 282–283.
163. C. Kilkenny, W. J. Browne, I. C. Cuthill, M. Emerson and D. G. Altman, *Osteoarthritis Cartilage*, 2012, **20**, 256–260.

CHAPTER 3

Lessons Learned from Recent Clinical Trials Targeting Diets to Treat and Prevent Obesity and Diabetes

N. S. WRIGHT*

Royal New Zealand College of General Practitioners, 50 Customhouse Quay, Wellington 6011, New Zealand
*Email: nicholas.samuel.wright@gmail.com

3.1 Introduction

Worldwide, obesity and diabetes are increasing in every country.[1,2] Obesity has increased sixfold from 1975 (100 million) to 2016 (671 million),[3,4] and results from chronic net positive energy balance (EB) (due to increased energy intake and/or decreased activity). Average US adult body weight increase is ~1/3 kg per year,[5] equivalent to the daily energy of 1 g fat, or a 5-minute walk.

Additionally, one in 11 adults has diabetes, although only half are diagnosed. Diabetes accounts for 10% of global healthcare costs.[6] Predictions for 2050 estimate 33% of US adults will have Type 2 diabetes (T2D).[7] T2D risk is strongly associated with increased body weight,[8] *e.g.* women with body mass

Nutritional Signaling Pathway Activities in Obesity and Diabetes
Food Chemistry, Function and Analysis No. 24
Edited by Zhiyong Cheng

Published by the Royal Society of Chemistry, www.rsc.org

index (BMI) > 35 compared with normal have a 93-fold risk.[9,10] Weight loss is the most effective treatment for T2D.[11,12]

Both environmental factors and individual differences contribute to EB (Figure 3.1). Isolated factors that influence EB are well researched, yet research-comprehensive models for consistent and sustained weight loss outcomes are lacking. Weight of food consumed tends to remain constant, so approaches to systematically reduce energy density (ED) (*e.g.* increased water or fibre, decreased fat) can assist weight loss and weight loss maintenance.[13–15] See Figure 3.2 for examples of differing ED in foods.

Most people seeking to lose weight attempt to do so more than once, and 80–90% do not sustain losses. Findings from one large cohort who did maintain an average weight loss of 30 kg for 5.5 years showed no other factors that associated this particular attempt with success other than greater commitment, stricter dieting and greater exercise. Weight loss maintainers: (1) undertook

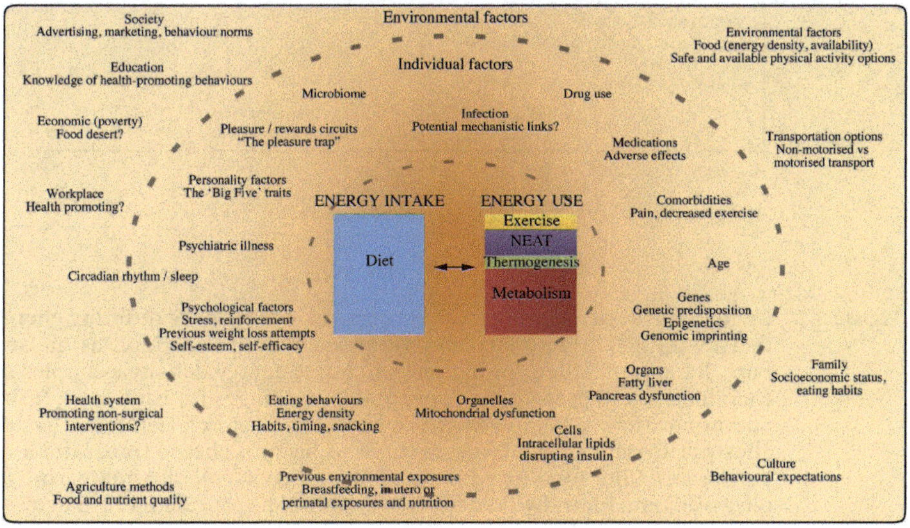

Figure 3.1 Overview of factors contributing to energy balance. This figure outlines many of the key causes of obesity. Factors here have been classified into external environmental factors, and individual factors unique to a person. External factors include wider social, political and systemic factors that impact on health choices. Individual factors include those from all stages of life, along with psychological factors, and medication use. Many factors have a combination of both environmental and individual components, *e.g. in utero* exposures, or the circadian rhythm. Interactions are present between all factors. All factors influence energy balance through either impacting energy intake or energy use. Dietary intake is generally accepted as the key component in regulating energy balance. Although exercise clearly plays a part in energy balance, the total impact of exercise is depicted here relative to the total energy use (approximately 15% of total energy use for an average adult), with the majority of energy use being metabolic processes. NEAT, non-exercise activity thermogenesis.

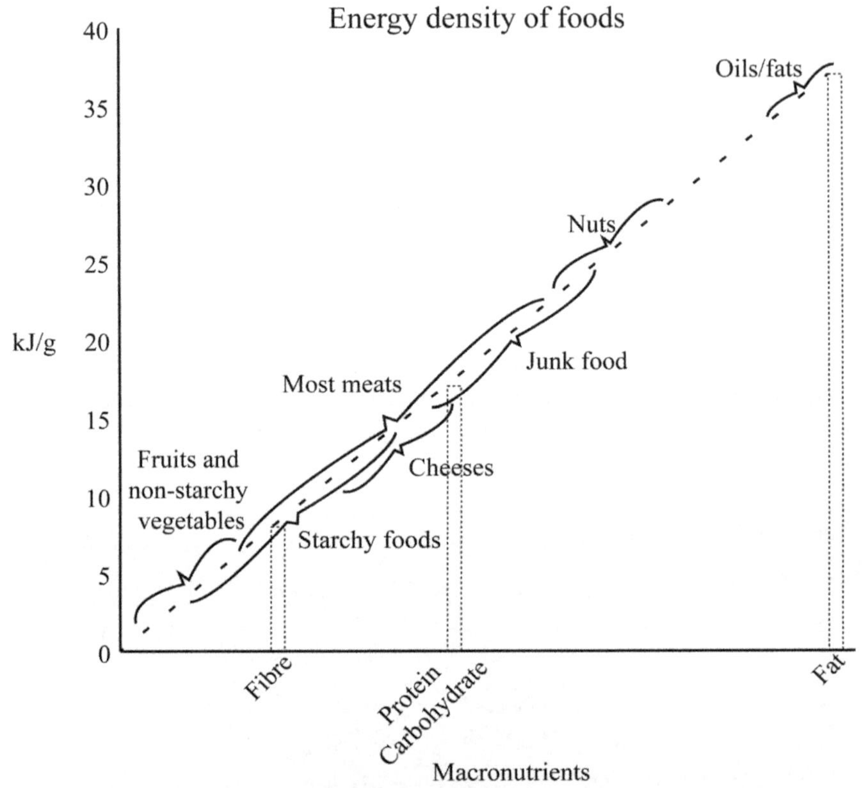

Figure 3.2 Energy density of different foods. Comparison of foods of differing energy density (kJ g^{-1}). Macronutrients are presented on the graph as dashed bars. Ranges provided for higher- and lower-energy density examples of each food group presented. Upper and lower ranges for each food: fruits and non-starchy vegetables (lettuce to avocado), starch-rich foods (water chestnut to white pasta), meats (tuna to bacon), cheese (mozzarella to parmesan), junk food (fried potato chips to cookies), oils (margarine to olive oil). Data for individual foods available at https://fdc.nal.usda.gov/

regular exercise; (2) ate a low-ED, low-fat diet (LFD) (average 24% total energy intake (E) from fat); (3) engaged in regular self-monitoring of food intake and body weight; (4) ate breakfast; and (5) were consistent in their approach in that there were few or no "cheat" days.[16]

3.2 Current Issues in Diabetes and Obesity Research

Diet research is abundant, and yet there is still conflicting data. If one approach is superior, what might this be, and what mechanisms are responsible? Extracting consistent lessons requires an in-depth look at the evidence, especially including real-world applicability. There is repeatedly poor adherence to long-term dietary changes, which is one of the major barriers to effective interventions. Which approach(es) offers better adherence, and why? This chapter aims to provide an overview of recent dietary clinical trials in obesity and diabetes prevention targeting diets, specifically looking at dietary patterns (rather than one food, or meal replacements), and discusses lessons learned from these to provide greater understanding and real-world applicability.

When discussing dietary interventions two primary approaches can be used to classify and dissect the data on nutrition outcomes: *either* patterns of eating (*e.g.* plant-based *vs.* paleolithic *vs.* Mediterranean) which can broadly reflect macronutrient intake *or* individual foods or groups (*e.g.* "effects of walnuts" or the "food groups" as per *Dietary Guidelines for Americans*). Because humans eat combinations of food rather than individual foods, studying individual nutrients does not reflect the complexity or any synergistic effects of the whole diet. The focus in this text is, therefore, on analysing commonly investigated dietary patterns.

In creating a representative sample of the data, a literature search was performed with PubMed for the following dietary approaches (combined using "OR"): Atkins; DASH (Dietary Approaches to Stop Hypertension); HEI (Healthy-Eating Index); high-fat; high-carbohydrate; ketogenic; low-fat; low-carbohydrate; Mediterranean; Nordic; paleolithic; Weight Watchers; vegetarian; vegan; AND diet AND obesity OR diabetes. Filters included: adults, full-text, English, from 2010–2020, AND RCTs, systematic reviews, or meta-analyses. High-protein was not used in a separate category due to the assumption these dietary patterns would be generally low in carbohydrates. Search was supplemented with Google and references review. Due to the sheer volume of literature, this search was intended to be representative rather than exhaustive, with results summarised in Table 3.1.

3.3 Results from Reviews on Obesity and Diabetes

The included review data from different diets broadly fell into either low-carbohydrate diet (LCD) or LFD. The reviews showed short-term (<1 year) improvements tended towards greater weight loss and improvements with diabetes markers utilising a LCD. Long-term (≥1 year) weight loss outcomes of any diet intervention compared to a control group ranged from 0 to 11 kg. There were strong consistent findings of no clear superiority of one diet pattern over another in either diabetes control or weight loss by 1 year, although there were less high-quality data for LCDs.

3.3.1 Limitations of Interpreting Review Data

Systematic reviews are considered a gold-standard measure; however, caution is needed when combining studies. Precisely categorising high- or low-fat diets for pooled data analysis presents difficulties due to differing definitions. An LFD can range from <30% E from fat or could be a no-added-oil whole food plant-based (WFPB) diet with as little as 7% E from fat. Likewise, LCD could be a very-low-carbohydrate ketogenic diet (VLCKD) with typically less than

Table 3.1 Representative sample of systematic reviews and meta-analyses for obesity and diabetes[†]

Obesity		
Objective	**Details**	**Results**
Churuangsuk et al. 2018 Low-carbohydrate diets for overweight and obesity: a systematic review of the systematic reviews[68]	12 reviews were included, which included patients with diabetes and other comorbidities	Several of the included systematic reviews have been discussed below (excluding several rated as very poor-quality evidence. Quality of reviews: 7 of 12 were critically low in quality; 3 moderate; 2 high. High-quality analyses reported little or no difference between diets in terms of weight loss. 3 of 4 meta-analyses supported by commercial organisations were rated critically low in quality and their pooled results showed superiority of LCD over LFD. The authors conclude: "Better quality reviews and RCTs are needed, before recommending low-carbohydrate diets as preferred to other approaches for energy restriction"
Anton et al. 2017 Effects of popular diets without specific calorie targets on weight loss outcomes[69]	16 RCTs, participant numbers not listed. Mean follow-up 12 months (range 3–24). Research included did not explicitly limit calories. Studies with structured physical activity (PA) component were excluded, but PA was allowed otherwise. The researchers examined 16 articles (numbers of studies listed) with evidence on popular diets: Atkins (10 studies); Dietary Approaches to Stop Hypertension (DASH) (1); Glycaemic Index (2); MedDiet (2); Ornish (2); paleolithic (1); and Zone (3)	Long-term weight loss results (in kg, ±95%[‡] CI where provided): Atkins range −2.1 (±4.8) to −11.4 (±3.4); DASH nil data; Glycaemic Index −3; MedDiet −7.4; Ornish range −2.2 to −3.3 (±7.3) (although this excluded Dr Ornish's studies); paleolithic −9.2; Zone range −1.5 to −3.2 (±6). The authors state: "high fat, low-carbohydrate diets are most advantageous for promoting long-term weight loss"
Snorgaard et al. 2017 Systematic review and meta-analysis of dietary carbohydrate restriction in patients with T2D[70]	10 RCTs with 1376 participants. Mean duration 12 months (range 3–24)	In short-term outcomes less than 1 year there was a greater effect of LCD over a standard LFD, and reduction in HbA_{1C} was correlated with the amount of restriction of carbohydrates. No effects on weight loss or HbA_{1C} seen at 1 year

Ma et al. 2017 Effects of weight loss interventions for adults who are obese on mortality, cardiovascular disease, and cancer[71]	54 RCTs, 30 206 participants. Follow-up mean 2.8 years (range 12–152 months). All but one trial included low-fat arm (usually low in saturated fat and < 30% E from fat), one trial used MedDiet, one intervention included ≤50 g day^{-1} carbohydrate, 2 low GI, 4 DASH, 8 US Diabetes Prevention Programme, 4 US Dietary Guidelines; all but 3 included exercise	Low-fat weight-reducing diets may reduce all-cause mortality in obesity. Mean weight difference compared to control changes at: 1 year (44 studies) −3.4 kg (±0.7); 2 years (20 studies) −2.5 (±0.9); 3 years (8 studies) −2.5 (±0.9)
Hooper et al. 2016 Effects of total fat intake on body weight[72]	32 RCTs, 54 000 participants (included children, however) and 25 cohorts. Looked at populations where there was no intention to lose weight but were randomised to a lower-fat diet intervention. Included RCTs >6 months' duration, and cohort studies >12 months' minimum duration	Consistent evidence in adults of decreased weight as a result of lower E from fat, even when not explicitly aiming to lose weight. Weight reductions in the order of −1.5 kg (±0.4) at end of follow-up
Mancini et al. 2016 Systematic review of the MedDiet for long-term weight loss[73]	5 RCTs, 998 participants, with RCTs on the MedDiet ≥12 months' duration (mean 25 months, range 12–48)	MedDiet produced similar weight loss to other comparator diets at ≥12 months, but showed greater weight loss than LFD in 3 RCTs. Glycaemic control and other improvements were seen in those with T2D
Mansoor et al. 2016 Effects of LCD *vs.* LFD on body weight and cardiovascular risk factors[74]	11 RCTs, 1369 participants. Studies included compared LCD with LFD. Inclusion and exclusion criteria made this publication quite specific. LCD was <20% E from carbohydrates, whereas LFDs were more liberal at <30% E from fat. Study mean length 12 months (range 6–24)	Weighted mean difference for LCD *vs.* LFD showed significant weight loss at −2.2 kg (±1.2); LDL cholesterol was increased in the LCD group
Huang et al. 2015 Vegetarian diets and weight reduction: a meta-analysis of randomized controlled trials[75]	12 RCTs, 1151 participants. Median duration 18 weeks. Trials included vegetarian group compared to non-vegetarian diets	In short-term data outcomes (<6 months median), vegetarian diet group (VDG) had significantly greater weight loss than non-VDG: WMD −2.0 kg (±0.8). Vegan subgroups' weight loss was greatest: −2.5 (±0.5), as were those whose diet was energy-restricted. Weight loss was greater under 1 year. >1 year: −1.1 (±0.9)

(*continued*)

Table 3.1 (*Continued*)

Obesity		
Objective	Details	Results
Atallah et al. 2014 Examine efficacy of Atkins, South Beach, Weight Watchers (WW), and Zone diets, focused on sustained (≥12 months) weight loss[76]	12 RCTs, 2559 participants. Mean duration 64 weeks (range 48–104). Ten *versus* usual care (5 of which studied Atkins, 4 WW and 1 South Beach) and 2 head-to-head (1 of Atkins, WW and Zone, and 1 of Atkins, Zone and control).	At 12 months, the 10 RCTs showed only WW had consistent weight reduction, at −5.6 to −6 kg compared to −0.8 to −3.5 kg weight loss for controls. Head-to-head data suggested modest weight loss for Atkins (−2.1 to −4.7 kg), WW (−3 kg), Zone (−1.6 to −3.2 kg) and control (−2.2 kg). No one diet was more beneficial than the others
Johnston et al. 2014 Comparison of weight loss among named diet programs in overweight and obese adults[77]	48 RCTs, 7286 participants. Comparing diet programme to control group(s) or usual care. 30 studies considered LCD, 17 moderate-fat diets, 6 LFD (several compared multiple diets). Total follow-up time was 12 months	LCD had greater weight loss at 6 months, but at 1 year LCD −7.3 kg (±2) and LFD −7.3 (±2) had equal weight loss. Differences between named diets were minimal. Authors support recommending any diet the patient will adhere to
Naude et al. 2014 Low carbohydrae *versus* isoenergetic balanced diets for reducing weight and cardiovascular risk[78]	19 RCTs, 3209 participants. Included T2D subgroups also. Considering trials with minimum-duration follow-up 12 weeks, outcomes were grouped by short-term (0–6 months) and long-term (1 and 2 years)	There were no mean differences between weight loss at 3–6 months or 1–2 years in LCD (range −2.9 to −12.3 kg in those without T2D) compared to isoenergetic balanced diet (range −3.5 to −10.9 kg in those without T2D). In summary, no differences in change between low- and balanced carbohydrate options at 2 years
Huo et al. 2014 Effects of Mediterranean-style diet on glycemic control, weight loss and cardiovascular risk factors among T2D individuals[79]	9 RCTs, 1178 patients. Mean duration 19 months (range 1–48)	Small reductions in % HbA_{1C} seen −0.3 (±0.2) at end of follow-up. Body weight reductions for the MedDiet were statistically but not clinically significant −0.3 kg (±0.3)

Bueno et al. 2013 VLCKD v. LFD for long-term weight loss[80]	13 RCTs, 1577 participants. Mean duration 16 months (range 12–24 months). Comparing studies with ≥12 months' follow-up, with <50 g day^{-1} compared to a conventional LFD (E < 30% fat)	Only 1 of 13 studies reported carbohydrate intake at the end of the follow-up to be consistent with the <50 g day^{-1} recommendation. For weight loss 7/13 studies showed non-significant trends of effects, with 1 showing a significant different for difference of weight loss. Total WMD weight loss −0.9 (±0.8) at end of follow-up, favouring VLCKDs. Subgroup analysis for 24-month follow-up showed no significant differences. No significant differences were seen in blood sugar, insulin or HbA_{1C}. Studies which included VLCKD interventions had favourable effects on body weight that, although statistically significant, "appear to be of little clinical significance"
Schwingshackl & Hoffmann 2013 Comparison of effects of long-term low-fat vs high-fat diets on blood lipid levels in overweight or obese patients[67]	32 RCTs, 8862 participants. Range of follow-up 1–6 years. Multiple dietary programmes were compared, each including at least one low-fat regimen	The WMD in body weight comparing low- and high-fat diets was not significant. Subgroup analysis including only hypocaloric diets showed a significantly greater decrease in body weight in the high-fat group: −0.6 kg (±0.5). LDL-C was improved with LFD compared with LCD, although this was not clinically significant
Esposito et al. 2011 MedDiet and weight loss[81]	16 RCTs, 3436 participants. Follow-up mean 64 weeks (range 6–260)	The MedDiet group mean difference compared to control diet was −1.7 kg (±1.1); this was greater with energy restriction or increased physical activity subgroup analysis
Nordmann et al. 2011 Meta-analysis comparing MedDiet to LFDs for modification of cardiovascular risk factors[82]	6 RCTs, 2650 participants. Studies include long follow-up periods, mean 33 months (range 24–72)	At 2-year follow-up MedDiet mean differences: body weight (compared to low-fat) was −2.2 kg (±1.6); along with slight improvements with fasting plasma glucose, and small reduction in C-reactive protein favouring the MedDiet interventions

(*continued*)

Table 3.1 (*Continued*)

Diabetes		
Objective	**Details**	**Results**
Qian et al. 2019 Association between plant-based dietary patterns and risk of Type 2 diabetes[83]	9 studies, 307 099 participants with 23 544 incident cases of T2D. All studies adjusted for BMI in their statistical analyses, therefore associations may be underestimated	All studies trended towards or showed that higher adherence to plant-based diet was associated with reduced T2D risk, with a pooled RR of 0.77 (95% CI 0.71–0.84). With subgroup analysis using data from four studies the association became stronger when healthful plant-based foods were included in defining plant-based diet patterns (fruits, vegetables, whole grains, legumes, nuts): RR 0.7 (95% CI 0.62–0.79)
Neuenschwander et al. 2019 Role of diet in T2D incidence[23]	53 publications were included, looking at risk associations with T2D and food behaviours. Numbers of studies considered for each dietary pattern and study participant numbers: AHEI 9 (numbers NA), vegetarian diet (numbers NA), DASH 7 (numbers NA), MedDiet 9 (n = 19 609), HEI 3 (numbers NA), LCD 4 (n = 8081)	153 diet behaviour or diet quality indices were included. Few associations were graded as high quality of evidence. Adjusted summary hazard ratios for different diets and T2D incidence: vegetarians *vs.* omnivores 0.67 (95% CI 0.54–0.84, low quality of evidence), AHEI high adherence *vs.* low 0.79 (0.73–0.85, moderate), MedDiet 0.85 (0.76–0.95, low), HEI 0.86 (0.79–0.93, low), LCD 1.17 (0.90–1.51, very low)
Papamichou et al. 2019 Effectiveness of studies ≥6 months length for LCD, macrobiotic, vegan, vegetarian, and MedDiets compared to LFD on diabetes[50]	20 RCTs, 2657 participants, mean 15 months (range 6–48 months). RCTs >6 months that investigated the effectiveness of various dietary patterns on T2D were included	Generally, 4 LCD showed greater glycaemic control over LFD but there were no significant differences in the majority. The MedDiet demonstrated greater reduction in body weight and HbA_{1C} levels and delayed requirement for diabetes medications. The vegan and macrobiotic diet demonstrated improved glycaemic control, while the vegetarian diet showed greater body weight reduction and insulin sensitivity. "Mounting evidence supports the view that vegan, vegetarian and Mediterranean dietary patterns should be implemented in public health strategies, in order to better control glycemic markers in individuals with T2DM"

Galbete et al. 2018 Evaluating MedDiet and risk of chronic disease in cohort studies: an umbrella review of meta-analyses[84]	14 publications reviewed 70 primary studies. Population being considered for analysis of T2D risk was 543 591	Those with highest adherence to the MedDiet showed inverse risk association for T2D: between RR 0.87 (±0.05) and RR 0.77 (±0.11)
Sainsbury et al. 2018 Examining differences in glycaemic control with carbohydrate-restricted diets compared to high carbohydrate diets[85]	25 RCTs, 2414 participants. Examined diet interventions only, with both studies on T1D and T2D. LCDs were broken down into further subgroups by % E from carbohydrate: low-carbohydrate <26%, moderate-carbohydrate <45%, high-carbohydrate >45%	Carbohydrate-restricted diets showed greater 3- and 6-month HbA_{1C} reductions in the LCD group, but no difference after accounting for differences in weight loss. There were no differences for moderate and high carbohydrates at any time point for HbA_{1C}. No significant differences at 12 or 24 months
Toumpanakis et al. 2018 Effectiveness of plant-based diets in promoting well-being in the management of T2D[86]	9 RCTs, 2 non-randomised trials, 606 participants. Intervention mean duration 28 weeks (range 3–74). Looked at a range of parameters to clearly define benefits on well-being of plant-based interventions	Pooled mean % HbA_{1C} differences were −0.6% in the intervention groups and −0.2% in the control groups. Pooled mean weight loss was 5.2 kg in the plant-based group compared with 2.8 kg in the control. Also seen were greater reductions in medications, total cholesterol (despite reduced lipid-lowering medications) and pain and improved quality of life and psychological well-being
Viguiliouk et al. 2018 Effect of vegetarian dietary patterns on cardiometabolic risk factors in diabetes[87]	9 trials, comparing 664 participants, mean duration of follow-up 12 weeks (range 4–74) comparing vegetarian dietary patterns with control diets, which included vegan through lacto-ovo-vegetarian diets. Vegetarian diets were higher in total % energy (% E) intake from carbohydrate (60% E (median) *vs.* 50% E in controls), fibre (28 g day^{-1} *vs.* 20 g day^{-1}), and were lower in: protein (15% E *vs.* 19% E), fat (25% E *vs.* 30% E) and saturated fat (5.1% E *vs.* 8.5% E). Studies included T1D or T2D	Vegetarian diets, compared to controls, showed significantly reduced HbA_{1C} −4.2 mmol mol^{-1} (±2.4) body weight −2.2 kg (±0.8), BMI −0.7 kg m^{-2} (±0.4). Significant improvements with the vegetarian group were also seen in waist circumference, non-HDL cholesterol, but not blood pressure or HDL-C
Huntriss et al. 2017 The interpretation and effect of a LCD in the management of T2D[88]	18 RCTs, 2204 participants. Follow-up mean duration 1 year (range 12 weeks–4 years). Comparing T2D interventions that used a low-carbohydrate approach with usual care or a control group	At 1 year nil significant differences for weight; low-carbohydrate interventions showed HbA_{1C} −0.3% (±0.2) compared to control. Consumption of <50 g day^{-1} carbohydrates appears unrealistic, but <130 g day^{-1} appears to be achievable

(*continued*)

Table 3.1 (*Continued*)

Diabetes		
Objective	Details	Results
van Wyk et al. 2016 A critical review of LCDs in people with T2D[20]	9 meta-analyses with 153 studies, and of these 12 high-quality studies were analysed. Mean duration 49 weeks (range 6–104). <45% E was used for LCD, but total carbohydrate intake was used in the outcome measures	Total energy intake predicts body weight, but no body weight differences were seen with LCD and LFD, except one study that favoured the LCD. The only study showing improved HbA_{1C} used a low-GI MedDiet with little difference in total carbohydrate: 16 g day^{-1}. LCD no different from LFD for metabolic markers and glycaemic control. Reduction in carbohydrate intake did not result in improved glycaemic control, even with difference in carbohydrate as much as 115 g day^{-1} between LCD and LFD with comparison to baseline. Study participants did not achieve LCD diet targets except in limited circumstances. Very LCD may not be sustainable long-term as carbohydrate intakes moderate over time (in diets aiming for <50 g day^{-1} intake ranged at 1 year between 132 and162 g), and lowest reported carbohydrate day^{-1} at 104 weeks was 95 g, although this was based on limited data
Carter et al. 2014 A MedDiet improves HbA1c but not fasting blood glucose compared to alternative dietary strategies[89]	8 RCTs, 1330 participants. Looking at outcomes for diabetes irrespective of weight loss examining articles that included at least one type of MedDiet as one intervention arm. Interventions considered: 9 MedDiets, 1 of which was low-carbohydrate, 1 national cholesterol programme diet, 1 paleolithic, 2 control. Study and follow-up mean 60 weeks (range 2–24 months)	Does not include weight loss outcomes. No superiority of one intervention over others in lowering glucose. MedDiet reduced HbA_{1C} significantly compared to usual care but not compared to the paleolithic diet (although this intervention group was small, at 14 participants, and brief, at 3 months)

† LCD, low-carbohydrate diet; LFD, low-fat diet; RCT, randomised controlled trial; MedDiet, Mediterranean diet; CI, confidence interval; T2D, Type 2 diabetes; Hb_{A1c}, haemoglobin A1c; E, total energy intake; GI, glycaemic index; WMD, weighted mean difference; VLCKD, very-low-carbohydrate ketogenic diet; LDL-C, low-density lipoprotein cholesterol; BMI, body mass index; RR, relative ratio; CI, confidence interval; AHEI, Alternative Healthy Eating Index; NA, not available; HEI, Healthy Eating Index; T2DM, Type 2 diabetes mellitus; T1D, Type 1 diabetes; HDL-C, high-density lipoprotein cholesterol.

‡ ± represents 95% confidence intervals.

20–50 g day^{-1} carbohydrate allowable (50 g day^{-1} is 10% E from carbohydrate for a 2000-kcal day^{-1} diet) or higher amounts, up to ≤45% E from carbohydrate. Both ends of the dietary spectrum can present additional difficulties for long-term adherence.

In comparing efficacy (how well does this work in the controlled setting?) *vs.* effectiveness (pragmatically, how well does this work in the real world?) we may see major differences. In assessing intervention effectiveness failure may be due to many factors, *e.g.* independently confounding variables that can be unique and difficult to predict, or participants' poor adherence. Many valid explanations are possible for poor dietary adherence in studies, not least of which is the difficulty in executing trials as per protocol. Because measuring, reporting and comparing data are problematic, standardised protocols for testing of adherence along with objective outcome tests will enable better reporting.

Data with poor dietary adherence leads to the potential for misclassifying an intervention. Classification of participants by randomisation allocation and not actual adherence to prescribed macronutrient intake at follow-up (which essentially all studies do) can make valid comparisons difficult. One example seen in the reviews was the Ornish diet, which was ~30% E from fat at follow-up in studies[17,18] rather than the recommended ~10%.[19] Data becomes further limited when there are very few data points. For example, the only long-term study of a paleolithic diet (PD) reported "poor adherence to the target protein intake (30% E) in the PD group" judged by urinary nitrogen, and the comparator did not reach carbohydrate targets. Do we continue to classify these as Ornish and paleolithic diets? Additionally, if a diet has undesirable consequences then an intention-to-treat analysis may reveal these through high dropout rates in the intervention compared to control, or poor adherence, which in turn may suggest the diet is too difficult to follow. Because of the aforementioned, prospective research misclassification of participants' data tends to be non-differential, resulting in the potential for underestimated true effect sizes.

When reporting adherence, % E from a macronutrient in conjunction with total energy intake will help avoid errors, *e.g.* in calorie-restricted diets comparing carbohydrates: a 1400-kcal LFD *versus* a 2400-kcal LCD could result in the LCD having more total g day^{-1} carbohydrates but very different % E from carbohydrates.[20]

The lesson here is to summarise all steps in research accurately and comprehensively, especially results data, which tends to be of variable quality. This will provide clarity rather than mislead the public, which is especially important given the high interest in weight loss research.[21]

3.3.2 Low-fat Diets

3.3.2.1 Clinical Trials

This discussion focuses on the Mediterranean diet (MedDiet), and the more generic term plant-based diet (PBD), which has become more popular in recent years for referring to healthy high-carbohydrate and low-fat dietary patterns. The MedDiet is perhaps the most studied diet pattern, high in fresh vegetables, but including allowance for some meat and animal products. Other diets that emphasise plant-based foods but not the exclusion of meats include the Alternative Healthy Eating Index and the Dietary Approaches to Stop Hypertension (DASH) diet. Also included in this low-fat category is a very-low-fat WFPB diet which typically heavily limits or excludes animal products (vegan, or strict vegetarian) and added oils. The Ornish, Eselstyn and McDougall diets are variations. The data is clear and consistent for weight loss on the MedDiet as well as vegetarian diets and PBDs.

The American College of Endocrinology states that a PBD is the optimal nutritional approach for those diagnosed with diabetes:

> All patients should strive to attain and maintain an optimal weight through a primarily plant-based meal plan high in polyunsaturated and monounsaturated fatty acids, with limited intake of saturated fatty acids and avoidance of *trans* fats.[22]

The discussion on PBDs represents a shift from traditional eating approaches focused on macronutrient quantities, and places more emphasis on dietary quality and whole plant foods. The main foods in a PBD (grains, cereal, fruits, vegetables) are all separately associated with positive health outcomes, and increased consumption of these foods leads to the corollary benefits of a reduction in red and processed meat and sugar-sweetened beverages.[23]

Observational data such as the Adventist Health Study showed vegan dietary patterns were associated with the lowest BMI of all dietary patterns, and a lower likelihood of diabetes,[24] a finding also seen in other observational data.[25] The EPIC-Oxford cohort data showed BMI increases with the amount of animal products consumed.[26] Other data suggests vegans gain less weight with ageing,[27] and a longer time following a vegan diet is related to lower BMI.[28] Data pooled from the Nurses' Health Study (NHS), NHS 2, and the Health Professionals Follow-Up Study found "plant-based diets, especially when rich in high-quality plant foods, are associated with substantially lower risk of developing T2D".[29] Using a nationally representative NHANES III data sample of >50-year-olds there was a link between increased protein consumption and increased diabetes mortality.[30] The authors suggest the link was seen only with animal protein but not plant-based protein.

Small intervention studies in those with T2D using a PBD have demonstrated improved glycaemic parameters independent of changes in BMI,[31] potentially related to loss of visceral fat.[32] Excess stored fat is strongly linked to high levels of circulating fats, and insulin resistance results from excess intracellular lipids in both muscle[33] and liver.[34,35] One randomised controlled trial (RCT) compared a PBD to the American Diabetes Association guidelines. Both groups achieved significant weight loss at 74 weeks with no significant between-groups difference in weight, yet the PBD had a significantly greater improvement in HbA_{1C}.[36] Significant weight loss can also occur without mandated exercise; using an ad-libitum WFPB diet there was -3.9 kg m^{-2} difference between intervention and standard Western diet control group at 6 months, along with a -5 mmol mol^{-1} difference in HbA_{1C}.[37]

3.3.2.2 *Potential Mechanisms of Low-fat Diets*

Potential key mechanisms (Figure 3.3) for the improvements conferred by PBDs include weight loss, increased fibre consumption impacting digestion and nutrient assimilation, differences in amino acid profiles and/or interactions with gut microbiota which in turn affect trimethylamine (TMA) and trimethylamine-*N*-oxide (TMAO) production. TMAO appears strongly related to cardiovascular disease and other chronic disease, and can be used as a prognostic marker in acute coronary syndrome.[38]

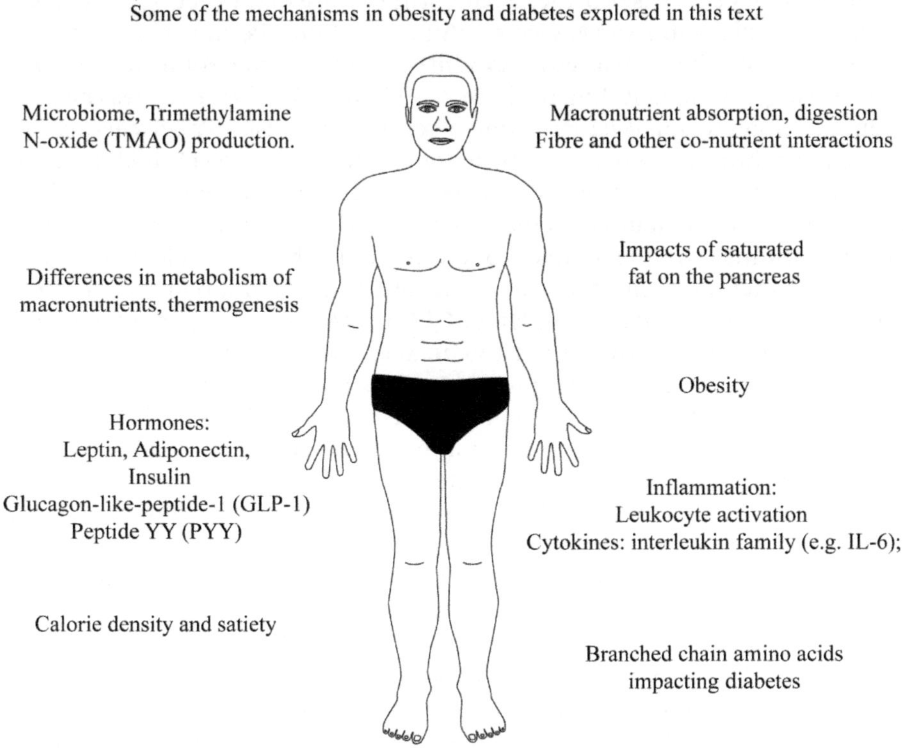

Figure 3.3 Mechanisms of obesity and diabetes. Dietary approaches for weight loss reduce intake of calories by reducing total energy consumed (*e.g.* portion control), and/or by consuming a lower-energy density diet. There is a strong relationship between obesity and Type 2 diabetes, and therefore most mechanisms are not exclusive to either disease state. The listed mechanisms have been discussed in the text as key or potentially key mechanisms; some need more research to elucidate their mechanistic role further.

A recent 4-week RCT showed a high saturated fat, animal protein diet compared with an isocaloric low-fat diet led to higher levels of both TMAO and the precursor, branched-chain amino acids (BCAAs: leucine, isoleucine and valine).[39] Higher intake of leucine (which appears to be the most deleterious BCAA in its effects) and serum BCAA levels increase the risk for insulin resistance, although findings suggest animal product sources as being of higher risk than plant product sources, as measured by homeostatic model assessment insulin resistance.[32,40,41] Higher serum BCAA levels and hyperlipidaemic states cause direct and continuous activation of the mammalian target of rapamycin complex 1 (mTORC1) signaling pathway, critical in cell metabolism and growth.[42,43] Continuous mTORC1 activation leads to serine phosphorylation of insulin receptor substrate 1 (IRS-1). The IRS-1/phosphatidylinositol (PI) 3-kinase/protein kinase B (Akt) pathway has decreased signal in insulin resistance, and pathognomic of T2D is the accompanying reduced uptake and utilisation of glucose.[43]

Intact plant cell walls provide the majority of fibre in the diet, and increased difficulty in digestion slows gastric emptying. Fibre is known to also have "anti-nutritive" effects whereby it inhibits energy-rich macronutrient absorption and digestion, including for both protein and fat.[44] Fibre slows the postprandial blood sugar level rise, which reduces reactive hypoglycaemia and subsequent hunger, and when fermented in the large bowel to short-chain fatty acids can improve the glucose response.[45] Increased fibre helps nutrients persist to the distal ileum (whereas they are usually absorbed prior) so nutrients here induce the "ileal brake" phenomenon, which is a release of satiety hormones including glucagon-like-peptide-1 (GLP-1) and peptide YY (PYY), both of which slow gastric emptying. GLP-1 also increases pancreatic β-cell growth.[46] When comparing vegan and meat meals with very similar macronutrients (both high-fat and protein), GLP-1 was significantly increased by the vegan meal.[41] Consistent patterns from diets rich in plant-based foods also show improved markers for inflammation (these markers are also typically elevated in obesity) like C-reactive protein (CRP) and interleukin-6 (IL-6),[47,48] but studies are less conclusive for tumour necrosis factor-α, adiponectin, leptin and resistin.[49]

While the individual studies and reviews included here have limitations, their combined effect creates a compelling and consistent picture that individuals consuming a PBD tend to have a lower BMI in meta-analyses of RCTs (when compared to control groups, but not necessarily resulting in greater weight loss when compared with an LCD). Large trials and reviews of the MedDiet have demonstrated real-world applicability. Diets rich in plant foods have taken over from the traditional focus of "high-carbohydrate" diets as quality of diet has great bearing on outcomes in addition to macronutrient intake. Furthermore, "there is a growing convergence of scientific evidence that an optimal diet is mostly plant-based", although further research is needed with regard to diabetes management.[50] Diets rich in plant foods reduce environmental demands and animal agriculture and are consistent with the EAT-Lancet Commission goals for sustainability.[51]

3.3.3 Low-carbohydrate Diets

3.3.3.1 Clinical Trials

LCDs include a spectrum from VLCKD through to a more moderate <45% E from carbohydrates. Although we tend to see a trend of greater weight loss (especially in the short term) for the VLCKD, this is not sustained to 1 year, where there is very limited evidence to support long-term use of a very-LCD. Many of the review articles cited in Table 3.1 commented on strict carbohydrate intakes, and, where addressed specifically, suggested this was not sustainable and that moderate carbohydrate restriction is a more realistic goal for those wishing to pursue LCDs.

In VLCKDs removing dietary glucose will lead to the use of stored glycogen and a lower level of blood sugar irrespective of weight loss, even in those with a diagnosis of T2D who are usually otherwise dependent on insulin. Sometimes this rapid normalisation of blood sugar level can occur within 24 hours.[52] Likewise, a review on LCD narrating evidence of T2D reversal[53] shows the potential of this approach but has been criticised for methodology and reporting of results.[54] Early weight loss in a ketosis-inducing diet results from glycogen (which combines with three to four parts water in liver, muscles and fat) stores being depleted as the body is forced into ketosis and gluconeogenesis.[55] This weight loss gives the *impression* of significant weight loss, but in reality can be very easily regained. Weight loss in the first 6 months of embarking on a dietary change is a predictor of long-term adherence,[16] so rapid results are potentially helpful, but weight cycling could be disheartening for patients, and initial glycogen weight loss is not a realistic representation of future weight loss speed (actual fat mass losses being far slower). Advocates of the VLCKD may suggest reducing carbohydrates as the most important variable in weight loss, as the concomitant increase in fat consumption is suggested in one study to create differential changes in metabolism leading to increased energy expenditure.[56] Limited findings here have not been replicated and have been criticised due to some inconsistencies in the data analysis, including how primary outcome data was calculated and presented.[57] If these effects are present this would be highly significant.

Another mechanism that has been suggested is fat and protein somehow having a greater effect on satiety (effect on subsequent nutrient intake).[58] However, these studies that implied protein > carbohydrates > fat as the most impactful on satiety did not consider palatability and ED; when these are similar between macronutrients there is little difference in satiety.[59] The body senses and responds to ingested calories to a *very* high degree of accuracy regardless of the macronutrient origin, which is intuitive – if calorie-regulating mechanisms were more than a fraction of a per cent erroneous, then large and rapid weight gain or loss would ensue. There is ample evidence that moving to a lower-ED diet promotes weight loss (see introduction). For those who consume either a calorie-restricted or lower-ED (not just ketogenic) LCD, weight loss will follow, and a net energy deficit is ultimately the mechanism for any weight loss approach. This seems a more likely mechanism for LCDs.

Review data indicates (as per Table 3.1) that LCD data shows evidence of effect in the short-term reduction of HbA_{1C} compared to either control diets or higher-carbohydrate diets, but this is inconclusive in longer-term management of T2D. Aside from the aforementioned putative mechanisms, there was no clear effect above and beyond simple calorie restriction causing weight loss. More research on both T2D and appetite–energy metabolism interactions would be of interest before drawing any further conclusions.

3.3.3.2 Potential Mechanisms of Low-carbohydrate Diets

Data on LCD and inflammatory markers, including IL-6 and CRP, is inconclusive.[47] Any significant weight loss can improve insulin sensitivity, but consuming an LCD and high-fat diet does not appear to promote greater insulin sensitivity in and of itself. In prediabetes, the liver accumulates fat and, as the toxic effects mount, triacylglycerol is secreted by the liver in very low-density lipoproteins and can then accumulate in the pancreas, causing pancreatic function to decline. Diets high in fat are often (but not always) higher by default in animal products and the concomitant increase in saturated fat in particular is linked to damaging health outcomes, which include: the onset of diabetes; harm to β-cell function;[60–64] and worse outcomes for cardiovascular health.[65,66] Cardiovascular disease is the leading cause of death in those with a diagnosis of T2D, and LCDs have statistically significant increases in low-density lipoprotein (LDL) cholesterol: weighted mean difference 0.08 mmol L^{-1}, along with increased high-density lipoprotein (HDL): weighted mean difference (WMD) 0.06 mmol L^{-1}, although these differences "do not allow for unequivocal recommendation" of either LCDs or LFDs given small total effect sizes.[67] Furthermore, the authors acknowledge HDL efflux capacity is not necessarily linked to concentration: "HDL particle functionality might be at least as important as HDL cholesterol plasma levels", and LDL, especially as more harmful dense particles, was not related to carbohydrate consumption, as had previously been considered. Therefore, findings suggest a small negative effect on cholesterol levels from LCD, but nothing that would be clinically significant. There was no difference between groups with hypocaloric diets. While these findings may appear to reduce harms from saturated fat it does not account for other biochemical processes (such as previously cited saturated fat causing increased TMAO).

3.4 Challenges with Diets as Interventions and Suggestions for the Future

There are many challenges with diet intervention research. The randomised trial format remains important, although a costly and resource-intensive approach to large-scale research. Blinding is practically impossible, and crossover interventions are often unsuitable, as knowledge cannot be removed. "Real-world" effectiveness research is the most useful, but all clinical trials fail to simulate this to some degree, with selection biases often exacerbating already poor external validity, with populations that are frequently retired, wealthy, highly adherent and literate – which does not represent the "average" person. Randomised trials often must use surrogate end-points rather than mortality as the cost and complexity of doing large RCTs with long-term follow-up render these not feasible.

Future research needs large RCTs and more standardisation, including guidance for reviewers and the ethical process, standardisation of methods for generating protocols in ways that allow for reproducibility and the management of biases or conflicts through far greater transparency. All data should be fully reported and easily accessible, and analysed using a gold-standard approach. Outcomes would be of higher reliability with greater use of objective biomarkers. Combining different dietary research in meta-analyses is particularly problematic when classifying data by initial allocation rather than follow-up dietary behaviour and allows for misclassifications of data. As guidelines move more towards eating of foods and are focused less on macronutrients, likewise research should continue to follow this same trend. The reductionist approach on macronutrient content rather than also including diet quality may be of only limited help, although reporting both macronutrient percentage and the actual foods consumed would be most helpful, and more information here is better.

Any of the mechanisms listed in Figure 3.3 could be more thoroughly investigated, given the many lingering questions. Research would be especially powerful if multiple hypotheses were tested simultaneously, and real-world adherence needs to be explored. Research could include more diverse opinions and experience, especially in the planning of interventions.

3.5 Summary

Despite differences in diets, nearly all agree that lower-ED, nutrient-rich whole plant foods are health-promoting for obesity and T2D. There was no clear short- or long-term superiority for weight loss or for diabetes control using LCD or LFD. Weight loss outcomes ranged from 0 to 11 kg at 1 year. There was a greater quantity of quality long-term data for low-fat approaches, yet still insufficient data to be conclusive around objective outcomes, despite some evidence of low-fat diets' superiority. All diets appear to have difficulties with long-term adherence. Classification of intervention participants by randomisation allocation and not final adherence outcomes can lead to erroneous conclusions, especially when few data points are present. Therefore, measuring, reporting and then comparing data are problematic. Standardisation would improve data.

PBDs have strong and consistent observational and experimental evidence to support their use for both diabetes and obesity, with the MedDiet, in particular, having strong evidence to support its use. This is reflected in reviews and now also guidelines. The focus on improving dietary quality, especially by consuming mainly or exclusively whole plant foods, represents a major and positive shift from diets focused solely on macronutrient percentage. Potential mechanisms for improvements with PBDs include: more plant foods reducing ED in the diet; greater fibre consumption, which increases postprandial satiety, slows gastric emptying and reduces nutrient absorption; amino acid profile and saturated fatty acid intake which impact on gut microbiota, including TMAO production, along with some impacts on inflammatory markers. This is important as whole-body inflammation affects obesity and diabetes. Diets rich in plant foods are more environmentally sustainable and do not pose the same ethical problems as diets rich in meat or animal products.

LCDs have ample short-term outcomes, the general trend being greater initial weight loss and improvements than high-carbohydrate diets with T2D, but this is not sustained to 1 year. Initial weight loss in VLCKDs is at least partly due to decreased glycogen stores that do not represent long-term losses. There is limited evidence to support long-term use of very-LCDs >1 year and very limited data >2 years. Moderate carbohydrate restriction appears more realistic than LCDs for most people wishing to consume an LCD. Despite increased fat intake the negative effects of LCD on total and LDL cholesterol are minimal in studies. Because these diets tend to be higher in foods that are otherwise demonstrated to be harmful, research might yet reveal harm outside of cholesterol. There are no clear mechanisms for improvements in LCDs beyond reduced total calorie intake, although more mechanistic studies could provide much greater insights into both low- and high-fat diets.

Acknowledgement

Thank you to Professor Cameron-Smith for input into planning this chapter and to Professor Cheng for his feedback and guidance.

References

1. D. Haslam, *Obes. Rev.*, 2007, **8**(s1), 31–36.
2. M. Blüher, *Nat. Rev. Endocrinol.*, 2019, **15**(5), 288–298.
3. J. Q. Purnell, Definitions, Classification, and Epidemiology of Obesity, in *Endotext*, ed. K. R. Feingold, B. Anawalt, A. Boyce, G. Chrousos, K. Dungan, A. Grossman, *et al.*, MDText.com, South Dartmouth, MA, (accessed November 18 2019).
4. Obesity and overweight, World Health Organization Key Facts, https://www.who.int/news-room/fact-sheets/detail/obesity-and-overweight (accessed November 18 2019).
5. D. Mozaffarian, T. Hao, E. B. Rimm, W. C. Willett and F. B. Hu, *N. Engl. J. Med.*, 2011, **364**(25), 2392–2404.
6. *IDF Diabetes Atlas*, 9th edn, International Diabetes Federation, Brussels, Belgium, 2019, https://diabetesatlas.org/ (accessed November 18 2019).
7. J. P. Boyle, T. J. Thompson, E. W. Gregg, L. E. Barker and D. F. Williamson, *Popul. Health Metr.*, 2010, **8**, 29.
8. F. B. Hu, J. E. Manson, M. J. Stampfer, G. Colditz, S. Liu, C. G. Solomon, *et al.*, *N. Engl. J. Med.* 2001, **345**(11), 790–797.

9. H. E. Bays, R. H. Chapman and S. Grandy, *Int. J. Clin. Pract.*, 2007, **61**(5), 737–747.
10. N. Gray, G. Picone, F. Sloan and A. Yashkin, *South. Med. J.*, 2015, **108**(1), 29–36.
11. J. P. H. Wilding, *Int. J. Clin. Pract.*, 2014, **68**(6), 682–691.
12. S. R. Kashyap, E. S. Louis and J. P. Kirwan, *Expert Rev. Endocrinol. Metab.*, 2011, **6**(4), 557–561.
13. B. J. Rolls, *J. Nutr.*, 2000, **130**(2S Suppl), 268S–271S.
14. J. A. Ello-Martin, L. S. Roe, J. H. Ledikwe, A. M. Beach and B. J. Rolls, *Am. J. Clin. Nutr.*, 2007, **85**(6), 1465–1477.
15. H. A. Raynor, E. L. Van Walleghen, J. L. Bachman, S. M. Looney, S. Phelan and R. R. Wing, *Eat. Behav.*, 2011, **12**(2), 119–125.
16. R. R. Wing and J. O. Hill, *Annu. Rev. Nutr.*, 2001, **21**, 323–341.
17. D. Ornish, L. W. Scherwitz, J. H. Billings, S. E. Brown, K. L. Gould, T. A. Merritt, *et al.*, *JAMA, J. Am. Med. Assoc.*, 1998, **280**(23), 2001–2007.
18. M. L. Dansinger, J. A. Gleason, J. L. Griffith, H. P. Selker and E. J. Schaefer, *JAMA, J. Am. Med. Assoc.*, 2005, **293**(1), 43–53.
19. D. Ornish, L. W. Scherwitz, J. H. Billings, S. E. Brown, K. L. Gould, T. A. Merritt, *et al.*, *JAMA, J. Am. Med. Assoc.*, 1998, **280**(23), 2001–2007.
20. H. J. van Wyk, R. E. Davis and J. S. Davies, *Diabetic Med.*, 2016, **33**(2), 148–157.
21. J. P. A. Ioannidis, *JAMA, J. Am. Med. Assoc.*, 2018, **320**(10), 969–970.
22. A. J. Garber, M. J. Abrahamson, J. I. Barzilay, L. Blonde, Z. T. Bloomgarden, M. A. Bush, *et al.*, *Endocr. Pract.*, 2019, **25**(1), 69–100.
23. M. Neuenschwander, A. Ballon, K. S. Weber, T. Norat, D. Aune, L. Schwingshackl, *et al.*, *BMJ*, 2019, **366**, l2368.
24. M. J. Orlich and G. E. Fraser, *Am. J. Clin. Nutr.* 2014, **100**(1), 353S–358S.
25. T. H. T. Chiu, H.-Y. Huang, Y.-F. Chiu, W.-H. Pan, H.-Y. Kao, J. P. C. Chiu, *et al.*, *PLoS One*, 2014, **9**(2), e88547.
26. A.-C. Vergnaud, T. Norat, D. Romaguera, T. Mouw, A. M. May, N. Travier, *et al.*, *Am. J. Clin. Nutr.*, 2010, **92**(2), 398–407.
27. M. Rosell, P. Appleby, E. Spencer and T. Key, *Int. J. Obes. (Lond.)*, 2006, **30**(9), 1389–1396.
28. T. Key and G. Davey, *BMJ*, 1996, **313**(7060), 816–817.
29. A. Satija, S. N. Bhupathiraju, E. B. Rimm, D. Spiegelman, S. E. Chiuve, L. Borgi, *et al.*, *PLoS Med.*, 2016, **13**(6), e1002039.
30. M. E. Levine, J. A. Suarez, S. Brandhorst, P. Balasubramanian, C.-W. Cheng, F. Madia, *et al.*, *Cell Metab.*, 2014, **19**(3), 407–417.
31. J. W. Anderson and K. Ward, *Am. J. Clin. Nutr.*, 1979, **32**(11), 2312–2321.
32. H. Kahleova, A. Tura, M. Hill, R. Holubkov and N. D. Barnard, *Nutrients*, 2018, **10**(2), 189.
33. J. Kuhlmann, C. Neumann-Haefelin, U. Belz, J. Kalisch, H.-P. Juretschke, M. Stein, *et al.*, *Diabetes*, 2003, **52**(1), 138–144.
34. A. Seppälä-Lindroos, S. Vehkavaara, A.-M. Häkkinen, T. Goto, J. Westerbacka, A. Sovijärvi, *et al.*, *J. Clin. Endocrinol. Metab.* 2002, **87**(7), 3023–3028.

35. D. K. Coletta and L. J. Mandarino, *Am. J. Physiol.: Endocrinol. Metab.*, 2011, **301**(5), E749–E755.
36. N. D. Barnard, J. Cohen, D. J. A. Jenkins, G. Turner-McGrievy, L. Gloede, A. Green, *et al.*, *Am. J. Clin. Nutr.*, 2009, **89**(5), 1588S–1596S.
37. N. Wright, L. Wilson, M. Smith, B. Duncan and P. McHugh, *Nutr. Diabetes*, 2017, 7(3), e256.
38. X. S. Li, S. Obeid, Z. Wang, B. J. Hazen, L. Li, Y. Wu, *et al.*, *Eur. Heart J.*, 2019, **40**(32), 2700–2709.
39. J. E. Park, M. Miller, J. Rhyne, Z. Wang and S. L. Hazen, *Nutr. Metab. Cardiovasc. Dis.*, 2019, **29**(5), 513–517.
40. G. Asghari, H. Farhadnejad, F. Teymoori, P. Mirmiran, M. Tohidi and F. Azizi, *J. Diabetes*, 2018, **10**(5), 357–364.
41. H. Kahleova, A. Tura, M. Klementova, L. Thieme, M. Haluzik, R. Pavlovicova, *et al.*, *Nutrients*, 2019, **11**(3), 486.
42. J. Ye, *Endocr., Metab. Immune Disord.: Drug Targets*, 2007, 7(1), 65–74.
43. B. Draznin, *Diabetes*, 2006, **55**(8), 2392–2397.
44. A. K. Hervik and B. Svihus, *J. Nutr. Metab.*, 2019, **2019**, 4983657.
45. M. McMacken and S. Shah, *J. Geriatr. Cardiol.*, 2017, **14**(5), 342–354.
46. M. L. Dreher, *Dietary Patterns and Whole Plant Foods in Aging and Disease*, Humana Press, 2018, https://www.springer.com/gp/book/9783319591797 (accessed January 1 2020).
47. A. Smidowicz and J. Regula, *Adv. Nutr.*, 2015, **6**(6), 738–747.
48. F. Haghighatdoost, N. Bellissimo, J. O. Totosy de Zepetnek and M. H. Rouhani, *Public Health Nutr.*, 2017, **20**(15), 2713–2721.
49. F. Eichelmann, L. Schwingshackl, V. Fedirko and K. Aleksandrova, *Obes. Rev.*, 2016, **17**(11), 1067–1079.
50. D. Papamichou, D. B. Panagiotakos and C. Itsiopoulos, *Nutr. Metab. Cardiovasc. Dis.*, 2019, **29**(6), 531–543.
51. W. Willett, J. Rockström, B. Loken, M. Springmann, T. Lang, S. Vermeulen, *et al.*, *Lancet*, 2019, **393**(10170), 447–492.
52. E. C. Westman, J. Tondt, E. Maguire and W. S. Yancy, *Expert Rev. Endocrinol. Metab.*, 2018, **13**(5), 263–272.
53. S. J. Hallberg, V. M. Gershuni, T. L. Hazbun and S. J. Athinarayanan, *Nutrients*, 2019, **11**(4), 766.
54. S. Joshi, T. Zaki, R. J. Ostfeld and M. McMacken, *Nutrients*, 2019, **11**(7), 1620.
55. S. N. Kreitzman, A. Y. Coxon and K. F. Szaz, *Am. J. Clin. Nutr.*, 1992, **56**(1 Suppl), 292S–293S.
56. C. B. Ebbeling, H. A. Feldman, G. L. Klein, J. M. W. Wong, L. Bielak, S. K. Steltz, *et al.*, *BMJ*, 2018, **363**, k4583.
57. K. D. Hall and J. Guo, *BMJ*, 2018, **363**, k4583.
58. R. Oh and K. R. Uppaluri, *Low Carbohydrate Diet*, StatPearls Publishing, Treasure Island, FL.
59. B. J. Rolls, *J. Nutr.* 2000, **130**(2S Suppl), 268S–271S.
60. R. Taylor, *Diabetologia*, 2008, **51**(10), 1781–1789.
61. G. I. Shulman, *J. Clin. Invest.*, 2000, **106**(2), 171–176.
62. R. Taylor, *Diabetic Med.*, 2013, **30**(3), 267–275.

63. P. M. Kris-Etherton, K. Petersen and L. V. Horn, *BMJ Nutr. Prevent. Health*, 2018, bmjnph-2018-000009.
64. F. Rosqvist, D. Iggman, J. Kullberg, J. Cedernaes, H.-E. Johansson, A. Larsson, *et al.*, *Diabetes*, 2014, **63**(7), 2356–2368.
65. M. Sacks Frank, A. H. Lichtenstein, J. H. Y. Wu, L. J. Appel, M. A. Creager P. M. Kris-Etherton, *et al.*, *Circulation,* 2017, **136**(3), e1–e23.
66. L. Hooper, N. Martin, A. Abdelhamid and G. Davey Smith, *Cochrane Database Syst. Rev.*, 2015, (6), CD011737.
67. L. Schwingshackl and G. Hoffmann, *J. Acad. Nutr. Diet,* 2013, **113**(12), 1640–1661.
68. C. Churuangsuk, M. Kherouf, E. Combet and M. Lean, *Obes. Rev.*, 2018, **19**(12), 1700–1718.
69. S. D. Anton, A. Hida, K. Heekin, K. Sowalsky, C. Karabetian, H. Mutchie, *et al.*, *Nutrients,* 2017, **9**(8), 822.
70. O. Snorgaard, G. M. Poulsen, H. K. Andersen and A. Astrup, *BMJ Open Diabetes Res. Care,* 2017, **5**(1), e000354.
71. C. Ma, A. Avenell, M. Bolland, J. Hudson, F. Stewart, C. Robertson, *et al.*, *BMJ*, 2017, **359**, j4849.
72. L. Hooper, A. Abdelhamid, D. Bunn, T. Brown, C. D. Summerbell and C. M. Skeaff, *Cochrane Database Syst. Rev.*, 2015, (8), CD011834.
73. J. G. Mancini, K. B. Filion, R. Atallah and M. J. Eisenberg, *Am. J. Med.,* 2016, **129**(4), 407–415.e4.
74. N. Mansoor, K. J. Vinknes, M. B. Veierød and K. Retterstøl, *Br. J. Nutr.*, 2016, **115**(3), 466–479.
75. R.-Y. Huang, C.-C. Huang, F. B. Hu and J. E. Chavarro, *J. Gen. Intern. Med.*, 2016, **31**(1), 109–116.
76. R. Atallah, K. B. Filion, S. M. Wakil, J. Genest, L. Joseph, P. Poirier, *et al.*, *Circ. Cardiovasc. Qual. Outcomes,* 2014, **7**(6), 815–827.
77. B. C. Johnston, S. Kanters, K. Bandayrel, P. Wu, F. Naji, R. A. Siemieniuk, *et al.*, *JAMA, J. Am. Med. Assoc.*, 2014, **312**(9), 923–933.
78. C. E. Naude, A. Schoonees, M. Senekal, T. Young, P. Garner and J. Volmink, *PLoS One*, 2014, **9**(7), e100652.
79. R. Huo, T. Du, Y. Xu, W. Xu, X. Chen, K. Sun, *et al.*, *Eur. J. Clin. Nutr.*, 2015, **69**(11), 1200–1208.
80. N. B. Bueno, I. S. V. de Melo, S. L. de Oliveira and T. da Rocha Ataide, *Br. J. Nutr.,* 2013, **110**(7), 1178–1187.
81. K. Esposito, C.-M. Kastorini, D. B. Panagiotakos and D. Giugliano, *Metab. Syndr. Relat. Disord.*, 2011, **9**(1), 1–12.
82. A. J. Nordmann, K. Suter-Zimmermann, H. C. Bucher, I. Shai, K. R. Tuttle, R. Estruch, *et al.*, *Am. J. Med.*, 2011, **124**(9), 841–851.e2.
83. F. Qian, G. Liu, F. B. Hu, S. N. Bhupathiraju and Q. Sun, *JAMA Intern Med.*, 2019 , **179**(10), 1335–1344.
84. C. Galbete, L. Schwingshackl, C. Schwedhelm, H. Boeing and M. B. Schulze, *Eur. J. Epidemiol.,* 2018, **33**(10), 909–931.
85. E. Sainsbury, N. V. Kizirian, S. R. Partridge, T. Gill, S. Colagiuri and A. A. Gibson, *Diabetes Res. Clin. Pract.,* 2018, **139**, 239–252.

86. A. Toumpanakis, T. Turnbull and I. Alba-Barba, *BMJ Open Diabetes Res. Care*, 2018, **6**(1), e000534.
87. E. Viguiliouk, C. W. Kendall, H. Kahleová, D. Rahelić, J. Salas-Salvadó, V. L. Choo, *et al.*, *Clin. Nutr.*, 2018, **38**(3), 1133–1145.
88. R. Huntriss, M. Campbell and C. Bedwell, *Eur. J. Clin. Nutr.*, 2017, **72**(3), 311–325.
89. P. Carter, F. Achana, J. Troughton, L. J. Gray, K. Khunti and M. J. Davies, *J. Hum. Nutr. Diet*, 2014, **27**(3), 280–297.

CHAPTER 4

Nutritional Regulation of Inflammation in Obesity and Diabetes

XIN WANG[a] AND JIAN LIU[a,b]*

[a]School of Food and Biological Engineering, Hefei University of Technology, Hefei, Anhui 230009, China
[b]Engineering Research Center of Bioprocess, Ministry of Education, Hefei University of Technology, Hefei 230009, China
*Email: liujian509@hfut.edu.cn

4.1 Introduction

In past decades, interest in nutrition has been rising, because of the tremendous increase and the global epidemic of obesity and related disorders.[1,2] Along with changes in lifestyle and dietary habits, the incidence of obesity and associated comorbidities has dramatically increased worldwide in both adults and children, and it is now alarming.[1–3] A growing number of epidemiologic studies suggest that even though the ability of adipose tissue to store excess calories was once advantageous when food access was intermittent, it now causes the development of obesity and associated comorbidities, particularly Type 2 diabetes, due to continuous overnutrition.[4–6]

In mammals, there are two classic types of adipose tissue: white adipose tissue (WAT) and brown adipose tissue (BAT)[7,8] (Figure 4.1). WAT located at

Nutritional Signaling Pathway Activities in Obesity and Diabetes
Food Chemistry, Function and Analysis No. 24
Edited by Zhiyong Cheng

Published by the Royal Society of Chemistry, www.rsc.org

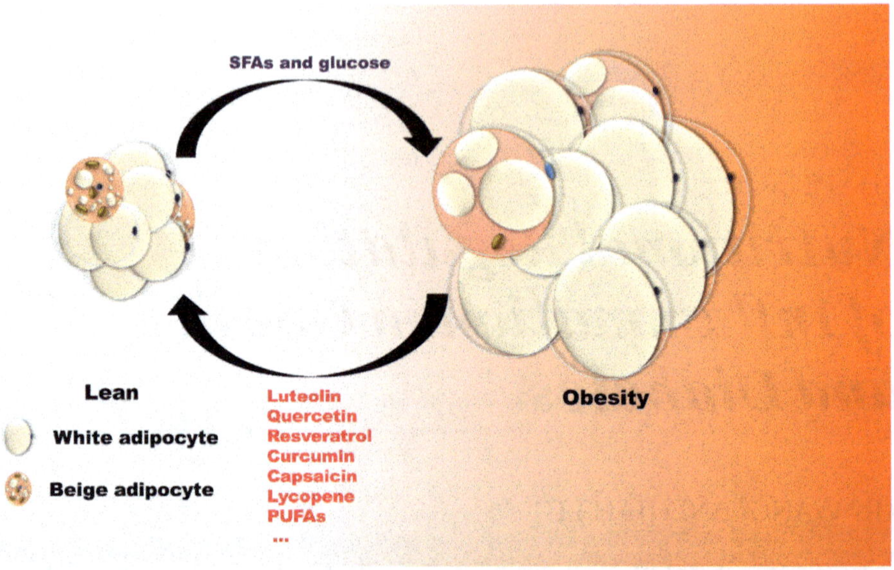

Figure 4.1 Nutritional regulation of adipocytes in the pathology of obesity during chronic overnutrition. Adipocytes become hypertrophied (increasing adipocyte size) and hyperplastic (increasing adipocyte number) accompanied by a decrease in the browning process, finally leading to adipose expansion and insulin resistance with long-term overnutrition. In contrast, dietary bioactive compounds, such as luteolin, quercetin, resveratrol, curcumin, capsaicin, lycopene, and polyunsaturated fatty acids (PUFAs), exert beneficial effects to inhibit the process of adipocyte hyperplasia and hypertrophy associated with energy storage and/or increase thermogenic activity *via* browning of adipose tissues, promoting energy dissipation in diet-induced adiposity. SFAs, saturated fatty acids.

subcutaneous and visceral level not only stores energy in the form of triglycerides when energy is surplus but also has the endocrine function of secreting adipokines and cytokines, regulating metabolism and inflammation.[9] In order to store excess energy, the processes of white adipocyte hyperplasia and hypertrophy are needed. BAT is found at the interscapular region of rodents as well as of human infants with high mitochondrial density, and it has highly specialized multilocular cells to dissipate stored chemical energy into heat *via* mitochondrial uncoupling protein 1 (UCP1).[8,10,11] Interestingly, besides white and brown adipocyte, beige adipocyte occurs in WAT depots and is induced by chronic cold exposure or adrenergic signaling activation.[12,13] Like brown adipocyte, beige adipocyte contains many mitochondria and multiple small lipid droplets and takes on a multilocular appearance. With the increasing expression of UCP1, these ordinarily energy-storing adipocytes produce heat efficiently.[13]

In adipose tissues, adipocytes are surrounded by a wide variety of cells, including endothelium, immune cells, fibroblasts, preadipocytes, and stem cells.[14] However, for decades, the roles of immune cells within the adipose tissue depots were unclear until recently; studies have increasingly revealed that these cells play central roles in maintaining local and whole-body metabolic homeostasis, especially under prolonged overnutrition[15–19] (Figure 4.2). Adipose tissue depots are the dominant sources of cytokines such as tumor necrosis factor-α (TNF-α), interleukin-1β (IL-1β), interleukin-6 (IL-6), and monocyte chemoattractant protein 1 (MCP1), creating a low-grade inflammatory environmental milieu that recruits the proinflammatory cells in response to overnutrition.[20] Rather than the adipocytes, many of these cytokines are produced by proinflammatory M1 macrophages (or "classically activated").[20] Other immune cells, including neutrophils, mast cells, B lymphocytes, dendritic cells, natural killer (NK) cells, and CD8+ and CD4+ T cells, also exert proinflammatory effects during obesity and Type 2 diabetes.[15,17,21–29] In contrast, M2 macrophages ("resident" or "alternatively activated"), eosinophils, innate lymphoid (ILC2) cells, and regulatory T cells

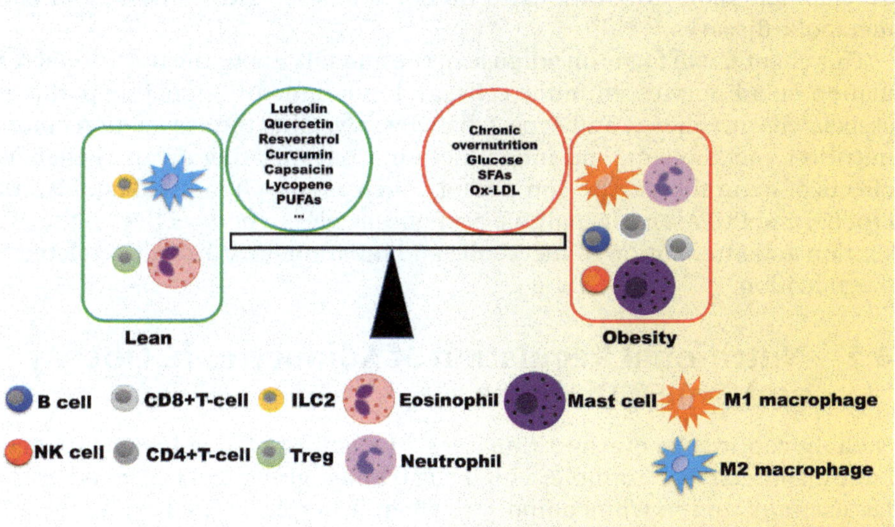

Figure 4.2 Nutrition regulation of adipose immune cells is associated with the initiation and integration of chronic low-grade inflammatory responses in leanness and obesity. In healthy adipose depots, there is a balance of pro- and anti-inflammatory mechanisms. The lean adipose depots contain many kinds of anti-inflammation immune cells, including M2 macrophages, eosinophils, innate lymphoid (ILC2) cells, and regulatory T cells (Tregs). Under persistent overnutrition, in adipose depots, there is an accumulation of proinflammatory cells, such as M1 macrophages, mast cells, B cells, natural killer (NK) cells, CD4+ and CD8+ T cells, and this contributes to the secretion of proinflammatory cytokines. Dietary bioactive compounds have potential roles in the treatment of obesity and Type 2 diabetes *via* inhibiting the secretion of proinflammatory cytokines (*e.g.*, tumor necrosis factor-α, interleukin-1β, and interleukin-6) and enhancing the accumulation of anti-inflammation cells (*e.g.*, ILC2, M2 macrophages, and Tregs). PUFAs, polyunsaturated fatty acids; SFAs, saturated fatty acids.

(Tregs), respectively, act to reduce inflammation in the development of obesity and Type 2 diabetes.[15,30–32]

When we talk about nutrition, generally, nutrients are classified into two groups: macronutrients that are in relatively high demand, including carbohydrates, fiber, fats, protein, and water, and micronutrients that include minerals, vitamins, and others, which are needed, although in much smaller quantities.[33] Together with dietary macronutrients and/or micronutrients, the most studied bioactive food components from fruits, vegetables, and seafood, including polyphenols (*e.g.*, curcumin, resveratrol, and flavonoids (such as quercetin and luteolin)), alkaloids (*e.g.*, capsaicin, caffeine), terpenoids (*e.g.*, lycopene), and n3-polyunsaturated fatty acids (PUFAs: *e.g.*, α-linolenic acid (ALA), docosahexaenoic acid (DHA), and eicosapentaenoic acid (EPA)), exert anti-oxidation and anti-inflammation properties *in vivo*.[34–38] They can reprogram metabolic and inflammatory signaling pathways, reduce the incidence of obesity and its associated complications, and bring about new developments and innovations in therapeutic strategies to these common metabolic diseases.[35–37]

This chapter will focus on adipose tissue and talk about the nutritional regulation on adipocytes, immune cells, and inflammatory signaling pathways associated with obesity and Type 2 diabetes, revealing the association among nutrition, metabolism, and inflammation. Concentrating on the beneficial effects of natural bioactive compounds, such as polyphenols, alkaloids, triterpene, and PUFAs, in the pathology of obesity gives a perspective on the prevention and attenuation of metabolic and inflammatory disorders related to overnutrition.

4.2 Nutritional Regulation of Adipocytes in Obesity and Type 2 Diabetes

Notably, continuous overnutrition is a significant risk factor leading to metabolic disorders.[4] For example, a high-sugar and high-fat diet can cause excess fat accumulation in white adipocytes when hyperplasia (increasing the adipocyte number) and hypertrophy (increasing the adipocyte size) occur,[39] leading to an imbalance between energy storage and expenditure, and secretion of proinflammatory cytokines.[40] Meanwhile, it can decrease the thermogenic activity of beige and brown adipocytes, disturbing energy dissipation,[41] causing obesity and its complications with systemic inflammation, glucose intolerance, and insulin resistance.[9,42] Therefore, to combat obesity and its associated complications such as Type 2 diabetes, it is a major emergency to develop pharmacological and nutritional strategies to inhibit the process of adipocyte hyperplasia and hypertrophy associated with energy storage and/or increasing thermogenic activity *via* browning of adipose tissues promoting energy dissipation.[7,16,43]

Fortunately, dietary bioactive compounds can alleviate obesity and insulin sensitivity *via* the inhibition of adipose tissue expansion and inflammation in both rodents and humans.[35,38,44] Dietary capsaicin markedly decreases visceral adipocyte size, along with reducing obesity and insulin resistance and hepatic steatosis in mice fed a high-fat diet (HFD).[45–47] Resveratrol, previously known to extend lifespan, impacts mitochondrial function and metabolic homeostasis, prevents diet-induced obesity, and promotes insulin sensitivity in obese mice and humans.[48] Interestingly, resveratrol improves adipose insulin sensitivity and reduces the inflammatory response, with adipocyte size decreasing in visceral, but not subcutaneous adipose, tissues of rhesus monkeys.[49] In addition, a low-dose dietary supply of luteolin, which is a flavonoid that has antioxidant and anti-inflammatory properties, can improve obesity and insulin resistance in diet-induced obese mice, and of course, reduce visceral adipocyte diameters.[50,51] Similarly, quercetin in the diet can improve HFD-induced obesity and promote insulin sensitivity when reducing adipocyte diameters in epididymal adipose tissues of mice.[52] Moreover, dietary curcumin has a potential benefit in preventing diet-induced obesity and insulin resistance by suppressing angiogenesis in adipose tissue.[53] Dietary n-3 PUFA intake can limit both hypertrophy and hyperplasia of fat cells, which reduce the accumulation of body fat in HFD-induced obese mice.[54]

On the other hand, dietary bioactive components can regulate whole-body metabolism *via* browning of beige and brown adipocytes to increase energy expenditure by catabolizing macronutrients, such as carbohydrate and saturated fatty acid.[55] For instance, through activation of mitochondrial UCP1 and mitochondrial biogenesis, caffeine induces browning features of adipocyte *in vitro* and *in vivo*.[56] Dietary capsaicin enhances browning of WAT and counters obesity by activating transient receptor potential vanilloid-1 (TRPV1) channel-dependent mechanisms.[46,57] When enabling AMP-activated protein kinase (AMPK) α_1, resveratrol promotes brown adipocyte formation and function in HFD-fed mice by increasing energy expenditure and reducing fat accumulation in adipose depots.[58,59] Also, quercetin can remodel white adipocytes to brown-like adipocytes, and the combination of resveratrol and quercetin induces browning in white adipose tissue and decrease of HFD-fed rats.[60,61] Moreover, dietary luteolin is able to reduce fat mass and increase energy expenditure when activating adipose tissue browning and thermogenesis in HFD-fed mice.[62] Additionally, dietary lycopene supplementation improves diet-induced obesity and insulin resistance by decreasing adipocyte hypertrophy and lipid accumulation together with the stimulation of thermogenic and mitochondrial functional genes in the adipose tissue.[63] Similarly, insufficient intake of n-3 PUFA is correlated with overweight and obesity,[64] and n-3 PUFA has beneficial effects on thermogenic function when promoting mitochondrial biogenesis and beige-like features to strengthen the recruitment of brown and beige adipocytes.[65,66]

These dietary bioactive compounds provide beneficial strategies in limiting both hypertrophy and hyperplasia of fat cells by reducing the accumulation of body fat, and on the other hand, increasing heat production efficiently *via* browning of adipose tissues and promoting energy dissipation in the long-term treatment of obesity and Type 2 diabetes.

4.3 Nutritional Regulation of Nutrition on Immune Cells in Obesity and Type 2 Diabetes

Immune cells act in cooperation with adipocytes to maintain the overall metabolic and inflammatory homeostasis when obesity disturbs the immune system and promotes inflammation, and this obesity-related inflammation potentially rescues a variety of chronic conditions and diseases.[16,67–70] As the primary source of cytokines in adipose tissue depots, macrophages can be observed histologically as "crown-like structures" surrounded by adipocytes that are dead or dying.[71] Two types of macrophages polarize in adipose depots: M1 and M2 macrophages.[72] Once activated, M1 macrophages exert a proinflammatory phenotype expressing and secreting proinflammatory cytokines, whereas alternatively activated M2 macrophages play a role in tissue remodeling and wound healing and secreting anti-inflammatory cytokines such as interleukin-10 (IL-10) in response to interleukin-4 (IL-4) and interleukin-13 (IL-13).[72] In diet-induced obese individuals, more M1 macrophages infiltrate into the adipose tissue depots, causing systemic inflammation and then leading to insulin resistance.[20] Intriguingly, alternative M2 macrophages do not diminish; on the contrary, they may increase in number, although there is a significant increment in the M1/M2 ratio favoring a proinflammatory status.[20 72,73]

Other innate and adaptive immune cells also play significant roles in setting the inflammatory status of the fat pad during chronic overnutrition;[5,19] in fact, almost all known kinds of immune cells are implicated in this process. Neutrophils,[21] mast cells,[22] B lymphocytes,[23,24] dendritic cells,[25] NK cells,[26,27] and CD8+ and CD4+ T lymphocytes[28,29] increase in abundance in the obese adipose tissue, and exert proinflammatory effects under persistent overnutrition. In contrast, eosinophils[30] and ILC2 cells[31] act to limit inflammatory response and maintain insulin sensitivity, and CD4+Foxp3+ Treg cells play a vital role in controlling other immune cells, including macrophages,[32] which are enriched in the normal visceral fat pad but sharply decreased under prolonged overnutrition (Figure 4.2).

Chronic, low-grade systemic inflammation plays a vital role in the development of obesity and associated disorders such as Type 2 diabetes.[4,15,74] Therefore, using natural bioactive products to regulate immune cells and reset the inflammatory status of the obese fat pad in diet-induced obesity is now becoming a research hotspot (Figure 4.2). For example, capsaicin, a spicy component of hot peppers, can suppress the adipose M1 macrophage infiltration that markedly decreases the expression of inflammatory cytokines as well as fasting glucose/insulin and triglyceride levels in both HFD-fed and genetically obese mice.[45,75,76] Similarly, resveratrol ameliorates obesity-associated inflammation by decreasing adipose tissue M1 macrophage infiltration and significantly inhibiting the secretion of proinflammatory cytokines.[49,77] Also, curcumin significantly improves obesity, diabetes, and associated inflammation that require reduction of macrophage infiltration of white adipose tissue in mouse models of HFD-induced or ob/ob obese mice.[53,78] Furthermore, luteolin can suppress inflammatory macrophage infiltration and polarization by activating AMPKα1 signaling and targeting the Toll-like receptor (TLR) signaling pathway in the epididymal adipose tissues of obese mice.[79] Moreover, lycopene inhibits HFD-induced obesity, inflammatory response, macrophage infiltration, and associated metabolic disorders as a result of downregulation in the phosphorylation levels of IκB, and p65 subunit.[63,80] Additionally, n-3 PUFA improved adipose tissue inflammation in HFD-induced obese or genetically diabetic mice, and completely prevented macrophage infiltration and changes in inflammatory gene expression, thereby dissecting obesity from adipose tissue inflammation.[81,82]

However, when it comes to other immune cells, there are only a few studies that have involved the anti-inflammatory properties of dietary bioactive components in adipose tissue depots. For example, quercetin can significantly suppress the accumulation and activation of immune cells that decrease in the ratio of CD4+ to CD8+ T cells, whilst inhibiting inflammatory macrophage infiltration and polarization, and then increase the expression of genes associated with mitochondrial oxidative phosphorylation.[83] Recently, mast cells have been observed to play crucial roles in obesity and diabetes, and their stabilizers can attenuate related metabolic disorders.[22] As a potent natural mast cell stabilizer in the diet, luteolin can reduce mast cell and macrophage infiltration and inflammatory cytokine production in epididymis adipose tissues to ameliorate diet-induced obesity and insulin resistance in mice.[50] The flaxseed-derived ALA exerts an antichemotactic and anti-inflammatory effect on CD8+ T-cell and adipocyte paracrine interactions, and has the potential to mitigate monocyte chemotaxis.[84] Unfortunately, studies of natural bioactive components on other immune cells (such as neutrophils, B lymphocytes, dendritic cells, NK cells, Tregs, eosinophils, and ILC2 cells) associated with inflammation in the adipose tissue depots are rarely found nowadays. Therefore, new developments and innovations of therapeutic strategies are needed that screen new molecules targeting the immune cells to limit inflammation and then restore insulin sensitivity of organs.

4.4 Nutritional Regulation of Inflammatory Signaling Pathways in Obesity and Type 2 Diabetes

Overnutrition activates several proinflammatory signaling pathways, leading to systemic chronic, low-grade inflammation in the context of obesity and Type 2 diabetes. Herein we will give an overview of the molecular mechanisms concerning the inflammatory signaling pathways in response to nutrition excess and the potential anti-inflammatory roles that dietary bioactive products have during the process (Figure 4.3).

4.4.1 5′ Adenosine Monophosphate-activated Protein Kinase (AMPK) as a Central Energy Sensor

As a central energy and nutrition sensor, AMPK is an evolutionally conserved serine/threonine kinase that plays a crucial role in the regulation of cellular and whole-body metabolism homeostasis and organism survival.[85–87]

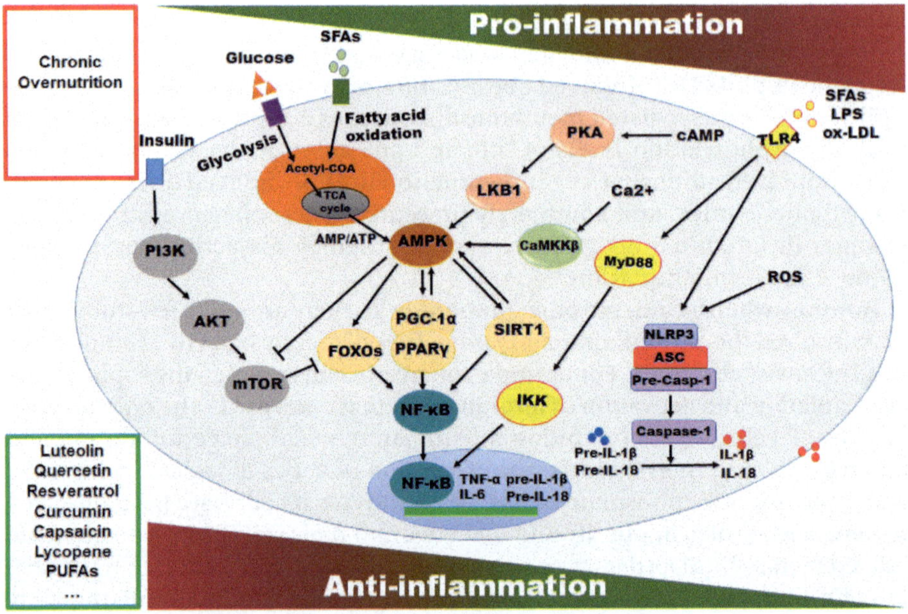

Figure 4.3 Simplified model outlining potential molecular mechanisms of nutrition and inflammation signaling. Obesity and Type 2 diabetes are characterized by hyperglycemia, hyperlipidemia, and hyperinsulinism, as well as systemic chronic, low-grade inflammation during persistent overnutrition. 5′ adenosine monophosphate-activated protein kinase (AMPK) is a central energy sensor also regulating inflammatory signaling through phosphorylation of downstream transcription factors such as silent information regulator T1 (SIRT1), peroxisome proliferator-activated receptor γ (PPARγ), PPARγ coactivator-1alpha (PGC-1α), and Forkhead box O family (FOXOs); all of them can suppress nuclear factor-κB (NF-κB) activities *via* inhibiting the transactivation capacity of p65 subunit. As an inflammatory switch involved in the activation of both the innate and adaptive immune systems, the activation of NF-κB thus stimulates the expression of various proinflammatory genes and leads to a reduction of insulin sensitivity. Glucose intake is broken down into pyruvate by glycolysis, subsequently translocated into the mitochondria and decarboxylated to acetyl-CoA, while saturated fatty acids (SFAs) form fatty acyl-CoA and transport into the mitochondria, forming acetyl-CoA through the β-oxidation pathway. Then the glucose or fat acid-derived acetyl-CoA enters the tricarboxylic acid (TCA) cycle, converting AMP or ADP to ATP when reducing equivalents to the electron transport chain and oxidative phosphorylation. A high-calorie diet with a decline in the ADP/ATP ratio downregulates AMPK activity. On the other hand, the protein kinase serine/threonine kinase 11, also known as liver kinase B1 (LKB1), and the Ca^{2+}/calmodulin-activated protein kinase kinases, particularly CaMKKβ, activate AMPK in responding to cAMP or Ca^{2+} levels respectively. In addition, insulin activates the phosphatidylinositol 3-kinase (PI3K), protein kinase B (Akt/PKB) to directly phosphorylate and activate the mechanistic target of rapamycin (mTOR) that enhances inflammation

Figure 4.3 (*Continued*)

through the inhibition of FOXOs. Toll-like receptor 4 (TLR4) and Nod-like receptor protein 3 (NLRP3) have been studied extensively in obesity-induced insulin resistance and inflammation during long-term overnutrition. The oxidized low-density lipoprotein (ox-LDL), SFAs, and lipopolysaccharide (LPS) can activate the MyD88/IKKβ/NF-κB signaling pathway as well as enhancing the NF-κB-mediated expression of NLRP3. Also, the activation of NLRs by saturated free fatty acids, lower activity of AMPK, or ROS exacerbates NLRP3 inflammasome that comprises NLRP3, procaspase-1, and ASC, leading to the maturation of caspase-1 and the processing of its substrates, IL-1β, favoring inflammation. As the crucial ingredient of nutrients, dietary bioactive compounds display anti-inflammatory properties and increased whole-body insulin sensitivity in diet-induced obesity and Type 2 diabetes targeting AMPK and downstream transcriptional factors that include SIRT1, PGC-1α, PPARγ, FOXOs as well as mTOR. Additionally, dietary bioactive compounds can improve TLR4 and NLRP3 signaling to alleviate adipose inflammatory response. PUFAs, polyunsaturated fatty acids.

Metabolism is fueled by the nutrients that have a direct impact on the energy status of the organism.[88] As a central energy sensor AMPK has a considerable number of downstream substrates; these are typically enzymes and proteins involved in the process of energy metabolic transcriptional regulation.[85] One significant role of AMPK is to respond to metabolic requirements either by stimulating catabolism, including glucose utilization, mobilization of lipid stores, and turnover of macromolecules by autophagy or by inhibiting anabolic pathways such as gluconeogenesis and synthesis of protein and fatty acids.[86] Impaired AMPK activity can cause several problems in many tissues, and it is not surprising that there are so many observations linking AMPK dysfunction with metabolic syndrome, including obesity and Type 2 diabetes.[85–87,89]

Adenosine triphosphate (ATP) is generally a substrate for energy currency that is broken down to adenosine diphosphate (ADP) and then adenosine monophosphate (AMP). On the one hand, increases in AMP/ATP or ADP/ATP ratios activate AMPK signaling pathways that enhance catabolism and decrease anabolism through the phosphorylation of essential proteins in multiple paths to replenish ATP levels and restore energy, whereas a reduction in AMPK activity occurs under prolonged overnutrition with a decline in ADP/ATP ratio.[85,90] On the other hand, two classic upstream kinases in mammals have been established as activators of AMPK: the protein kinase (liver kinase B1) LKB1 complex[91] and the Ca^{2+}/calmodulin-activated protein kinase kinases (CaMKK), particularly CaMKKβ. The LKB1 complex phosphorylates AMPK in response to changes in the AMP/ATP ratio and/or cyclic adenosine monophosphate (cAMP) level,[92,93] whereas CAMKKβ phosphorylates AMPK in response to an increase in intracellular Ca^{2+} concentration.[87,94]

Accumulation of lipids in adipocytes along with the ectopic storage of fat in the liver, muscle, and pancreas stimulates the innate immunity defense, leading to the secretion of proinflammatory cytokines and subsequently to the recruitment of immune cells.[4,5,95] Extensive studies have identified that chronic low-grade inflammation plays a crucial role in the development of metabolic syndrome.[4,96] With long-term overnutrition, the decrement of AMPK activity enhances inflammatory responses *via* inhibition of the downstream activation of nuclear factor-κB (NF-κB), a key regulator of innate immunity and inflammation.[87]

4.4.2 NF-κB Signaling Pathways as an Inflammatory Switch

The inhibitor of the κB kinase-β (IKKβ)/NF-κB pathway is the best-known evolutionarily conserved inflammatory signaling that has been studied extensively over the last two decades.[97] During chronic overnutrition, metabolic diseases such as obesity and Type 2 diabetes are accompanied by hyperlipidemia and hyperglycemia, both of which activate the inflammatory IKKβ/NF-κB pathway in adipocytes and macrophages, stimulating the expression of various genes associated with inflammation, including those encoding proinflammatory cytokines, chemokines, adhesion molecules, and many others, as well as leading to a reduction of insulin sensitivity.[97,98]

Although AMPK has several direct phosphorylation substrates; it suppresses NF-κB signaling indirectly through its downstream mediators such as silent information regulator T1 (SIRT1), Forkhead box O (FOXO) family, and peroxisome proliferator-activated receptor γ (PPARγ)/PPARγ coactivator-1α (PGC-1α), as well as the mechanistic target of rapamycin (mTOR) pathway, which subsequently inhibits the expression of inflammatory genes.[87,95] The interaction of genetic, dietary, lifestyle, and environmental factors favors a chronic positive energy balance and leads to increased body fat mass, causing obesity that facilitates the development of metabolic and inflammatory dysfunction. Indeed, studies have demonstrated that a growing number of natural bioactive products derived from fruits and vegetables exert the beneficial effects of antiobesity and antidiabetes.

4.4.3 AMPK-SIRT1-NF-κB Signaling Pathway

SIRT1, widely expressed in mammals, downstream of AMPK, is an evolutionarily conserved protein that has coordinated functions in metabolism and inflammation.[99–101] SIRT1 responds to increases and decreases in nutrient availability and/or energy expenditure, exerting an anti-inflammatory effect, and improves insulin sensitivity when activated.[102] Briefly, as a nicotinamide adenine dinucleotide (NAD) positive-dependent protein deacetylase, SIRT1 can directly interact with and deacetylate the p65 subunit of NF-κB, resulting in a reduction of the transactivation capacity of p65 subunit and consequently suppressing the transcription of the NF-κB-dependent gene expression and inflammation.[100,101]

An enormous number of natural products are involved in AMPK/SIRT1. For instance, through a mechanism including AMPKα1/SIRT1, quercetin can reduce obesity-associated ATM infiltration and inflammation in mice.[50]

Similarity, dietary luteolin can activate AMPK activity in macrophage to inhibit their inflammatory polarization and enhance insulin sensitivity in adipocytes.[103] Also, dietary capsaicin reduces obesity-related inflammation and metabolic disorders such as insulin resistance by suppressing the expression of proinflammatory cytokines (*e.g.*, TNF-α, MCP1, IL-6) and reducing macrophage infiltration in fat pads.[45,57,75,76] Further, resveratrol activates AMPK and SIRT1 and then modulates PGC-1α to improve mitochondrial function and enhances brown adipocyte formation, leading to improved adipose insulin sensitivity and reduced chronic low-grade-inflammation in mice and rhesus monkeys on a HFD.[48,49,58] Curcumin increases the expression of AMPK, leading to a diminishment of NF-κB protein in spontaneous obesity and Type 2 diabetes mice, even though it cannot modify the expression of PGC-1α or SIRT1.[104] n-3 PUFAs exert anti-inflammatory functions that antagonize macrophage inflammation *via* activation of AMPK/SIRT1 when DHA increases expression, phosphorylation, and activity of the major isoform AMPKα1 and then leads to SIRT1 overexpression.[82] On the other hand, DHA mimics the effect of SIRT1 on deacetylation of the p65 subunit of NF-κB and inhibits its downstream cytokine expression.[37] Similarly, EPA has been reported to have beneficial effects in obesity-associated metabolic disorders. Dietary EPA intake can prevent and reverse insulin resistance in diabetes or diet-induced obese mice *via* modulation of adipose tissue inflammation.[105,106]

4.4.4 AMPK-FOXOs-NF-κB Signaling Pathway

In mammals, FOXOs proteins are a family of transcription factors, downstream of AMPK, that act at the interconnections between metabolic and inflammatory pathways inducible by many important signal transducers and mediators, including p53, Ikkβ, NF-κB, and others.[107] FOXO1 appears to play a crucial role in adipocyte differentiation, acting as an inhibitor of adipogenesis at the early differentiated phase, thus regulating energy storage and expenditure through regulation of adipocyte size and adipose tissue-specific gene expression in WAT and BAT after excessive calorie intake.[108,109] Further, FOXO3 can suppress NF-κB activation by inhibiting p65 DNA-binding activity, thus restraining NF-κB activation. Overexpression of FOXO3a decreases lipid accumulation and the expression of adipocyte markers during adipocyte differentiation; moreover, it enhances the expression of thermogenic genes accompanied by increased mitochondrial mass in C3H10T1/2 adipocytes.[110,111] Therefore, FOXOs activation presents anti-inflammatory potential specifically *via* the inhibition of NF-κB. Also, as a downstream mediator of insulin signaling, FOXOs activity is modulated by phosphorylation catalyzed by protein kinase B (PKB, also known as Akt) and other kinases.[112] Besides, immunologically, the activation of FOXO facilitates the balances between T helper (Th) cells such as Th1/2/17 and Treg cells; meanwhile, FOXOs cooperatively enhance the early differentiation of Foxp3+Treg cells and then promote anti-inflammatory proceeding.[113]

As a result, FOXOs can be very promising targets for the cure for obesity and Type 2 diabetes. Resveratrol has been proved to alleviate obesity and Type 2 diabetes, and it upregulates FOXO1, alongside SIRT1 and adiponectin in human visceral adipocytes. Also, dietary curcumin intake improves obesity-associated inflammation and diabetes.[34,78] In murine models of insulin-resistant obesity, it significantly increases perigonadal adipose expression of the genes encoding the transcription factor FOXO1 and the anti-inflammatory adipokine adiponectin.[78] Curcumin regulates lipid accumulation by increasing the expression of two lipid transport genes, the fatty acid transporter (FAT/CD36), and the fatty acid-binding protein 4 (FABP4), while enhancing the activity of FOXO3a, thus leading to increased lipid levels in monocytes/macrophages.[114] Further, chronic intake of quercetin reduces oxidative stress and raises the expression of FOXO3-mediated antioxidant enzymes in visceral adipose tissues;[115] on the other hand, quercetin has previously been demonstrated to enhance insulin sensitivity in mice and improve β-cell mass and function in fructose-induced hyperinsulinemia by preserving pancreatic Akt/FOXO1 activation.[116]

Furthermore, lycopene has been shown to inhibit HFD-induced obesity, inflammation, and associated metabolic disorders by reducing the phosphorylation levels of IκB, and p65-related inflammatory responses, while lycopene-rich tomato oleoresin can modulate plasma adiponectin concentration and levels of adiponectin, SIRT1, and FOXO1 in adipose tissue of obese rats. Moreover, the dietary supplement of DHA reduces the expression of FOXO1 and FOXO3 in the adipose tissue of pigs as well as decreases their expressions in human and porcine primary adipocytes.[117]

4.4.5 PGC-1α/PPARγ-NF-κB Signaling Pathway

PPAR-γ, a highly expressed nuclear regulator in adipose tissues, is an essential regulator of adipogenesis as well as a crucial modulator of inflammatory and immune responses.[118] Meanwhile, PGC-1α is a member of a family of transcription coactivators that interacts specifically with PPAR-γ.[119] Moreover, they have been extensively described as inducible coregulators of nuclear receptors to stimulate the transcription of genes concerning the control of cellular energy expenditure and inflammatory response that have been associated with obesity and Type 2 diabetes.[120] Incidentally, AMPK can directly interact with and phosphorylate PGC-1α, while SIRT1 can deacetylate and regulate PGC-1α activity in adipose depots as well.[120–122]

During long-term overnutrition, PGC-1α is downregulated, leading to obesity and insulin resistance accompanied by systemic inflammation.[19], [123] Also, adipose PPARγ expression is upregulated after exposure of mice to a HFD.[124] Conversely, as a negative regulator for macrophage activation, PPAR-γ inhibits gene expression in part by antagonizing the activities of the transcription factors NF-κB.[125,126] PPARγ is highly expressed in F4/80hi adipose tissue macrophages and dampens adipose tissue inflammation.[127] On the other hand, PPARγ activation promotes infiltration of alternatively activated macrophages into adipose depots,[128] and PPARγ agonists inhibit the production of monocyte inflammatory cytokines.[129]

While the PGC-1α and PPARγ systems play a vital role in the control of cellular energy expenditure and inflammatory response,[130–132] in the meantime, many nutritional bioactive components that target PGC-1α and/or PPARγ have been studied and describe promising strategies to combat obesity and Type 2 diabetes. For instance, it has been demonstrated that resveratrol can decrease PGC-1α acetylation and hence PGC-1α activity *via* activating SIRT1, thus repressing PPARγ increases *UCP1* expression in adipose depots.[48,59] Also, dietary curcumin interacts with nuclear transcription factors, including PPARγ as well as NF-κB, downregulating the inflammatory cytokines in adipocytes and macrophages and then reversing hyperglycemia, hyperlipidemia, insulin resistance, and other inflammatory symptoms associated with obesity and associated metabolic diseases.[133,134] Moreover, by inhibiting the transactivation of PPARγ luteolin inhibits adipocyte differentiation and enhances insulin sensitivity.[135,136] Furthermore, lycopene improves diet-induced obesity, insulin resistance, and systemic inflammation by blocking adipocyte hypertrophy and lipid accumulation when modulating the expression of thermogenic and/or mitochondrial functional genes such as PPARγ, PGC-1α, and SIRT1 in the adipose tissue.[63] In addition, n-3 PUFAs are potent PPARγ activators leading to the increased expression of NF-κB while reducing proinflammatory cytokine productions and macrophage infiltration in adipose tissue depots, which improves whole-body lipid metabolism, glucose homeostasis, insulin sensitivity, and inflammation during the development of obesity and Type 2 diabetes.[81,82,137,138]

4.4.6 mTOR Signaling Pathway

The mTOR protein is a serine/threonine-protein kinase that belongs to the phosphatidylinositol 3-kinase (PI3K)-related kinase family to form two distinct complexes named mTOR complex (mTORC1 and mTORC2), which is an inhibitory insulin regulatory protein.[139] mTOR is one of the well-known signaling modulators that integrates environmental nutrients, cellular energy status, and whole-body immune responses, and plays a critical role in the regulation of adipose tissue biology and function, including adipogenesis, lipid metabolism, glucose homeostasis, thermogenesis, and inflammation.[139–142] Dysregulation in the mTOR signaling pathway is associated with obesity and insulin resistance under persistent overnutrition.[139] When it comes to the mTOR signaling pathway, briefly, the upstream growth factors such as insulin and insulin-like growth factor 1 (IGF1) stimulate the PI3K, thus enabling the effector kinases–PKB to directly phosphorylate and activate the mTOR inhibitor–tuberous sclerosis complex-1/2 (TSC1 and TSC2) and IKKβ, resulting in the activation of mTOR complexes. Activated by mTOR complex, S6 kinase 1 (S6K1) directly phosphorylates the insulin receptor substrate (IRS) and promotes IRS degradation, resulting in increased feedback phosphorylation of Akt.[139–141] In addition, specific mTORC1 signaling disruption in macrophage and other innate immune cells reduces inflammation and increases insulin sensitivity of obese mice.[142,143]

Moreover, nutrients and other factors that influence energy status, leading to a rise in intracellular AMP levels, which activate AMPK, exert inhibitory effects on mTORC1 through activation of TSC2 and phosphorylation. For example, the HFD results in adipose depots mediated by vascular dysfunction and remodeling through the downregulation of AMPK with concurrent upregulation of mTOR.[144] Besides, phosphorylated FOXO1 and FOXO3a can restrain mTORC2 activities through serum- and glucocorticoid-induced protein kinase 1 (SGK1) protein and mTORC2 negatively regulates TLR4-mediated inflammatory response *via* FOXO1.[145]

Since mTOR signaling plays a crucial role in the control of whole-body metabolism, nutrition, and inflammation, it is becoming a new potential therapeutic target for the development of drugs curing obesity and Type 2 diabetes, and enormous natural products in the diet. Most frequently, resveratrol has been observed to bring beneficial effects in obesity and Type 2 diabetes mice on a high-calorie diet by inhibiting the activity of the mTOR signaling proteins and increasing AMPK and LKB1 activity, which can suppress mTOR signaling.[146,147] Besides, quercetin can attenuate obesity and inflammation through the mechanism inhibiting mTOR signaling that downregulates AKT and PI3K, which are essential factors upstream of mTOR, and decreasing mTOR and ribosomal protein S6 kinase beta-1 (S6K1, also known as p70S6K) protein in obesity-induced mice.[148] Other natural products, such as capsaicin, curcumin, lycopene, and n-3 PUFA, also have been proved to regulate the Akt/PI3K/mTOR signaling pathway in cancer or fatty liver disease,[149–151] and the regulation of inflammation and insulin sensitivity in adipose depots needs further study.

4.4.7 The Roles of TLRs and NLRPs in Inflammation

Emerging evidence has suggested that obesity and Type 2 diabetes are characterized by hyperglycemia, hyperlipidemia, and hyperinsulinism as well as systemic chronic, low-grade inflammation, indicating the interaction between nutrition, metabolism, and inflammation.[5,6] Innate immune cells such as macrophages initiate inflammatory immune responses *via* the recognition of pattern recognition receptors (PRRs) that include TLRs at the cell surface and intracellular NOD-like receptors (NLRs).[6,152,153] Recently, TLR4 and NOD-like receptor protein 3 (NLRP3) have been studied extensively in obesity-induced insulin resistance and inflammation during long-term overnutrition.[152–156] Briefly, oxidized low-density lipoprotein (ox-LDL), saturated free fatty acids, or lipopolysaccharide (LPS) can activate the myeloid differentiation primary response gene 88(MyD88)-IKKβ-NF-κB signaling pathway, leading to proinflammatory cytokine secretion as well as enhancing the NF-κB-mediated expression of NLRP3.[156,157] On the other hand, the activation of NLRs by saturated free fatty acids, lower activity of AMPK, or reactive oxygen species (ROS) exacerbates NLRP3 inflammasome that comprises an intracellular sensor

NLRP3, the precursor procaspase-1 and the adaptor-apoptosis-associated speck-like protein containing a CARD (ASC) leading to the activation of caspase-1 and then processing the maturation and secretin of its substrates, IL-1β, and interleukin-18 (IL-18)[153] (Figure.4.3).

Now that TLR and NLR signaling pathways play crucial roles in the regulation of inflammation during the pathogenesis of diet-induced obesity and Type 2 diabetes, many dietary bioactive compounds have been proved to disturb the process. For instance, in the adipose tissue of diabetic mice, resveratrol inhibits ROS-associated mitochondrial fission by upregulating dynamin-related protein 1 (Drp1) phosphorylation *via* AMPK, alleviating inflammation, and reducing cell apoptosis through NLRP3 inflammasome activation.[158] And luteolin effectively inhibits the production of ROS and expression of NLRP3, ASC, caspase-1, IL-18, and IL-1β proteins in RAW264.7 cells incubated with LPS, whereas luteolin pretreatment can enhance the expression of M2 macrophage marker arginase 1 (Arg-1) and IL-10 and decrease the expression of M1 macrophage marker TNF-α, IL-6, and inducible nitric oxide synthase (iNOS).[159] Besides, curcumin inhibits caspase-1 activation, and IL-1β secretion in the LPS priming inflammasome-activated mouse bone marrow-derived macrophages, and then through the inhibition of the NLRP3 inflammasome activation, curcumin prevents inflammation, reducing proinflammatory cytokine IL-1β secretion and improving HFD-induced insulin resistance in mice.[160] Furthermore, even though dietary saturated fatty acids activate NF-κB signaling *via* TLR4 and prime the NLRP3 inflammasome activation in diet-induced insulin resistance,[161] the PUFAs prevent NLRP3 inflammasome-dependent inflammation from the metabolic syndrome in a HFD-induced Type 2 diabetes model and thus suppress adipose macrophage activation.[162,163]

4.5 Conclusion

With progress in modern technology and agriculture of our society so that we have stable and excess nutrition supplies, a growing number of metabolic dysfunction disorders such as obesity and Type 2 diabetes occur with long-term overnutrition. Long-term overnutrition exposure, when excess energy is stored in adipose tissue depots, can disturb metabolic and physiological homeostasis, leading to whole-body metabolism dysfunction such as obesity and Type 2 diabetes, characterized by a chronic, low-grade inflammatory status with immune cell infiltration and the activation of inflammatory signaling pathways. Over the last two decades, studies have observed that with long-term overnutrition adipose tissue depots display characteristic features. First of all, adipocytes become dysfunctional and hypertrophy and hyperplasia occur, along with the whitening of brown adipocytes. The second feature is that almost every cluster of innate immune cells is associated with the initiation and integration of chronic low-grade inflammatory responses. Thirdly, the energy-intensive metabolic and inflammatory signaling pathways that include the molecules and networks integral to inflammatory responses integrate metabolism, inflammation, and nutrition intimately with the development of metabolic syndrome, particularly in obesity and Type 2 diabetes.

Fortunately, mother nature gives us a treasure containing an enormous variety of natural bioactive components in our diets with antioxidative and anti-inflammatory properties. The most studied natural products, such as polyphenols, alkaloids, flavonoids, triterpene, and PUFAs, commonly found in fruits, vegetables, and seafood, are bringing about new developments and innovations in therapeutic strategies through reprogramming metabolic and inflammatory pathways in both adipocytes and immune cells of adipose tissue depots, which can help to reduce the incidence of obesity and Type 2 diabetes.

References

1. M. Ezzati and E. Riboli, *N. Engl. J. Med.*, 2013, **369**, 954–964.
2. C. M. Apovian, *N. Engl. J. Med.*, 2016, **374**, 177–179.
3. WHO, https://www.who.int/news-room/fact-sheets/detail/obesity-and-overweight, 2018.
4. G. S. Hotamisligil, *Nature*, 2006, **444**, 860–867.
5. G. S. Hotamisligil and E. Erbay, *Nat. Rev. Immunol.*, 2008, **8**, 923–934.
6. S. M. Reilly and A. R. Saltiel, *Nat. Rev. Endocrinol.*, 2017, **13**, 633–643.
7. M. Giralt and F. Villarroya, *Endocrinology*, 2013, **154**, 2992–3000.
8. A. Pfeifer and L. S. Hoffmann, *Annu. Rev. Pharmacol. Toxicol.*, 2015, **55**, 207–227.
9. J. K. Sethi and A. J. Vidal-Puig, *J. Lipid Res.*, 2007, **48**, 1253–1262.
10. M. E. Lidell, M. J. Betz, O. D. Leinhard, M. Heglind, L. Elander, M. Slawik, T. Mussack, D. Nilsson, T. Romu, P. Nuutila, K. A. Virtanen, F. Beuschlein, A. Persson, M. Borga and S. Enerbäck, *Nat. Med.*, 2013, **19**, 631–634.
11. K. A. Virtanen, M. E. Lidell, J. Orava, M. Heglind, R. Westergren, T. Niemi, M. Taittonen, J. Laine, N. J. Savisto, S. Enerback and P. Nuutila, *N. Engl. J. Med.*, 2009, **360**, 1518–1525.
12. A. Vegiopoulos, K. Muller-Decker, D. Strzoda, I. Schmitt, E. Chichelnitskiy, A. Ostertag, M. Berriel Diaz, J. Rozman, M. Hrabe de Angelis, R. M. Nusing, C. W. Meyer, W. Wahli, M. Klingenspor and S. Herzig, *Science*, 2010, **328**, 1158–1161.
13. J. Wu, P. Bostrom, L. M. Sparks, L. Ye, J. H. Choi, A. H. Giang, M. Khandekar, K. A. Virtanen, P. Nuutila, G. Schaart, K. Huang, H. Tu, W. D. van Marken Lichtenbelt, J. Hoeks, S. Enerback, P. Schrauwen and B. M. Spiegelman, *Cell*, 2012, **150**, 366–376.
14. T. Tchkonia, D. E. Morbeck, T. Von Zglinicki, J. Van Deursen, J. Lustgarten, H. Scrable, S. Khosla, M. D. Jensen and J. L. Kirkland, *Aging Cell*, 2010, **9**, 667–684.
15. J. R. Brestoff and D. Artis, *Cell*, 2015, **161**, 146–160.
16. F. Villarroya, R. Cereijo, J. Villarroya, A. Gavalda-Navarro and M. Giralt, *Cell Metab.*, 2018, **27**, 954–961.
17. H. S. Schipper, B. Prakken, E. Kalkhoven and M. Boes, *Trends Endocrinol. Metab.*, 2012, **23**, 407–415.

18. A. W. Ferrante, Jr., *Diabetes Obes. Metab.*, 2013, **15** (Suppl 3), 34–38.
19. S. S. Choe, J. Y. Huh, I. J. Hwang, J. I. Kim and J. B. Kim, *Front. Endocrinol. (Lausanne)*, 2016, 7, 30.
20. C. N. Lumeng, J. L. Bodzin and A. R. Saltiel, *J. Clin. Invest.*, 2007, **117**, 175–184.
21. S. Talukdar, D. Y. Oh, G. Bandyopadhyay, D. Li, J. Xu, J. McNelis, M. Lu, P. Li, Q. Yan, Y. Zhu, J. Ofrecio, M. Lin, M. B. Brenner and J. M. Olefsky, *Nat. Med.*, 2012, **18**, 1407–1412.
22. J. Liu, A. Divoux, J. Sun, J. Zhang, K. Clement, J. N. Glickman, G. K. Sukhova, P. J. Wolters, J. Du, C. Z. Gorgun, A. Doria, P. Libby, R. S. Blumberg, B. B. Kahn, G. S. Hotamisligil and G. P. Shi, *Nat. Med.*, 2009, **15**, 940–945.
23. D. A. Winer, S. Winer, L. Shen, P. P. Wadia, J. Yantha, G. Paltser, H. Tsui, P. Wu, M. G. Davidson, M. N. Alonso, H. X. Leong, A. Glassford, M. Caimol, J. A. Kenkel, T. F. Tedder, T. McLaughlin, D. B. Miklos, H. M. Dosch and E. G. Engleman, *Nat. Med.*, 2011, **17**, 610–617.
24. J. DeFuria, A. C. Belkina, M. Jagannathan-Bogdan, J. Snyder-Cappione, J. D. Carr, Y. R. Nersesova, D. Markham, K. J. Strissel, A. A. Watkins, M. Zhu, J. Allen, J. Bouchard, G. Toraldo, R. Jasuja, M. S. Obin, M. E. McDonnell, C. Apovian, G. V. Denis and B. S. Nikolajczyk, *Proc. Natl. Acad. Sci. U. S. A.*, 2013, **110**, 5133–5138.
25. A. Bertola, T. Ciucci, D. Rousseau, V. Bourlier, C. Duffaut, S. Bonnafous, C. Blin-Wakkach, R. Anty, A. Iannelli, J. Gugenheim, A. Tran, A. Bouloumie, P. Gual and A. Wakkach, *Diabetes*, 2012, **61**, 2238–2247.
26. D. O'Shea and A. E. Hogan, *Cancers (Basel)*, 2019, **11**, 573.
27. S. Boulenouar, X. Michelet, D. Duquette, D. Alvarez, A. E. Hogan, C. Dold, D. O'Connor, S. Stutte, A. Tavakkoli, D. Winters, M. A. Exley, D. O'Shea, M. B. Brenner, U. von Andrian and L. Lynch, *Immunity*, 2017, **46**, 273–286.
28. S. Nishimura, I. Manabe, M. Nagasaki, K. Eto, H. Yamashita, M. Ohsugi, M. Otsu, K. Hara, K. Ueki, S. Sugiura, K. Yoshimura, T. Kadowaki and R. Nagai, *Nat. Med.*, 2009, **15**, 914–920.
29. S. Winer, Y. Chan, G. Paltser, D. Truong, H. Tsui, J. Bahrami, R. Dorfman, Y. Wang, J. Zielenski, F. Mastronardi, Y. Maezawa, D. J. Drucker, E. Engleman, D. Winer and H. M. Dosch, *Nat. Med.*, 2009, **15**, 921–929.
30. D. Wu, A. B. Molofsky, H. E. Liang, R. R. Ricardo-Gonzalez, H. A. Jouihan, J. K. Bando, A. Chawla and R. M. Locksley, *Science*, 2011, **332**, 243–247.
31. A. B. Molofsky, J. C. Nussbaum, H. E. Liang, S. J. Van Dyken, L. E. Cheng, A. Mohapatra, A. Chawla and R. M. Locksley, *J. Exp. Med.*, 2013, **210**, 535–549.
32. M. Feuerer, L. Herrero, D. Cipolletta, A. Naaz, J. Wong, A. Nayer, J. Lee, A. B. Goldfine, C. Benoist, S. Shoelson and D. Mathis, *Nat. Med.*, 2009, **15**, 930–939.
33. C. D. Berdanier and L. A. Berdanier, *Advanced Nutrition: Macronutrients, Micronutrients, and Metabolism*, CRC Press, Boca Raton, 2015.
34. A. Asai and T. Miyazawa, *J. Nutr.*, 2001, **131**, 2932–2935.

35. G. A. Mohamed, S. R. M. Ibrahim, E. S. Elkhayat and R. S. El Dine, *Bull Faculty Pharm Cairo Univ*, 2014, **52**, 269–284.
36. S. Wang, M. H. Pan, W. L. Hung, Y. C. Tung and C. T. Ho, *Food Funct.*, 2019, **10**, 1263–1279.
37. N. Alves, S. Valdes, C. Silveira, H. Duarte-Martino, F. I. Milagro-Yoldi, M. J. Moreno-Aliaga and S. Ribeiro, *Open Nutraceut. J.*, 2012, **5**, 193–206.
38. M. Meydani and S. T. Hasan, *Nutrients*, 2010, **2**, 737–751.
39. J. Jo, O. Gavrilova, S. Pack, W. Jou, S. Mullen, A. E. Sumner, S. W. Cushman and V. Periwal, *PLoS Comput. Biol.*, 2009, **5**, e1000324.
40. E. D. Rosen and B. M. Spiegelman, *Nature*, 2006, **444**, 847–853.
41. A. Bartelt and J. Heeren, *Nat. Rev. Endocrinol.*, 2014, **10**, 24–36.
42. L. Lionetti, M. P. Mollica, A. Lombardi, G. Cavaliere, G. Gifuni and A. Barletta, *Nutr. Metab. Cardiovasc. Dis.*, 2009, **19**, 146–152.
43. V. Pellegrinelli, S. Carobbio and A. Vidal-Puig, *Diabetologia*, 2016, **59**, 1075–1088.
44. P. Paoli, A. Caselli, G. Camici and P. Cirri, *Curr. Med. Chem.*, 2015, **22**, 80–111.
45. J. H. Kang, T. Goto, I. S. Han, T. Kawada, Y. M. Kim and R. Yu, *Obesity (Silver Spring)*, 2010, **18**, 780–787.
46. P. Baskaran, V. Krishnan, J. Ren and B. Thyagarajan, *Br. J. Pharmacol.*, 2016, **173**, 2369–2389.
47. G.-R. Lee, M. K. Shin, D.-J. Yoon, A.-R. Kim, R. Yu, N.-H. Park and I.-S. Han, *Obesity*, 2012, **21**, 115–122.
48. M. Lagouge, C. Argmann, Z. Gerhart-Hines, H. Meziane, C. Lerin, F. Daussin, N. Messadeq, J. Milne, P. Lambert, P. Elliott, B. Geny, M. Laakso, P. Puigserver and J. Auwerx, *Cell*, 2006, **127**, 1109–1122.
49. Y. Jimenez-Gomez, J. A. Mattison, K. J. Pearson, A. Martin-Montalvo, H. H. Palacios, A. M. Sossong, T. M. Ward, C. M. Younts, K. Lewis, J. S. Allard, D. L. Longo, J. P. Belman, M. M. Malagon, P. Navas, M. Sanghvi, R. Moaddel, E. M. Tilmont, R. L. Herbert, C. H. Morrell, J. M. Egan, J. A. Baur, L. Ferrucci, J. S. Bogan, M. Bernier and R. de Cabo, *Cell Metab.*, 2013, **18**, 533–545.
50. N. Xu, L. Zhang, J. Dong, X. Zhang, Y. G. Chen, B. Bao and J. Liu, *Mol. Nutr. Food Res.*, 2014, **58**, 1258–1268.
51. E. Y. Kwon, U. J. Jung, T. Park, J. W. Yun and M. S. Choi, *Diabetes*, 2015, **64**, 1658–1669.
52. J. Dong, X. Zhang, L. Zhang, H. X. Bian, N. Xu, B. Bao and J. Liu, *J. Lipid Res.*, 2014, **55**, 363–374.
53. A. Ejaz, D. Wu, P. Kwan and M. Meydani, *J. Nutr.*, 2009, **139**, 919–925.
54. J. Ruzickova, M. Rossmeisl, T. Prazak, P. Flachs, J. Sponarova, M. Veck, E. Tvrzicka, M. Bryhn and J. Kopecky, *Lipids*, 2004, **39**, 1177–1185.
55. M. Okla, J. Kim, K. Koehler and S. Chung, *Adv. Nutr.*, 2017, **8**, 473–483.
56. K. Velickovic, D. Wayne, H. A. L. Leija, I. Bloor, D. E. Morris, J. Law, H. Budge, H. Sacks, M. E. Symonds and V. Sottile, *Sci. Rep.*, 2019, **9**, 9104.
57. S. K. Panchal, E. Bliss and L. Brown, *Nutrients*, 2018, **10**, 630.

58. S. Wang, X. Liang, Q. Yang, X. Fu, M. Zhu, B. D. Rodgers, Q. Jiang, M. V. Dodson and M. Du, *Mol. Nutr. Food Res.*, 2017, **61**, 1600746.
59. J. M. Andrade, A. C. Frade, J. B. Guimaraes, K. M. Freitas, M. T. Lopes, A. L. Guimaraes, A. M. de Paula, C. C. Coimbra and S. H. Santos, *Eur. J. Nutr.*, 2014, **53**, 1503–1510.
60. Y. Zhao, B. Chen, J. Shen, L. Wan, Y. Zhu, T. Yi and Z. Xiao, *Oxid. Med. Cell Longev.*, 2017, **2017**, 1459497.
61. L. Zhao, F. Cen, F. Tian, M. J. Li, Q. Zhang, H. Y. Shen, X. C. Shen, M. M. Zhou and J. Du, *Exp. Ther. Med.*, 2017, **14**, 5942–5948.
62. X. Zhang, Q. X. Zhang, X. Wang, L. Zhang, W. Qu, B. Bao, C. A. Liu and J. Liu, *Int. J. Obes. (*Lond *.)*, 2016, **40**, 1841–1849.
63. J. Wang, Y. Suo, J. Zhang, Q. Zou, X. Tan, T. Yuan, Z. Liu and X. Liu, *J. Nutr. Biochem.*, 2019, **69**, 63–72.
64. M. Micallef, I. Munro, M. Phang and M. Garg, *Br. J. Nutr.*, 2009, **102**, 1370–1374.
65. R. A. Ghandour, C. Colson, M. Giroud, S. Maurer, S. Rekima, G. Ailhaud, M. Klingenspor, E. Z. Amri and D. F. Pisani, *J. Lipid Res.*, 2018, **59**, 452–461.
66. L. M. Laiglesia, S. Lorente-Cebrian, P. L. Prieto-Hontoria, M. Fernandez-Galilea, S. M. Ribeiro, N. Sainz, J. A. Martinez and M. J. Moreno-Aliaga, *J. Nutr. Biochem.*, 2016, **37**, 76–82.
67. K. Sun, C. M. Kusminski and P. E. Scherer, *J. Clin. Invest.*, 2011, **121**, 2094–2101.
68. J. Lu, J. Zhao, H. Meng and X. Zhang, *Front. Immunol.*, 2019, **10**, 1173.
69. K. Makki, P. Froguel and I. Wolowczuk, *ISRN Inflamm.*, 2013, **2013**, 139239.
70. D. Mathis and S. E. Shoelson, *Nat. Rev. Immunol.*, 2011, **11**, 81.
71. S. P. Weisberg, D. McCann, M. Desai, M. Rosenbaum, R. L. Leibel and A. W. Ferrante, Jr., *J. Clin. Invest.*, 2003, **112**, 1796–1808.
72. N. Ouchi, J. L. Parker, J. J. Lugus and K. Walsh, *Nat. Rev. Immunol.*, 2011, **11**, 85–97.
73. C. N. Lumeng, S. M. Deyoung, J. L. Bodzin and A. R. Saltiel, *Diabetes*, 2007, **56**, 16–23.
74. Y. S. Lee, J. Wollam and J. M. Olefsky, *Cell*, 2018, **172**, 22–40.
75. J. H. Kang, C. S. Kim, I. S. Han, T. Kawada and R. Yu, *FEBS Lett.*, 2007, **581**, 4389–4396.
76. J. X. Song, H. Ren, Y. F. Gao, C. Y. Lee, S. F. Li, F. Zhang, L. Li and H. Chen, *Front. Physiol.*, 2017, **8**, 602.
77. B. T. Jeon, E. A. Jeong, H. J. Shin, Y. Lee, D. H. Lee, H. J. Kim, S. S. Kang, G. J. Cho, W. S. Choi and G. S. Roh, *Diabetes*, 2012, **61**, 1444–1454.
78. S. P. Weisberg, R. Leibel and D. V. Tortoriello, *Endocrinology*, 2008, **149**, 3549–3558.
79. E. Y. Kwon and M. S. Choi, *Nutrients*, 2018, **10**, 1415.
80. S. Fenni, H. Hammou, J. Astier, L. Bonnet, E. Karkeni, C. Couturier, F. Tourniaire and J. F. Landrier, *Mol. Nutr. Food Res.*, 2017, **61**, 1601083.

81. J. Todoric, M. Loffler, J. Huber, M. Bilban, M. Reimers, A. Kadl, M. Zeyda, W. Waldhausl and T. M. Stulnig, *Diabetologia*, 2006, **49**, 2109–2119.
82. B. Xue, Z. Yang, X. Wang and H. Shi, *PLoS One*, 2012, 7, e45990.
83. M. Kobori, Y. Takahashi, M. Sakurai, Y. Akimoto, T. Tsushida, H. Oike and K. Ippoushi, *Mol. Nutr. Food Res.*, 2016, **60**, 300–312.
84. J. M. Monk, D. M. Liddle, M. J. Brown, L. Zarepoor, A. A. De Boer, D. W. Ma, K. A. Power, D. W. L. Ma and L. E. Robinson *J. Nutr.*, 2016, **60**, 621–630.
85. D. G. Hardie, F. A. Ross and S. A. Hawley, *Nat. Rev. Mol. Cell Biol.*, 2012, **13**, 251–262.
86. S. Herzig and R. J. Shaw, *Nat. Rev. Mol. Cell Biol.*, 2018, **19**, 121–135.
87. N. B. Ruderman, D. Carling, M. Prentki and J. M. Cacicedo, *J. Clin. Invest.*, 2013, **123**, 2764–2772.
88. R. J. DeBerardinis and C. B. Thompson, *Cell*, 2012, **148**, 1132–1144.
89. G. R. Steinberg and D. Carling, *Nat. Rev. Drug Discov.*, 2019, **18**, 527–551.
90. D. G. Hardie, D. Carling and S. J. Gamblin, *Trends Biochem. Sci.*, 2011, **36**, 470–477.
91. S. A. Hawley, J. Boudeau, J. L. Reid, K. J. Mustard, L. Udd, T. P. Makela, D. R. Alessi and D. G. Hardie, *J. Biol.*, 2003, **2**, 28.
92. M. J. Sanders, P. O. Grondin, B. D. Hegarty, M. A. Snowden and D. Carling, *Biochem. J.*, 2007, **403**, 139–148.
93. S. P. Collins, J. L. Reoma, D. M. Gamm and M. D. Uhler, *Biochem. J.*, 2000, **345**, 673–680.
94. A. Woods, K. Dickerson, R. Heath, S. P. Hong, M. Momcilovic, S. R. Johnstone, M. Carlson and D. Carling, *Cell Metab.*, 2005, **2**, 21–33.
95. A. Salminen, J. M. Hyttinen and K. Kaarniranta, *J. Mol. Med. (Berl.)*, 2011, **89**, 667–676.
96. Y.-H. Lee and R. E. Pratley, *Curr. Diabetes Rep.*, 2005, **5**, 70–75.
97. R. G. Baker, M. S. Hayden and S. Ghosh, *Cell Metab.*, 2011, **13**, 11–22.
98. L. Catrysse and G. van Loo, *Trends Cell Biol.*, 2017, **27**, 417–429.
99. T. T. Schug and X. Li, *Ann. Med.*, 2011, **43**, 198–211.
100. A. Kauppinen, T. Suuronen, J. Ojala, K. Kaarniranta and A. Salminen, *Cell Signal.*, 2013, **25**, 1939–1948.
101. N. B. Ruderman, X. J. Xu, L. Nelson, J. M. Cacicedo, A. K. Saha, F. Lan and Y. Ido, *Am. J. Physiol.: Endocrinol. Metab.*, 2010, **298**, E751–E760.
102. T. Yoshizaki, J. C. Milne, T. Imamura, S. Schenk, N. Sonoda, J. L. Babendure, J. C. Lu, J. J. Smith, M. R. Jirousek and J. M. Olefsky, *Mol. Cell Biol.*, 2009, **29**, 1363–1374.
103. L. Zhang, Y. J. Han, X. Zhang, X. Wang, B. Bao, W. Qu and J. Liu, *Diabetologia*, 2016, **59**, 2219–2228.
104. L. M. Jimenez-Flores, S. Lopez-Briones, M. H. Macias-Cervantes, J. Ramirez-Emiliano and V. Perez-Vazquez, *Molecules*, 2014, **19**, 8289–8302.
105. M. Figueras, M. Olivan, S. Busquets, F. J. Lopez-Soriano and J. M. Argiles, *Obesity (Silver Spring)*, 2011, **19**, 362–369.

106. N. S. Kalupahana, K. Claycombe, S. J. Newman, T. Stewart, N. Siriwardhana, N. Matthan, A. H. Lichtenstein and N. Moustaid-Moussa, *J. Nutr.*, 2010, **140**, 1915–1922.
107. M. Watroba, D. Maslinska and S. Maslinski, *Adv. Med. Sci.*, 2012, **57**, 183–195.
108. D. N. Gross, A. P. van den Heuvel and M. J. Birnbaum, *Oncogene*, 2008, **27**, 2320–2336.
109. J. Nakae, Y. Cao, M. Oki, Y. Orba, H. Sawa, H. Kiyonari, K. Iskandar, K. Suga, M. Lombes and Y. Hayashi, *Diabetes*, 2008, **57**, 563–576.
110. J. H. Choi, N. J. Song, A. R. Lee, D. H. Lee, M. J. Seo, S. Kim, S. H. Chang, D. K. Yang, Y. J. Hwang, K. A. Hwang, T. S. Ha, U. J. Yun and K. W. Park, *Int. J. Mol. Sci.*, 2018, **20**, 26.
111. J. W. Hwang, S. Rajendrasozhan, H. Yao, S. Chung, I. K. Sundar, H. L. Huyck, G. S. Pryhuber, V. L. Kinnula and I. Rahman, *J. Immunol.*, 2011, **187**, 987–998.
112. R. K. Yadav, A. S. Chauhan, L. Zhuang and B. Gan, *Semin. Cancer Biol.*, 2018, **50**, 65–76.
113. C. M. Evans and R. G. Jenner, *Brief Funct. Genomics*, 2013, **12**, 499–511.
114. J. M. Zingg, S. T. Hasan, D. Cowan, R. Ricciarelli, A. Azzi and M. Meydani, *J. Cell Biochem.*, 2012, **113**, 833–840.
115. M. Kobori, Y. Takahashi, Y. Akimoto, M. Sakurai, I. Matsunaga, H. Nishimuro, K. Ippoushi, H. Oike and M. Ohnishi-Kameyama, *J. Funct. Foods*, 2015, **15**, 551–560.
116. J. M. Li, W. Wang, C. Y. Fan, M. X. Wang, X. Zhang, Q. H. Hu and L. D. Kong, *J. Evidence-Based Complementary Altern. Med.*, 2013, **2013**, 303902.
117. Y.-J. Chen, C.-C. Chen, T.-K. Li, P.-H. Wang, L.-R. Liu, F.-Y. Chang, Y.-C. Wang, Y.-H. Yu, S.-P. Lin, H. J. Mersmann and S.-T. Ding, *J. Nutr. Biochem.*, 2012, **23**, 1609–1616.
118. E. Fuentes, L. Guzman-Jofre, R. Moore-Carrasco and I. Palomo, *Mol. Med. Rep.*, 2013, **8**, 1611–1616.
119. P. Puigserver, Z. Wu, C. W. Park, R. Graves, M. Wright and B. M. Spiegelman, *Cell,* 1998, **92**, 829–839.
120. C. Canto and J. Auwerx, *Curr. Opin. Lipidol.*, 2009, **20**, 98–105.
121. Z. Zhang, H. Zhang, B. Li, X. Meng, J. Wang, Y. Zhang, S. Yao, Q. Ma, L. Jin, J. Yang, W. Wang and G. Ning, *Nat. Commun.*, 2014, **5**, 5493.
122. J. T. Rodgers, C. Lerin, W. Haas, S. P. Gygi, B. M. Spiegelman and P. Puigserver, *Nature,* 2005, **434**, 113.
123. K. Ruschke, L. Fishbein, A. Dietrich, N. Kloting, A. Tonjes, A. Oberbach, M. Fasshauer, J. Jenkner, M. R. Schon, M. Stumvoll, M. Bluher and C. S. Mantzoros, *Eur. J. Endocrinol.*, 2010, **162**, 515–523.
124. A. Vidal-Puig, M. Jimenez-Linan, B. B. Lowell, A. Hamann, E. Hu, B. Spiegelman, J. S. Flier and D. E. Moller, *J. Clin. Invest.*, 1996, **97**, 2553–2561.
125. M. Ricote, A. C. Li, T. M. Willson, C. J. Kelly and C. K. Glass, *Nature*, 1998, **391**, 79–82.

126. N. R. Gough, *Sci. Signaling*, 2010, **3**, ec296–ec296.
127. J. Bassaganya-Riera, S. Misyak, A. J. Guri and R. Hontecillas, *Cell Immunol.*, 2009, **258**, 138–146.
128. J. I. Odegaard, R. R. Ricardo-Gonzalez, M. H. Goforth, C. R. Morel, V. Subramanian, L. Mukundan, A. Red Eagle, D. Vats, F. Brombacher, A. W. Ferrante and A. Chawla, *Nature*, 2007, **447**, 1116–1120.
129. C. Jiang, A. T. Ting and B. Seed, *Nature*, 1998, **391**, 82–86.
130. P. Puigserver and B. M. Spiegelman, *Endocr. Rev.*, 2003, **24**, 78–90.
131. B. N. Finck and D. P. Kelly, *J. Clin. Invest.*, 2006, **116**, 615–622.
132. M. Ahmadian, J. M. Suh, N. Hah, C. Liddle, A. R. Atkins, M. Downes and R. M. Evans, *Nat. Med.*, 2013, **19**, 557–566.
133. A. Shehzad, T. Ha, F. Subhan and Y. S. Lee, *Eur. J. Nutr.*, 2011, **50**, 151–161.
134. Z. Song, X. Revelo, W. Shao, L. Tian, K. Zeng, H. Lei, H. S. Sun, M. Woo, D. Winer and T. Jin, *Obesity (Silver Spring)*, 2018, **26**, 547–558.
135. H. S. Park, S. H. Kim, Y. S. Kim, S. Y. Ryu, J. T. Hwang, H. J. Yang, G. H. Kim, D. Y. Kwon and M. S. Kim, *Biofactors*, 2009, **35**, 373–379.
136. L. Ding, D. Jin and X. Chen, *J. Nutr. Biochem.*, 2010, **21**, 941–947.
137. C. M. Mejia-Barradas, B. E. Del-Rio-Navarro, A. Dominguez-Lopez, R. Campos-Rodriguez, M. Martinez-Godinez, S. Rojas-Hernandez, E. Lara-Padilla, E. Abarca-Rojano and A. Miliar-Garcia, *Endocrine*, 2014, **45**, 98–105.
138. E. Oliver, F. McGillicuddy, C. Phillips, S. Toomey and H. M. Roche, *Proc. Nutr. Soc.*, 2010, **69**, 232–243.
139. M. Laplante and D. M. Sabatini, *Cell*, 2012, **149**, 274–293.
140. H. Cai, L. Q. Dong and F. Liu, *Trends Pharmacol. Sci.*, 2016, **37**, 303–317.
141. J. J. Howell and B. D. Manning, *Trends Endocrinol. Metab.*, 2011, **22**, 94–102.
142. H. Jiang, M. Westerterp, C. Wang, Y. Zhu and D. Ai, *Diabetologia*, 2014, **57**, 2393–2404.
143. T. Weichhart, G. Costantino, M. Poglitsch, M. Rosner, M. Zeyda, K. M. Stuhlmeier, T. Kolbe, T. M. Stulnig, W. H. Horl, M. Hengstschlager, M. Muller and M. D. Saemann, *Immunity*, 2008, **29**, 565–577.
144. L. Ma, S. Ma, H. He, D. Yang, X. Chen, Z. Luo, D. Liu and Z. Zhu, *Hypertens. Res.*, 2010, **33**, 446–453.
145. M. Laplante and D. M. Sabatini, *J. Cell Sci.*, 2009, **122**, 3589–3594.
146. J. A. Baur, K. J. Pearson, N. L. Price, H. A. Jamieson, C. Lerin, A. Kalra, V. V. Prabhu, J. S. Allard, G. Lopez-Lluch, K. Lewis, P. J. Pistell, S. Poosala, K. G. Becker, O. Boss, D. Gwinn, M. Wang, S. Ramaswamy, K. W. Fishbein, R. G. Spencer, E. G. Lakatta, D. Le Couteur, R. J. Shaw, P. Navas, P. Puigserver, D. K. Ingram, R. de Cabo and D. A. Sinclair, *Nature*, 2006, **444**, 337–342.
147. A. L. Widlund, J. A. Baur and O. Vang, *Expert Rev. Mol. Med.*, 2013, **15**, e10.
148. M. J. Seo, Y. J. Lee, J. H. Hwang, K. J. Kim and B. Y. Lee, *J. Nutr. Biochem.*, 2015, **26**, 1308–1316.
149. Y. Safdari, M. Khalili, M. A. Ebrahimzadeh, Y. Yazdani and S. Farajnia, *Pharmacol. Res.*, 2015, **93**, 1–10.

150. S. M. Johnson, P. Gulhati, I. Arrieta, X. Wang, T. Uchida, T. Gao and B. M. Evers, *Anticancer Res.*, 2009, **29**, 3185–3190.
151. S. Shin, K. Jing, S. Jeong, N. Kim, K. S. Song, J. Y. Heo, J. H. Park, K. S. Seo, J. Han, J. I. Park, G. R. Kweon, S. K. Park, T. Wu, B. D. Hwang and K. Lim, *Biomed. Res. Int.*, 2013, **2013**, 568671.
152. B. Vandanmagsar, Y. H. Youm, A. Ravussin, J. E. Galgani, K. Stadler, R. L. Mynatt, E. Ravussin, J. M. Stephens and V. D. Dixit, *Nat. Med.*, 2011, **17**, 179–188.
153. H. Wen, J. P. Ting and L. A. O'Neill, *Nat. Immunol.*, 2012, **13**, 352–357.
154. G. M. Tannahill and L. A. O'Neill, *FEBS Lett.*, 2011, **585**, 1568–1572.
155. H. Shi, M. V. Kokoeva, K. Inouye, I. Tzameli, H. Yin and J. S. Flier, *J. Clin. Invest.*, 2006, **116**, 3015–3025.
156. L. A. Velloso, F. Folli and M. J. Saad, *Endocr. Rev.*, 2015, **36**, 245–271.
157. Y. Qiao, P. Wang, J. Qi, L. Zhang and C. Gao, *FEBS Lett.*, 2012, **586**, 1022–1026.
158. A. Li, S. Zhang, J. Li, K. Liu, F. Huang and B. Liu, *Mol. Cell Endocrinol.*, 2016, **434**, 36–47.
159. B. C. Zhang, Z. Li, W. Xu, C. H. Xiang and Y. F. Ma, *Am. J. Transl. Res.*, 2018, **10**, 265–273.
160. H. Yin, Q. Guo, X. Li, T. Tang, C. Li, H. Wang, Y. Sun, Q. Feng, C. Ma, C. Gao, F. Yi and J. Peng, *J. Immunol.*, 2018, **200**, 2835–2846.
161. C. M. Reynolds, F. C. McGillicuddy, K. A. Harford, O. M. Finucane, K. H. Mills and H. M. Roche, *Mol. Nutr. Food Res.*, 2012, **56**, 1212–1222.
162. Y. Yan, W. Jiang, T. Spinetti, A. Tardivel, R. Castillo, C. Bourquin, G. Guarda, Z. Tian, J. Tschopp and R. Zhou, *Immunity*, 2013, **38**, 1154–1163.
163. L. Shen, Y. Yang, T. Ou, C. C. Key, S. H. Tong, R. C. Sequeira, J. M. Nelson, Y. Nie, Z. Wang, E. Boudyguina, S. V. Shewale and X. Zhu, *J. Lipid Res.*, 2017, **58**, 1808–1821.

CHAPTER 5

The Interaction of Nutrition with Nuclear Receptors in Obesity and Diabetes

CLAUDIA TOVAR-PALACIO,[a*†] LILIA G. NORIEGA,[b†] IVAN TORRE VILLALVAZO,[b†] ANDREA DÍAZ-VILLASEÑOR[c†] AND BERENICE PALACIOS-GONZÁLEZ[d†]

[a]División de Nutrición, Instituto Nacional de Ciencias Médicas y Nutrición "Salvador Zubirán"
[b]Departamento de Fisiología de la Nutrición, Instituto Nacional de Ciencias Médicas y Nutrición "Salvador Zubirán"
[c]Instituto de Investigaciones Biomédicas, Universidad Nacional Autónoma de México
[d]Unidad de Vinculación Científica de la Fac. de Medicina UNAM en Instituto Nacional de Medicina Genómica
*Email: claudia.tovarp@incmnsz.mx, tovarpal@gmail.com

5.1 Introduction

All the energy that is required by the body to perform its daily functions is supplied by the nutrients present in food; thus, these nutrients should be readily available. However, food intake is intermittent and varies in nutrient composition from meal to meal. Energy-demanding tissues, such as adipose tissue, liver, muscle, and kidney, need a constant supply of energetic substrates to sustain their functions. Organs with a rapid cell turnover rate, such as the

† All authors contributed equally to this work.

Nutritional Signaling Pathway Activities in Obesity and Diabetes
Food Chemistry, Function and Analysis No. 24
Edited by Zhiyong Cheng

Published by the Royal Society of Chemistry, www.rsc.org

digestive system, need a continuous supply of nutrients to be used as building blocks for cellular proliferation. To achieve such a constant supply of nutrients despite daily variations in nutrient availability, the human body possesses fine-tuned mechanisms for energy storage after feeding and of energy release during periods of food deprivation.

All cellular processes involving the transport, storage, and breakdown of nutrients for ATP generation or synthesis of structural components are tightly controlled at the transcriptional level by the activity of specific transcription factors. Transcription factors bind to specific regions named response elements in the promoter region of genes to activate or repress transcription. Some transcription factors are activated by binding to specific molecules named ligands. These transcription factors are called nuclear receptors and exist as monomers, homodimers, or heterodimers.

Nuclear receptors were first identified nearly 50 years ago as intracellular receptors for steroid hormones.[1,2] Fat-soluble hormones and intermediary metabolites are particularly well suited to mediate intracellular communication because they can pass freely through the cell membrane lipid bilayer.[3,4] Interestingly, although hydrophobic hormones are chemically diverse, their sizes are quite similar. In principle, this regulatory system does not require additional second messenger systems or protein cascades.[3] Thus far, it is known that nuclear receptors are a large superfamily consisting of 48 ligand-regulated transcription factors in humans that are activated by steroid hormones, metabolites such as fatty acids, bile acids (BAs), oxysterols, and xeno- and endobiotics.[5] These molecules act as intracellular receptor ligands that are capable of crossing the plasma membrane.[6,7] Nuclear receptors are ligand-activated transcription factors that regulate target gene expression in a wide variety of physiological pathways such as metabolic processes.[8–11] Ligand binding induces conformational changes to nuclear receptors, allowing them to bind to their respective DNA response element, releasing corepressors and recruiting coactivators to promote the transcription of target genes involved in a variety of cellular processes, including metabolism.[6]

Nuclear receptors share a common structure, comprising a highly variable amino-terminal domain that includes several distinct transactivation regions: the A/B domain, also referred to as AF1 for activation function 1, a central conserved DNA-binding domain that includes two Zn fingers (C domain), a short region responsible for nuclear localization (D domain), and a large fairly well-conserved carboxy-terminal ligand-binding domain (E domain, or LBD) that also contributes to interactions of the subset of nuclear receptors that form heterodimers.[10,11] Some also possess a highly variable carboxy-terminal tail (F domain) that, in most cases, has unknown functions (Figure 5.1A).

5.2 Metabolic Nuclear Receptors

The notion that nuclear receptors integrate complex metabolic homeostasis is supported by systemic analysis of tissue-, time-, and stimulus-dependent nuclear receptor gene expression.[12,13] The nuclear receptors act as metabolic

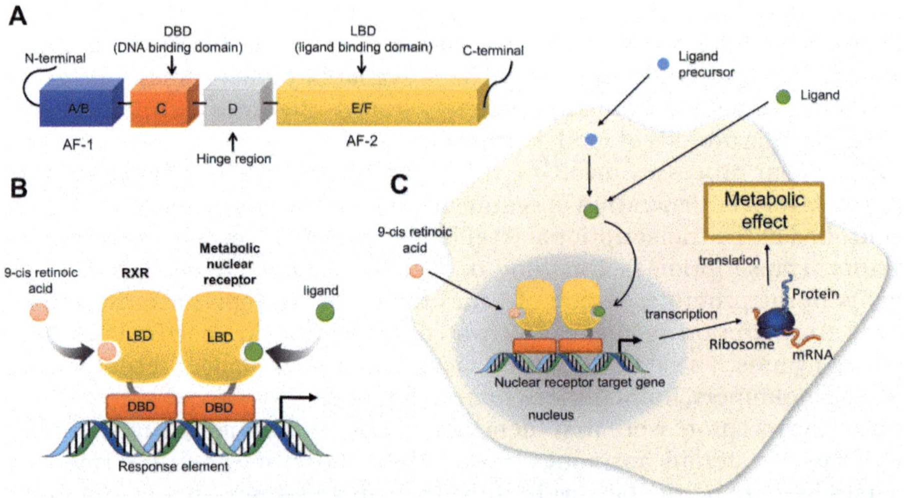

Figure 5.1 Structure and activation of metabolic nuclear receptors. (**A**) Metabolic nuclear receptors are single polypeptides containing four major functional/structural domains: the N-terminal A/B domain, containing a ligand-independent transactivation function (AF)-1 region that binds coactivator proteins; the C region containing the DNA-binding domain (DBD), which binds to specific sequences of DNA in the promoter regions of target genes *via* two zinc finger motifs; the D domain composed of a flexible hinge region that connects the DBD and ligand-binding domains (LBDs); and the E/F or LBD, which contains the hydrophobic pocket for ligand binding and the AF-2 region that binds coregulators. (**B**) Metabolic nuclear receptors function by forming obligate heterodimers with the RXR. In the absence of a ligand, the AF-2 domain is associated with corepressor proteins. Ligand binding induces a conformational modification of the LBD such that the AF-2 domain becomes capable of dimerizing with RXR and recruiting transcriptional coactivators. Metabolic nuclear receptors bind to short DNA sequences named response elements, typically containing the AGGTCA motif. (**C**) The ligands of nuclear receptors are a broad range of small lipophilic molecules that diffuse through the plasma and nuclear membranes. Ligand binding induces conformational changes in nuclear receptors, allowing heterodimerization, DNA binding, and recruitment of coactivators and the transcriptional machinery. Translated genes leave the nucleus to be translated. Common target genes for metabolic nuclear receptors are enzymes and transporters involved in the uptake, synthesis, modification, and degradation of nutrients.

and toxicological sensors, enabling the organism to quickly adapt to environmental changes by inducing metabolic genes and pathways. Moreover, the importance of this family of nuclear receptors in metabolic diseases is exemplified using nuclear receptor ligands for the treatment of diabetes, dyslipidemia, hypercholesterolemia, and other metabolic abnormalities. Genetic studies in humans and rodents support the notion that nuclear receptors control a wide variety of metabolic processes by regulating the expression of genes encoding key enzyme transporters and other proteins involved in metabolic homeostasis. Recognition of the importance of nuclear receptors as master regulators of genes involved in metabolic control has resulted in the intensive search for novel ligands for these receptors that might be used as preventive and therapeutic strategies to combat common diseases such as diabetes and obesity.

Nuclear receptors are typically classified according to the structure of their ligands. This classification groups nuclear receptors into seven families. Additionally, nuclear receptors are classified according to their participation in the homeostatic control of: (1) energy and glucose metabolism through peroxisome proliferator-activated receptor gamma (PPARγ); (2) fatty acid, triglyceride, and lipoprotein metabolism *via* peroxisome proliferator-activated receptor alpha (PPARα), peroxisome proliferator-activated receptor beta/delta (PPARβ/δ), and peroxisome proliferator-activated receptor gamma (PPARγ); (3) reverse cholesterol transport and cholesterol absorption through the liver X receptors (LXRs); and (4) BA metabolism through the farnesoid X receptor (FXR) and LXRs.[5,13] For this reason, these nuclear receptors are considered to be metabolic nuclear receptors, as defined by Francis *et al.*, because they are activated by nutrients, diet metabolites, or drugs and act as metabolic and toxicological sensors.[5,14] Ligands for these metabolic receptors are present in high concentrations in the diet compared with classical nuclear receptor ligands. In fact, nuclear receptors have low affinity for their ligands, so high concentrations of ligands are required to activate the receptors. The transcriptional control of these metabolic circuits requires coordination between nuclear receptors and other transcription factors and coregulators.

5.2.1 The Interaction of Nutrition with Nuclear Receptors

Dietary factors deserve special attention because they have a major role in metabolism. These daily necessary factors are modifiable regarding their quantity, quality, and periodicity of consumption. Additionally, it is important to emphasize that crops by themselves can have seasonal nutritional variations. The dietary energy density and its nutrient composition determine the amount of substrates available for catabolic and anabolic pathways in different organs. Moreover, food-derived molecules exert important metabolic actions, which do not have well-defined mechanisms, thus they comprise a fruitful research area.[15,16] These molecules include plant polyphenols such as catechins and isoflavones, omega-3 and omega-6 polyunsaturated fatty acids (PUFAs), short-chain fatty acids (SCFAs), and sulfur-containing compounds, among others. The therapeutic potential of these bioactive dietary molecules has been evaluated in several basic and clinical studies revealing important beneficial health effects on metabolic diseases, such as obesity and diabetes.[17,18]

These nutrients and their metabolites bind nuclear receptors, as steroid hormones do, adding a new layer to our understanding of metabolic regulation. Nutrient availability elicits metabolic signals that are sensed by nuclear receptors activating cellular signaling pathways (Figure 5.1B). All living organisms are continuously challenged to adapt their physiology to changes in environmental conditions, such as nutrient availability. Therefore, high-energy chronic intake challenges the capacity of the organism to maintain homeostasis, and when this fails, obesity-derived chronic diseases appear. More research is needed to better understand obesity pathogenesis to develop novel dietary approaches to treat these conditions and their comorbidities.

5.2.2 Interaction of Nutrients with Metabolic Nuclear Receptors in Different Tissues During Obesity and Diabetes

Although much has been learned about the physiology of certain nuclear receptors in a limited number of well-studied target tissues, little is known about the function of these receptors throughout the body as a whole. Furthermore, relatively little is known about the complexity of interactions that may exist between nuclear receptors and the different nutrients that act as ligands, which can activate or repress these receptors. Indeed, the ability of nuclear receptors to cross-react with each other's DNA binding sites and to govern similar regulatory cascades implies the existence of higher-order transcriptional regulatory networks and dynamic physiological parameters (Figure 5.1C).

The chapter contains information about the five most important organs involved in nutrient metabolism, as well as in the physiopathology of obesity and diabetes. The chapter includes the intestine, the liver, adipose tissue, skeletal muscle, and kidney, arranged in the following sections in a physiological nutrient order.

5.3 Intestine

The intestine is formed by long, small-diameter segments divided into the duodenum, jejunum, and ileum and by a short, large-diameter part divided into the cecum, colon, rectum, and anus. The functions of the intestine are mainly fourfold: (1) maintenance of fluid and electrolyte homeostasis;[19] (2) digestion and absorption of nutrients;[19] (3) hormone secretion;[20] and (4) a reservoir and place of interaction with the microbiota.[21] These functions are precisely modulated by nuclear receptors in several portions of the intestine.[22] The anatomical profiling of nuclear receptors performed by Bookout *et al.* revealed that the intestine expresses high levels of FXR, hepatocyte nuclear factor 4 gamma (HNF4γ), retinoic acid receptor-related orphan nuclear receptor gamma (RORγ), vitamin D receptor (VDR), PPARβ/δ, retinoid X receptor alpha (RXRα), estrogen-related receptor alpha (ERRα), and glucocorticoid receptor (GR) in duodenum, jejunum, ileum, and colon.[13] Although not highly expressed, other nuclear receptors have also been described to be relevant to maintaining the functions of the intestine, such as PPARγ and LXR alpha (LXRα). Therefore, the roles of PPARs, FXR, and LXRs regarding intestine functions are described below as well as how nutrients positively and negatively affect intestinal modulation.

5.3.1 Physiological Role of FXR on Bile Acid (BA) Homeostasis and Hormone Secretion in the Intestine

BAs modulate lipid absorption in the intestine through activation of FXR. The endogenous ligands of FXR are the primary BAs chenodeoxycholic acid (CDCA) and cholic acid (CA), the secondary BAs deoxycholic acid (DA) and lithocholic acid (LA), and conjugated BA taurocholic acid (TA). When BAs are absorbed by the enterocytes, they activate FXR.[23] The activation of FXR increases the expression of fibroblast growth factor (FGF) 15/19 and organic solute and steroid transporter alpha (OSTα), organic solute and steroid transporter beta (OSTβ), and ileal bile acid binding protein (IBABP). FGF 15/19 is secreted to the portal circulation and activates fibroblast growth factor receptor 4 (FGFR4) in the liver, where it downregulates BA synthesis.[24] In addition to the role of FXR in BA homeostasis, FXR modulates ceramide production[25] and glucagon-like peptide-1 (GLP1) secretion. In fact, FXR interferes with the transcription factor carbohydrate-responsive element vinding protein (ChREBP) and GLP1 secretion by glycolysis inhibition.[26] Additionally, FXR inhibits the transcriptional activity of cAMP response element-binding protein (CREB) in enteroendocrine L-cells, inhibiting the production of GLP1.[27] However, the role of FXR in GLP1 secretion is still controversial since intestine-specific FXR-deficient mice presented augmented GLP1 secretion. Notably, nutrient modulation of FXR pathways during obesity and diabetes is highly impacted by the composition of the microbiota, a mechanism that will be addressed below.

5.3.2 Physiological Role of LXRs on Reverse Cholesterol Transport in the Intestine

Dietary cholesterol is absorbed by the enterocyte through the Niemann-Pick C1-Like 1 (NPC1L1) transporter. Once in the enterocyte, cholesterol can: (1) be excreted back to the intestinal lumen by ATP binding cassette subfamily G member 5 and 8 (ABCG5/ABCG8) and then secreted in feces; or (2) enter the lymph in the form of chylomicrons, mediated by acyl-coenzyme A:cholesterol acyltransferase 2 (ACAT2), or in the form of high-density lipoprotein (HDL) through the ATP-binding cassette transporter (ABCA1) receptor.[28] A diet with an excess of cholesterol increases the oxysterol content in enterocytes activating LXR. Activated LXR promotes the downregulation of NPC1L1 and an increase in ABCG5/ABCG8 and ABCA1 promoting the loss of cholesterol in feces.[29] In fact, activation of LXR in the intestine avoids the development of atherosclerosis in low-density lipoprotein receptor (LDLr)-deficient mice that are prone to atherosclerosis. During obesity and diabetes, there is a reduction in fecal cholesterol elimination,[30] possibly associated with defective LXR signaling.

5.3.2.1 Nutrient Modulation of LXR Signaling in the Intestine

Further research is needed to establish whether nutrients can modulate the LXR pathway in the intestine since this would represent a potential therapeutic approach. For instance, a high glucose concentration increases cholesterol uptake by a mechanism involving a higher expression of NPC1L1 and a decrease in the scavenger receptor class B type 1 (SR-B1), suggesting a reduced activity of LXR signaling; however, the expression of LXR is surprisingly increased.[31,32] Nevertheless, higher expression of LXR does not necessarily equal higher activity, which indicates the importance of evaluating LXR activity together with LXR expression. In addition to glucose, intestinal cholesterol absorption is modulated by PUFAs.[33] PUFAs decrease the expression of NPC1L1 and ABCA1 in human enterocytes, but whether this occurs through an LXR-dependent mechanism is unknown. The potential therapeutic use of LXR activation has been demonstrated by the use of an LXR agonist. In fact, GW3965 or T0901317, both LXR agonists, reduced chylomicron secretion by modifying the localization of SR-B1 and decreasing triglyceride absorption.[34] The work performed in zebrafish models confirms that LXR activation in the intestine regulates the rate of absorption of dietary lipids by promoting a lipid intestinal droplet storage depot and protects animals from hypercholesterolemia and hepatic steatosis induced by an HFD.[35,36]

Bioactive compounds could play an important role in modulating the LXR pathway. In fact, a garlic extract increases LXR expression in the intestine, decreasing cholesterol and triglycerides in serum.[37] Soy protein isoflavones modify the expression of LXR target genes, increasing total fecal BA excretion and insulin sensitivity through an LXR-dependent mechanism since soy protein isoflavones have no effect on LXR-deficient mice.[38]

5.3.3 Dysfunctional Interplay Between the Intestine and Microbiota During Obesity and Diabetes Through FXR and PPARs

The intestine houses more than 100 trillion microbes. The duodenum has fewer bacteria, and the density increases throughout the intestine, reaching the highest number in the cecum and colon. Most intestinal bacteria belong to four phyla: Firmicutes, Bacteroidetes, Actinobacteria, and Proteobacteria.[39] The diet composition is the main determinant of the bacterial population. However, there is a tight interplay between the intestine and microbiota involving: (1) the metabolites produced by the bacteria; (2) the hormones or nutrients released by the enterocytes and the enteroendocrinal cells; and (3) the immune system. Louis *et al.* have extensively reviewed the effect of the different dietary nutrients on microbiota composition.[40] Here, we emphasize the nutrients necessary for producing nuclear receptor ligands once metabolized by bacteria. Ligands can be divided into two groups: a short fatty acid producer that leads to the activation of PPARs and secondary BA producers that activate FXR. Both mechanisms are important regulators of the immune response, which will also be addressed.

5.3.3.1 PPARs

The microbiota species predominant in the intestine are obligate anaerobes that produce SCFAs, such as butyrate, acetate, and propionate, limit the production of nitrate, and increase oxygen consumption by the intestinal epithelial cells. The latter effects are mediated by PPARγ. Butyrate enters enterocytes and activates PPARγ promoting oxidation and repressing the production of inducible nitric oxide synthase (iNOS),[41] as well as reducing colonic paracellular permeability.[42] A decrease in PPARγ induces an inflammatory response and glycolytic metabolism and increases the production of nitrate that is accompanied by the growth of pathogenic bacteria. Notably, HFD consumption induces a decrease in PPARγ accompanied by an increase in dense microbiota in the intervillous zone of the ileum (Figure 5.2). The microbiota composition changes by an expansion of Firmicutes, Proteobacteria, and Verrucomicrobia,[43] through modulation of the mucosal surface of the ileum.

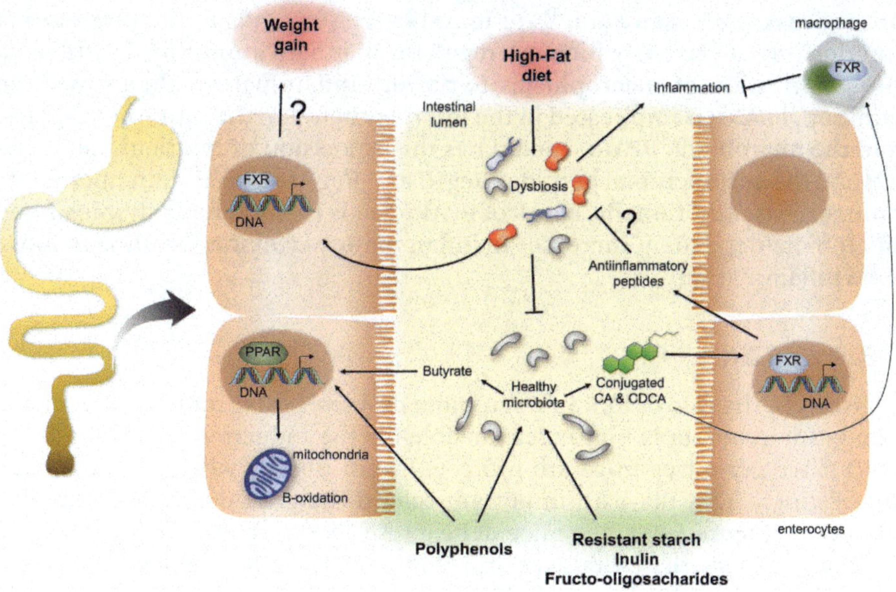

Figure 5.2 A high-fat diet induces microbiota dysbiosis, reducing butyrate production and PPARγ activation, which has two major consequences: a decrease in β-oxidation and an increase in inflammation. Notably, dietary fibers such as resistant starch, inulin, or fructo-oligosaccharides and polyphenols promote healthy microbiota, reestablishing butyrate levels and therefore PPARγ activation, preventing the induction of oxidation. Conversely, the role of FXR in the interaction among the gut microbiota, enterocytes, and nutrients is far more controversial. FXR is necessary for the weight gain induced by a high-fat diet. However, independent studies have demonstrated that FXR activation increases antimicrobial peptides and decreases macrophage-mediated inflammation which could decrease microbiota dysbiosis.

Determining why an HFD reduces PPARγ signaling in the intestine is a subject that requires further research. It is most likely a result of a decrease in butyrate production observed during HFD consumption,[44] since the use of a PPARγ agonist, such as rosiglitazone, or the use of dietary fiber that reestablishes butyrate levels, is able to reverse the altered phenotype observed during HFD consumption, including the reestablishment of the mucosal surface and the microbiota density.[43] In fact, butyrate supplementation improves insulin sensitivity and increases energy expenditure in mice,[45] and regulates G protein-coupled receptors and gut microbiota.[46] In addition, PPARγ controls the expression of the lactase gene as shown in both *in vivo* and *in vitro* analyses in Caco-2 cells (human epithelial colorectal adenocarcinoma cell line) treated with PPAR agonist. Thus, PPARγ modulation in the intestine may constitute a new important target in the treatment of lactase deficiency that leads to lactose malabsorption and intolerance.[47]

Moreover, intestinal PPARγ activation or inhibition modulates the inflammatory response to allow the growth of commensal bacteria and to restrict that of pathogenic bacteria. Studies performed in humans within this field are reviewed by Hasan *et al.*[48] For example, PPARγ binds to the promoter of interleukine 10 (IL-10), which is a regulator of innate immunity, by inducing the polarization of macrophages to an anti-inflammatory phenotype.[49] In addition, PPARα is implicated in the interplay between the immune response and the microbiota. PPARα modulates the expression of interleukine 22 (IL-22) and the antimicrobial peptides RegIII and RegIIIδ to maintain microbiota homeostasis. Additionally, a lack of PPARα leads to microbiota dysbiosis and microbiota-dependent increases in inflammatory cytokines, leading to intestinal inflammation.[50]

5.3.3.2 FXR

Conversely, the role of FXR in microbiota dysbiosis is far more controversial. The gut microbiota is highly responsible for the induction of obesity by an HFD since germ-free mice are protected from obesity when fed an HFD.[51] Interestingly, FXR plays an important role in the gut microbiota induction of obesity (Figure 5.2) since mice deficient in FXR have less weight gain and hepatic steatosis than wild-type mice on an HFD.[52] In line with this, consumption of an HFD increases FXR expression by a reduction in the FXR promoter methylation.[53] Furthermore, the accumulation of intestinal tauro-B-muricholic acid (T-B-MCA), an antagonist of FXR, by tempol, an antioxidant that induces microbiome remodeling characterized by an increase in the genus *Lactobacillus* and its bile salt hydrolase, decreases diet-induced obesity, and an intestine-specific FXR-deficient mouse model had a similar phenotype.[54] These studies suggest that FXR inhibition could represent an important therapeutic venue. An independent model is also associated with the hypothesis that FXR inhibition protects against diet-induced obesity. The intestinal-specific sirtuin1-deficient mice have reduced intestinal FXR signaling and reduced OST and apical sodium-dependent bile acid transporter

(ASBT) and, therefore, decreased absorption of ileal BAs. Additionally, these mice had a reduced accumulation of BAs and were protected from liver damage under a lithogenic diet.[55]

In contrast to the above information, activation of FXR only in the intestine by the use of a gut-restricted FXR agonist called fexaramine reduces weight gain, inflammation, and glucose production associated with an increase in thermogenesis and browning of white adipose tissue.[56] Hence, until now it was difficult to suggest whether tissue-selective activation or inhibition of FXR could be a therapeutic approach to treat obesity, and further research is needed to determine the status of FXR signaling upon HFD consumption.

FXR activity is critical for mucosal immune homeostasis and is decreased during chronic intestinal inflammation.[57] Briefly, higher concentrations of unconjugated BAs in the ileum activate FXR in intestinal endothelial cells, dendritic cells, and macrophages, maintaining a controlled inflammatory response. In intestinal endothelial cells, FXR increases the production of antimicrobial peptides (Figure 5.2) and suppresses the assembly of ATP-dependent Clp protease proteolytic subunit-related protein 3 (chloroplastic) (CLPR3)-containing inflammasomes.[58] In addition, FXR activation by the pharmacologic agonist INT-747 represses Toll-like receptor 4 (TLR4)-induced proinflammatory gene expression in intestinal endothelial and dendritic cells.[59,60] FXR activation represses nuclear factor kappa light-chain-enhancer of activated B cells (NF-κB) activity in macrophages (Figure 5.2),[61,62] and increases Treg cells in the intestine.[63] Taken together, these studies demonstrate the role of FXR in limiting the mucosal inflammatory responses of the intestine, especially in the ileum.

5.3.4 Restoration of Intestine Dysfunction by Nutrients and Phytochemicals Through Attenuation of Gut Dysbiosis

Nutrients that are not digested and absorbed in the small intestine become available to the bacteria in the large intestine. These nutrients are non-digestible carbohydrates such as resistant starch, plant cell wall material, and oligosaccharides, as well as some dietary proteins and polyphenolic compounds ingested in the diet.[64]

5.3.4.1 Carbohydrates

The addition of resistant starch, inulin, and fructo-oligosaccharides to the diet[65] increased the production of butyrate and the increase in *Lactobacillus* and *Bifidobacterium*[66,67] with the potential to activate PPARγ and reestablished microbiota homeostasis upon HFD consumption (Figure 5.2). Additionally, polydextrose, a soluble fiber with lipid-lowering effects, upregulates FXR expression in the intestine and increases the abundance of *Allobaculum*, *Bifidobacterium,* and Coriobacteriaceae in HFD-mice.[68]

5.3.4.2 Proteins

High-protein diets are known to promote weight loss.[69] Much of the effect observed by dietary protein depends on the type consumed because each source provides different amino acids, and other macronutrients such as fibers (legumes), saturated fat (red meat), or calcium (dairy).[70] For example, soy protein prevents weight gain and insulin resistance upon an HFD consumption.[71,72] However, as with the non-digestible carbohydrates, much research is needed to determine which effects are mediated by nuclear receptors. PPARs and FXR are the most likely candidates since the type of protein consumed can modify the production of BAs and butyrate. For example, rats fed with a fish protein hydrolysate rich in taurine and glycine could produce more secondary BAs conjugated with these amino acids, modifying the activity of FXR.

5.3.4.3 Polyphenols

The interaction between dietary polyphenols and microbiota has been vastly reviewed.[73,74] However, future research is needed to determine which of the beneficial effects observed by polyphenols are mediated by nuclear receptors. PPARs have already been described to be at the intersection between dietary polyphenols and microbiota.[75] For instance, a cranberry extract, without modifying weight gain, reversed hepatic steatosis and insulin resistance associated with an upregulation of PPARα, and genes involved in lipid catabolism, as well as by a drastic expansion of *Akkermansia muciniphila* and *Bamesiella* spp.[76] A crude extract of camu camu, an Amazonian fruit rich in phytochemicals, prevented weight gain, fat accumulation, metabolic inflammation, and endotoxemia in mice fed a high-fat/high-sucrose diet. This phenotype was associated with an alteration in the plasma BA pool size, an increase in *A. muciniphila,* and a reduction in *Lactobacillus*, which suggests a modification in FXR signaling. The gut microbiota of both treated mice recapitulated the phenotype in germ-free mice. The changes in microbiota induced the increase in BAT thermogenesis through the activation of membrane BA G-protein-coupled bile acid receptor, Gpbar1 (TGR5).[76] Moreover, epigallocatechin (EGCG), the most abundant polyphenol in green tea, reduced the effects of the FXR agonist CDCA, and it was associated with a reduction in diet-induced obesity and insulin resistance.[77] Barberine, a compound extracted from the roots of *Rhizoma coptidis*, inhibits *Clostridium* spp. and bile salt hydrolase, increasing the tricarboxylic acid cycle (TCA) and FXR activation[78] and preventing diet-induced obesity.

5.4 The Liver

The liver is the central node of lipid, carbohydrate, and protein metabolism in the body. This organ synthesizes, stores, and distributes nutrients to other metabolic tissues according to their particular energetic needs.[79] After a meal, dietary carbohydrates and amino acids are transported directly to the liver from the intestine through the portal vein. This anatomical feature prevents a sharp increase in circulating glucose during a high-carbohydrate meal, avoiding acute hyperglycemia and brain damage. Moreover, the liver is able to act as a glucose damp, releasing enough glucose into the circulation to be available for peripheral organs at a tolerable concentration and storing excess glucose as glycogen. The rate of hepatic glucose uptake is determined primarily by the abundance of membrane-bound glucose transporter 2 (GLUT2) and the activity of the high-K_m enzyme glucokinase. After different organs absorb dietary glucose, the circulating glucose concentration declines. To avoid hypoglycemia, the liver (as well as the kidney) releases glucose into the circulation to maintain its availability for glucose-dependent tissues and cells such as the brain and erythrocytes.[80] Thus, in periods of energy needs such as fasting or physical activity, the liver degrades glycogen and augments gluconeogenesis in order to release glucose to the circulation.

5.4.1 Physiological Role of PPARα, HNF4α, LXR, FXR, and PXR Receptors in the Liver

All these metabolic activities are tightly controlled by the metabolic nuclear receptors PPARα, HNF4α, LXR, FXR, and pregnane X receptor (PXR). The activity of these receptors in the liver is directly regulated by the intracellular concentration of specific nutrients and nutrient-derived metabolites that act as ligands, and indirectly the activity of these receptors is regulated by hormonal signaling.[81] Thus, the activity of the metabolic nuclear receptors in the liver is directly coupled to fasting/feeding cycles and, interestingly, to the composition of nutrients in the diet.

5.4.1.1 Liver Glucose and Lipid Metabolism are Regulated by PPARα and HNF4

Glucose metabolism in the liver is transcriptionally controlled by the nuclear receptors PPARα and HNF4α. These nuclear receptors are abundantly expressed in the liver and regulate the rate of glycolysis, glycogen synthesis, and gluconeogenesis through the modulation of enzyme content.[82] PPARα binds to its peroxisome proliferator-activated response element (PPRE) in the promoter region of enzymes such as phosphoenolpyruvate carboxykinase (PEPCK), glucose 6 phosphatase (G-6-Pase), pyruvate dehydrogenase kinase 1 (PDK1), glycerol kinase (GK), and lactate dehydrogenase (LDH).[83] This nuclear receptor has an uncommon ligand-binding domain (LBD) that enables PPARα to bind several molecules with different structures.

Since hepatocytes have a physical constraint that limits the storage of glycogen molecules, all surplus dietary glucose entering the liver is converted to fatty acids and stored as triglycerides. These triglycerides along with dietary or endogenous cholesterol are packed in sequential density lipoproteins with different contents of triglycerides and cholesterol. Hepatic lipoprotein metabolism is governed primarily by PPARα and HNF4. Many of the genes coding for lipid-synthesizing enzymes, apolipoproteins, lipid transfer proteins, and lipoprotein receptors have a PPRE in their promoter regions. These genes include the fatty acid binding proteins (FABPs), long-chain-fatty-acid-CoA ligases (ACSLs), acylglycerol-3-phosphate acyltransferases (AGPATs), fatty acid desaturase (FADS), fatty acid elongases (ELOVs), Patatin-like phospholipase domain-containing protein (PNPLA) (also known as adiponutrin (ADPN) or acylglycerol O-acyltransferase (MOGAT)), apolipoprotein A1 (ApoA1), lipoprotein lipase (Lip), microsomal triglyceride transfer protein (MTTP), very-low-density lipoprotein receptor (VLDLr), and the HDL receptor SR-B1.[83] Several genes that are under the transcriptional control of PPARα are also targets of HNF4.[84]

Triglycerides, along with dietary or endogenous cholesterol, are packed into very-low-density lipoproteins (VLDL) to be released into circulation. Peripheral organs, such as skeletal muscle, adipose tissue, and the heart, release Lip to their surrounding endothelial surfaces to release free fatty acids from the lipoproteins that can be stored or oxidized for ATP production. The resulting triglyceride-poor but cholesterol-rich low-density lipoprotein (LDL) is absorbed by the liver *via* the LDLr to be recycled. The liver also determines the abundance of HDL in the circulation.[85] This is of paramount importance since HDL collects cholesterol from peripheral tissues and cells, such as endothelial macrophages, to deliver it to the liver in a loop named reverse cholesterol transport. Since HDL is not synthesized and secreted in the liver (as occurs with VLDL), hepatocytes secrete ApoA1, which is the substrate for HDL formation during its passage through the circulation. The cholesterol-rich HDL particles are absorbed by hepatocytes through the receptor SR-B1.[86]

The flexibility of PPARα in its binding to different molecules has obscured for many years the identification of its endogenous ligands. However, in 2009, Chakravarthy *et al.*[87] identified the phospholipid 1-palmitoyl-2-oleoyl-sn-glycerol-3-phosphocholine (16:0/18:1 GPC) as a physiologically relevant PPARα ligand in the liver.[87] After this initial discovery, several endogenous molecules have been identified that bind and activate this nuclear receptor in the liver. These molecules are lipids or lipid derivatives synthesized directly in the hepatocytes[88] or derived from adipose tissue, such as the lipokine palmitoleate.[89] HNF4 is also a fatty acid receptor, and its ligands have been identified as medium- and long-chain saturated fatty acyl-CoAs (C14:0 to C:16:0 CoA).[90]

In the liver, the crosstalk between metabolic nuclear receptors is a common feature, exerting tight control over vital metabolic pathways (Figure 5.3).

5.4.1.2 Liver Cholesterol Metabolism is Regulated by LXRs and FXR

To synthesize BAs, the liver uses cholesterol from lipoprotein fractions. After a meal BAs are released into the duodenum to favor intestinal lipid solubilization and absorption. BAs return to the liver through specific transporters in the enterocytes and hepatocytes, all of which are under strict transcriptional control. The nuclear receptors controlling BA synthesis and transport are LXR

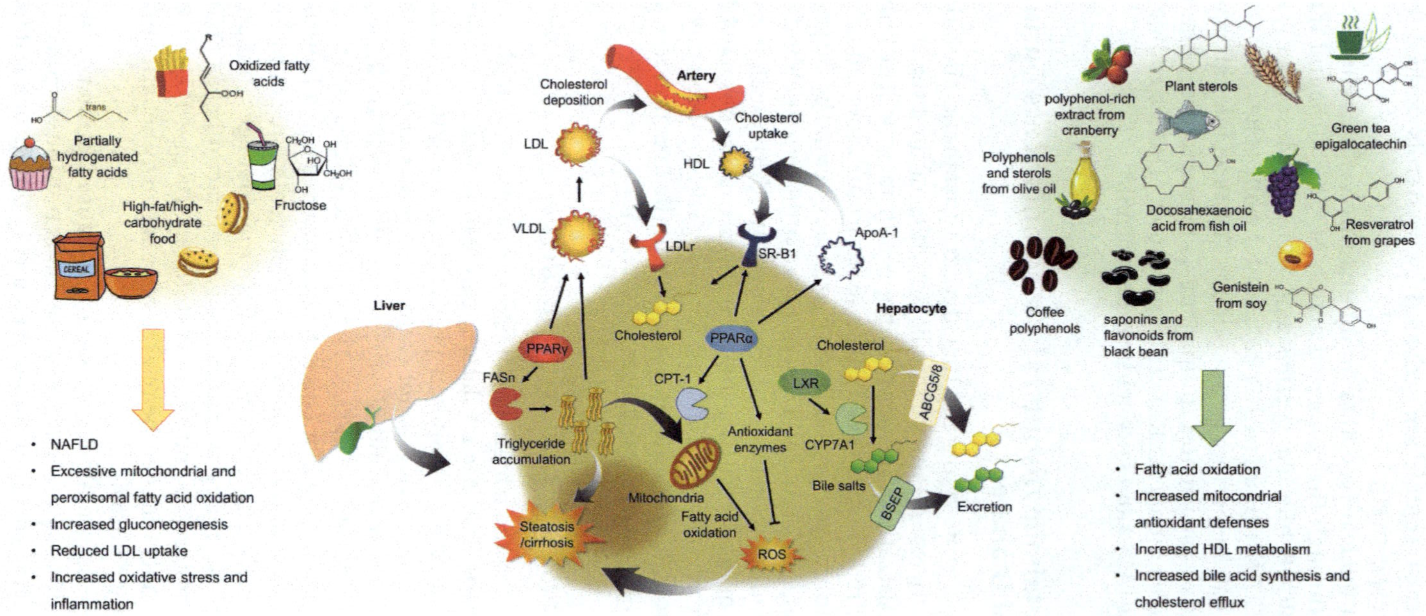

Figure 5.3 Dietary components that modulate the activity of metabolic nuclear receptors in the liver to promote or prevent NAFLD. High-fat/HCHDs, along with particular dietary components such as oxidized and partially hydrogenated fatty acids and fructose, induce hepatic fatty acid synthesis, triglyceride accumulation, VLDL hypersecretion and, consequently, increased circulating LDL. These dietary components also induce uncontrolled mitochondrial and peroxisomal fatty acid oxidation, leading to the generation of reactive oxygen species and oxidative stress. All these alterations are mediated by aberrant activation of the nuclear receptors PPARα, PPARγ, and LXR. Conversely, polyphenols, sterols, and PUFAs are common constituents found in fruits and vegetables (polyphenols and sterols) and fish (PUFAs). The intake of diets rich in these foods can revert hepatic steatosis and prevent the progression to cirrhosis and fibrosis by modulating the activity of PPARα and LXR.

and FXR. These nuclear receptors are activated by oxysterols derived from cholesterol and BAs, respectively.

The activity of both nuclear receptors maintains intracellular cholesterol and BA homeostasis through the concerted action of the cholesterol transporters ABCG5 and ABCG8, bile acid synthesizing 7α hydroxylase (CYP7) and bile acid synthesizing 27α hydroxylase (CYP27), and the bile salt transporters sodium/taurocholate cotransporting polypeptide (NTCP) and bile salt export pump (BSEP).[91] LXR also participates in lipogenesis through direct binding to the liver X response elements (LXRE) in the promoter of genes coding for acetyl-CoA carboxylase (ACC) and fatty acid synthase (FAS).[92]

5.4.2 Hepatic Metabolic Disorders by Macronutrients During Obesity and Diabetes

The liver is one of the primary organs that develop metabolic alterations during obesity and diabetes. All the metabolic activities of the liver are progressively impaired. Alterations in hepatic lipoprotein metabolism induce atherosclerosis. Impaired glucose metabolism leads to sustained gluconeogenesis, leading to chronic hyperglycemia. An increase in ketone body synthesis leads to ketoacidosis and reduced rates of urea synthesis, which induces hyperammonemia, leading to hepatic encephalopathy. All these alterations are the origin of many of the life-threatening conditions that are observed during the progression from obesity to diabetes.[93,94] Along with reduced physical activity, all the metabolic derangements that occur in the liver of obese subjects are caused by long-term dietary choices.

5.4.2.1 High-fat and High-carbohydrate Diets

Chronic intake of an HFD and a high-carbohydrate diet (HCHD) (high carbohydrate increases hepatic lipid accumulation) can lead to hepatic steatosis. This condition is completely reversible if proper dietary modifications are followed. However, if the condition is not early diagnosed and treated early, it progresses to hepatic cirrhosis and fibrosis, a chronic condition known as nonalcoholic fatty liver disease (NAFLD). The lipotoxic effects of excessive lipid accumulation in hepatocytes[95] mediate NAFLD progression. Lipotoxicity induces cellular damage, leading to parenchymal cell death and activation of Kupffer cells. Kupffer cells, also known as stellate macrophages, initiate an inflammatory response, secreting transforming growth factor beta 1 (TGF-β) and chemoattractant molecules that induce cytotoxic CD8 and Tregs.[96] This chronic inflammatory state induces the release of collagen by hepatic stellate cells, leading to end-stage fibrosis.[97]

The elevated worldwide prevalence of NAFLD is caused primarily by the increased intake of industrialized energy-dense foods previously called junk foods and currently referred to as ultraprocessed foods. Ultraprocessed foods are made from a vast assortment of ingredients; however, common constituents are present in almost all of their formulations, such as simple sugars/

high-fructose corn syrup, palm/hybrid palm oil and oxidized/trans fatty acids.[98] These ultraprocessed foods include sugar-sweetened beverages, packaged breads, cookies, savory snacks, candies, ice cream, breakfast cereals, and pre-prepared frozen meals. Additional studies also confirm a direct relationship between the consumption of ultraprocessed food and higher fasting glucose, higher total and LDL cholesterol and the risk of hypertension.[98]

These products have in common the fact that they are industrial formulations, elaborated from food processing, synthetic constituents, and food additives.[99] In the United States, ultraprocessed foods provide 58% of energy intake and 89% of added sugars in the diet. This elevated consumption of ultraprocessed food is directly associated with excess weight, particularly among women.[100] Data from the Brazilian Longitudinal Study of Adult Health (ELSA-Brasil) cohort revealed a clear association between body weight and waist circumference and ultraprocessed foods intake.[101] Alarmingly, this positive association between the consumption of ultraprocessed foods and body fat is observed not only in adults but also in children and adolescents.[102] Longo-Silva *et al.* observed that approximately 75% of preschool children had received at least one or more ultraprocessed foods daily in their diet.[103] Thus, there is an imperative urgency to understand the biochemical mechanisms that induce the metabolic derangements caused by ultraprocessed food intake in order to develop new strategies to curtail this threat to life.

The associations between ultraprocessed foods and higher body weight are independent of total energy intake.[101] This finding clarifies that the metabolic derangements induced from ultraprocessed food are mediated by the direct actions of its components on metabolic tissues rather than its energy density alone. It is increasingly recognized that the development and progression of NAFLD are mediated by a maladaptive activation of hepatic metabolic nuclear receptors. In addition to their elevated energy content, HFDs and HCHDs include several lipid species or precursors for lipid synthesis that act as non-common ligands for PPARα and LXR in the liver, inducing aberrant transcriptional activities.

5.4.2.2 Dietary Components Involved in the Progression of NAFLD Through Actions on Metabolic Nuclear Receptors in the Liver

Numerous nutrients and non-nutritive compounds found in foods can bind to PPARα, LXRs, and FXR in the liver, increasing or repressing their transcriptional activities, leading to beneficial or detrimental outcomes. Studies in rodents are very accurate models that resemble human liver diseases. Basic studies in rats have shown that chronic intake of HFDs/HCHDs induces obesity and NAFLD.[104] Obesity in humans is also the result of chronic intake of HFDs/HCHDs, containing ultraprocessed foods, which include ingredients that are able to modify the physiological activity of hepatic nuclear receptors.

Feeding mice with high fructose/glucose in drinking water for 16 weeks induces hepatic steatosis and cirrhosis along with an increase in the hepatocyte nuclear abundance of PPARα, FXR, and LXRs.[105] Glucose has been identified as a direct LXR ligand[106] but there is no evidence that glucose or fructose binds PPARα. However, derivatives from glucose or fructose act as endogenous ligands for PPARα in liver.[88] Feeding mice with palm oil or hybrid palm oil for 8 weeks induces hepatic steatohepatitis associated with augmented expression of many PPARα target genes, indicating a direct effect of these oils on this nuclear receptor activity.[107] Many ultraprocessed foods, such as donuts, potato chips, and French fries, are subjected to deep-frying. In addition to the surplus calories from fat, the deep-frying process oxidizes the vegetable oils used for frying. These oxidized vegetable oils have been directly associated with liver damage. Feeding rats with oxidized soybean oil increases PPARα target genes such as acyl-CoA oxidase (ACOX) and cytochrome P450 4A (CYP4A) involved in peroxisomal fatty acid oxidation.[108] Partially hydrogenated or trans fatty acids are produced by food companies by hydrogenation of vegetal oils rich in PUFAs to increase their melting temperature and shelf life. Several foods, such as cookies, bread, margarine, snack food, packaged baked goods, and frying oil, contain trans fatty acids. Intake of trans fatty acids induces NAFLD and dyslipidemia, increasing the risk of diabetes and cardiovascular disease.[109,110] Part of the detrimental effects of trans fatty acids in liver metabolism is mediated by the activation of nuclear receptors.[111]

Since PPARα induces fatty acid oxidation and LXR favors cholesterol excretion into bile, it could be thought that PPARα and LXR activation is advantageous to prevent excessive lipid accumulation in the liver. However, the metabolic outcome of nuclear receptor activation depends on the nature of the ligand that binds to it. Ligand structure induces a particular tridimensional conformation of the nuclear receptor, leading to a specific recruitment/dissociation of coactivators and corepressors.[112] The activities of nuclear receptors are also modulated by their interaction with other nuclear receptors, such as PPARα and HNF4α.[113] Thus, the activation of PPARα and LXR in the liver by ultraprocessed food components could activate an aberrant transcriptional program leading to excessive mitochondrial and peroxisomal fatty acid oxidation.[114] Elevated mitochondrial and peroxisomal fatty acid oxidation increases the release of reactive oxygen species (ROS) to the cytosol. Elevated ROS levels induce cellular oxidative stress, DNA damage, protein peroxidation, and disruption of lipid bilayers. All of these alterations cause oncogenic proliferation, inflammation, and apoptosis. In fact, there is solid evidence of the involvement of free radicals and oxidative stress in the progression from simple steatosis to non-alcoholic steatohepatitis and to cirrhosis and cancer.[115]

The activation of PPARα and LXR by HFDs and HCHDs increases the lipogenic pathway and lipoprotein assembly. These dietary regimens also induce the expression of PPARγ in the liver.[111] This nuclear receptor is the master transcription factor present in adipose tissue; however, its atypical expression in the liver leads to an increase in fatty acid synthesis, triglyceride esterification, and lipoprotein secretion.[116] Consequently, aberrant activation of PPARα, PPARγ, and LXRs in the liver in response to excessive carbohydrate and fat intake is the main cause of NAFLD, dyslipidemia, and atherosclerosis in obesity and diabetes[116,117] (Figure 5.3).

5.4.3 Restoration of Hepatic Metabolic Homeostasis by Dietary Components Through the Activation of Metabolic Nuclear Receptors

5.4.3.1 Proteins

Despite there being no evidence to date of the direct activation of nuclear receptors by peptides or single amino acids, it is clear that dietary protein influences the activity of metabolic nuclear receptors such as PPARα, HNF4α, and LXR.[118] Soy protein intake reduces hepatic lipid accumulation and the inflammatory response in Zucker diabetic fatty rats[119] and in rats fed an HFD.[72] These effects are mediated in part by the activation of PPARα and LXRα.[120] The mechanisms involved in the effects of dietary soy protein on metabolic nuclear receptors are probably indirect, mediated by the modification of the insulin–glucagon ratio.[121] Interestingly, the protein content in the diet also modulates the activity of hepatic nuclear receptors. A low-protein diet activates PPARα, which in turn represses HNF4α transcriptional activity. This trans-repression increases the expression of genes involved in fatty acid oxidation and simultaneously reduces the expression of amino acid enzymes. In turn, an increase in protein intake is accompanied by a reduction in PPARα function and an increase in HNF4α activity, reducing fatty acid oxidation enzyme abundance and increasing amino acid-degrading enzymes. Thus, when protein availability is scarce, the liver avoids amino acid catabolism by increasing fatty acid oxidation, preventing amino acid degradation. Conversely, a high-protein diet increases amino acid catabolism, reducing lipid oxidation. This metabolic switch between metabolic nuclear receptors allows continuous energy production through alternating substrate utilization in the liver depending on nutrient availability.[113]

5.4.3.2 Fatty Acids

Fat intake provides more than double the calories of carbohydrate or protein per gram. However, the type of fat consumed exerts very distinct metabolic effects on hepatic energy metabolism. SCFAs such as butyric acid and medium-chain fatty acids such as lauric acid are derived from dietary fiber fermentation by the intestinal microbiota and from coconut oil, respectively. These fatty acids bind and activate PPARα and even PPARγ in the liver, increasing lipid mobilization and mitochondrial fatty acid oxidation. The effects elicited by these fatty acids through PPARs induce a metabolic switch from lipid synthesis to utilization, preventing hepatic steatosis even when fed a high-fat diet.[122] Long-chain fatty acids are also dietary PPAR ligands. Saturated fatty acids such as palmitic acid, monounsaturated fatty acids such as oleic, and PUFAs such as linoleic and α-linolenic acid bind and activate PPARα *in vitro* and *in vivo*.[123,124] However, each fatty acid induces differential effects on PPARα activity and thus dissimilar metabolic outcomes. This is why there are many contradictory or unclear results in different studies. The most reproducible beneficial effects of long-chain fatty acids in hepatic energy metabolism are those involving long-chain ω3 PUFAs such as docosahexanoic acid (DHA) and eicosapentanoic aicd (EPA), which are abundant in fish and algae.[125–130]

Thus, the impact of fat intake on lipid metabolism in the liver is not totally determined by its energy content but mainly by its fatty acid composition. Consequently, dietary habits that include fish, algae, coconuts, fatty seeds, and dietary fiber provide beneficial PPARα and PPARγ ligands that prevent lipid accumulation and inflammation.

5.4.3.3 *Polyphenols*

Several studies have demonstrated the ability of different plant polyphenols to bind and modulate PPARα activity. As an example, studies with reporter gene assays and time-resolved fluorescence energy transfer analyses along with surface plasmon resonance experiments revealed that cyanidin (a pigment found in many red berries) binds directly to PPARα with great affinity, activating the transcriptional activity of this nuclear receptor. These molecular events translate into evaluable metabolic effects. Cyanidin reduces cellular lipid accumulation in lipid-loaded hepatocytes.[131] A polyphenol-rich extract from cranberry reverses hepatic steatosis in mice fed a high-fat/high-sucrose diet in association with an upregulation of PPARα target genes.[132] In mice fed an HFD, coffee extract induces hepatic gene expression of PPARα and ACOX1 associated with reduced liver fat content.[133] Saponins and flavonoids extracted from black bean seed coats stimulate the expression of hepatic PPARα and some of its target genes in mice fed a 0.5% cholesterol diet.[134] Naringenin (a flavonoid present in grapefruit) reduces triglycerides and total cholesterol in the plasma and liver of Long–Evans hooded rats. These biochemical determinations are associated with increased PPARα protein expression in the liver along with elevated expression of some of its target genes.[135] Administration of the flavonoid kaempferol to rats fed an HFD for 8 weeks produces similar effects as those exerted by fenofibrate (a synthetic PPARα ligand) with respect to body weight gain, plasma lipid levels, coronary artery risk, and atherogenic indices. Kaempferol also reduced the hepatic triglyceride and cholesterol contents associated with an increased expression of PPARα and its target genes ACOX and CYP4A1.[136] Several studies have also shown beneficial effects of different plant polyphenols on hepatic lipid and glucose metabolism in rodents fed an HFD or HCHD. Despite there being no direct experimental evidence that the metabolic benefits of dietary polyphenols are mediated directly by nuclear receptor activation, the metabolic features of polyphenol-administered mice and rats are similar to those observed when administered known PPARα ligands such as wy14643 or fibrates.[137,138] Some of these polyphenols that are able to prevent fatty liver in animals fed an HFD are isorhamnetin glycosides extracted from *Opuntia ficus-indica*,[139] the soy isoflavone genistein,[140] and saponins extracted from *Agave salmiana*.[141] (Figure 5.3).

5.4.3.4 Plant Sterols

In addition to endogenous cholesterol metabolites, plant sterols and stanols are potent dietary LXR ligands. Plant stanols increase hepatic cholesterol efflux and reduce inflammatory LXR activity, increasing the expression of its target genes Abca1 and Abcg1.[142,143] Stigmasterol and β-sitosterol prevent the progression of NAFLD in mice fed a high-fat Western diet for 33 weeks, which is associated with a reduction in hepatic cholesterol.[144] Curcumin also reduces hepatic steatosis and dyslipidemia in mice fed a high-fat/high-fructose diet through the activation of LXRs.[145] Soy protein isoflavones can regulate LXR isoforms to modulate lipid metabolism and cholesterol transport in the liver of mice fed a 0.5% cholesterol diet for 28 days.[38] Saponins and flavonoids from black bean seed coats increase BA secretion in mice fed a 0.5% cholesterol diet by increasing the expression of hepatic LXR target genes, such as the reverse cholesterol transporters ABCG5/ABCG8 and bile acid synthesizing enzymes 7A1 hydroxylase (CYP7A1)[134] (Figure 5.3).

5.4.4 Dietary Supplements and Liver Damage

However, despite all of the above evidence indicating the beneficial effects of plant polyphenols in hepatic lipid and carbohydrate metabolism, the use of natural medicine to treat liver problems must be prescribed very cautiously.[146] Herbal supplements are increasing in popularity and are perceived as safe because they are supposedly "natural." However, Schoepfer *et al.* and Chen *et al.* reported several cases of acute liver injury implicating herbal products.[147,148]

The detrimental effects of dietary supplements on liver integrity and function are related to the chronic administration of highly concentrated extracts. Thus, while the intake of fruits and vegetables has beneficial health effects due to their low content of polyphenols, the consumption of dietary supplements composed of fruit extracts or purified polyphenols could have contrary effects. This biphasic response to different amounts of a substance is termed hormesis, *i.e.*, the biological and toxicological concept that small quantities have opposite effects from large quantities.[149] Thus, to maintain the effects of dietary ligands for metabolic nuclear receptors in the hormetic zone, the best prescription is consumption of fresh, unprocessed foods for a healthy liver.

5.5 White and Brown Adipose Tissue

White adipose tissue is the organ responsible for storing energy mainly in the form of triglycerides for energy demand periods. Depending on its localization, white adipose tissue is divided into subcutaneous and visceral adipose tissue, resembling different metabolic functions.[150] The metabolism of white adipocytes is highly regulated by insulin, specifically lipid synthesis and release, as well as glucose uptake. In response to different stimuli, the mass of adipose tissue can increase. This increase can occur through hypertrophy or through hyperplasia with the formation of new adipocytes (derived from preadipocytes that are present in the adipose tissue) by a process called adipogenesis.

Adipose tissue hypertrophy induces several alterations in its functions, such as insulin resistance, alterations in adipokine secretion, the release of inflammatory markers, low lipid storage capacity, and glucose intolerance, among others. All these characteristics lead to metabolically unhealthy obesity. Moreover, since the reduction in the esterification capacity leads to increased fatty acid release into the circulation, the accumulation of fatty acids in tissues such as liver, pancreas, skeletal muscle, and heart induces insulin resistance and other metabolic alterations due to lipotoxicity. In contrast, when fat mass increases by hyperplasia, new insulin-sensitive adipocytes are formed, generating a metabolically healthy obesity.[151,152]

Thus, inhibiting adipogenesis exacerbates hepatic and skeletal muscle lipid accumulation,[153] as adipogenesis is necessary to avoid fatty acid spillover to non-adipose tissue such as the liver and skeletal muscle.[154]

In contrast, brown adipose tissue is composed of rich mitochondria and multivacuolar smaller adipocytes that are positive for the expression of uncoupling protein-1 (UCP-1), which is responsible for thermogenesis and the generation of heat rather than ATP from the oxidation of fatty acids.

5.5.1 Physiological Role of PPARs and LXR in White and Brown Adipose Tissue

PPARs are transcription factors that upon activation by ligands are heterodimerized with retinoid X receptor (RXR) and translocate into the nucleus to bind to DNA sequences called PPARE in the promoter of its target genes involved in lipid and glucose metabolism in adipose tissue.[155–157]

The PPAR family includes three isotypes: PPARα, PPARβ/δ, and PPARγ. Different fatty acids or their derivatives called eicosanoids act as ligands of PPARγ and PPARα, both of which play important roles in white and brown adipose tissue lipid metabolism[158,159] (Figure 5.4).

In particular, PPARγ is most highly expressed in white adipose tissue and is the master regulator of adipogenesis, which is also required for lipogenesis and survival in mature adipocytes. In particular, PPARγ promotes adipocyte differentiation and maturation, stimulates triglyceride synthesis as well as the storage of fatty acids in adipocytes, enhances insulin sensitivity, regulates glucose homeostasis by increasing the expression of GLUT4 as well as numerous secreted factors that affect insulin sensitivity, such as adiponectin, resistin, leptin, and tumor necrosis factor α (TNFα).[159–161]

To regulate all of these metabolic processes, PPARγ controls the transcription of a number of genes, such as fatty acid binding protein (aP2), Lip, adiponectin, leptin, resistin, adipsin, perilipin, phosphoenolpyruvate carboxykinase (PEPCK), acyl CoA synthase (ACS), GK, aquaporin 7 (AQP7), and cluster of differentiation 36 (CD36).

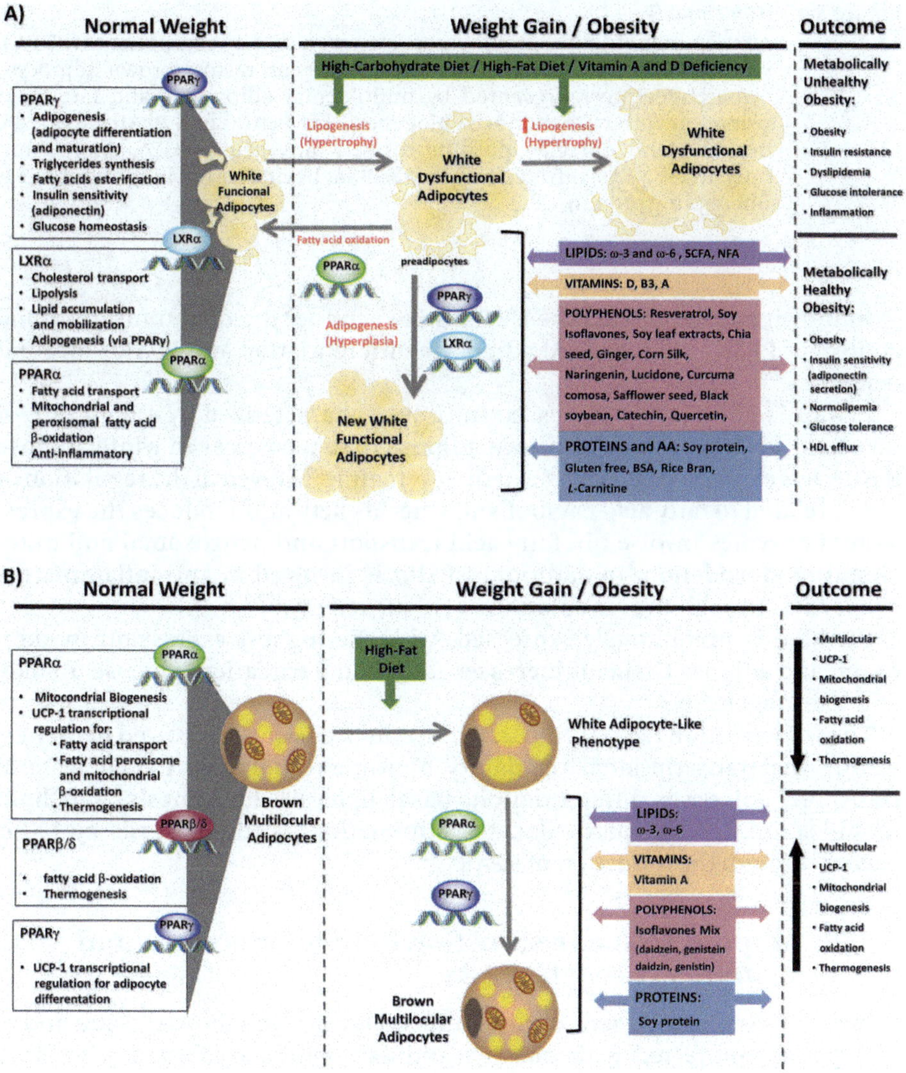

Figure 5.4 (**A**) White adipose tissue: High-carbohydrate diet, high-fat diet, and/or vitamin A and D deficiencies induce weight gain or obesity by lipogenesis (triglyceride synthesis and fatty acid esterification), gradually generating hypertrophic and dysfunctional white adipocytes. However, adipogenesis can be induced by dietary components (lipids, vitamins, polyphenols, and proteins), generating new white functional adipocytes from preadipocytes *via* PPARγ and LXRα. In addition, dietary components can induce fatty acid oxidation *via* PPARα in hypertrophic adipocytes, reducing its triglyceride content and, in turn, its size. Two metabolic outcomes can occur depending on whether obesity takes place by hypertrophy (metabolically unhealthy obesity) or by hyperplasia (metabolically healthy obesity) induced by dietary components. (**B**) Brown adipose tissue: a high-fat diet induces weight gain or obesity by generating white-like adipocytes with a reduced number of mitochondria and less fatty acid oxidation, consequently with

(*continued*)

Figure 5.4 (*Continued*) an increment in lipid droplet area. However, dietary components (PUFAs, vitamin A, isoflavone mix, and soy protein) can restore brown adipocyte cytoarchitecture represented by multilocular adipocytes and fatty acid oxidation with mitochondrial biogenesis *via* PPARα and PPARγ. w-3, w-6, omega-3 and omega-6 polyunsaturated fatty acids; SCFA, short-chain fatty acids; NFAs, nitrated fatty acids; BSA, bovine serum albumin; UCP-1, uncoupling protein-1.

In brown adipose tissue, the UCP-1 gene is under the dual transcriptional control of PPARγ in relation to adipocyte differentiation and PPARα for lipid oxidation.[162]

PPARα is abundantly expressed in tissues characterized by a high rate of fatty acid oxidation, such as brown adipose tissue, whereas in white adipose tissue, it is expressed to a lower extent. It is mainly involved in the regulation of genes related to fatty acid catabolism, since its activation induces the expression of enzymes involved in fatty acid transport and peroxisomal and mitochondrial β-oxidation. In addition, PPARα is involved in anti-inflammatory properties through the modulation of NF-κB activity.[157,160]

PPARβ/δ is ubiquitously expressed, with the highest expression in skin, brain, and adipose tissue. It increases fatty acid β-oxidation, glucose uptake, and thermogenesis.[163]

The transcription factor LXRα in white adipose tissue is involved in orchestrating the transcriptional regulatory network responsible for cholesterol transport, adipocyte differentiation, basal lipolysis, lipogenesis, and lipid mobilization.[164–167] In particular, the stimulation of adipogenesis by LXRα occurs through the induction of PPARγ.[168]

5.5.2 Adipose Tissue Dysfunction by Macronutrients and Vitamins Through PPARs

PPARγ is a core component of diet-induced obesity in adipose tissue and is highly interconnected with obesity-candidate genes associated with lipid metabolism and inflammation.[169]

5.5.2.1 High-fat Diet

In obesity, PPARγ is downregulated.[170] In rodents fed an HFD, the gene expression of PPARγ in epididymal white adipose tissue is reduced, together with an increase in adipocyte area and an increase in leptin secretion.[72] This effect is even observed in epididymal white adipose tissue of obese Zucker (*fa/fa*) rats, where an excess of coconut or soybean oil as a dietary fat source decreases the gene expression of PPARγ and PPARα, as well as their target genes, GLUT4 and UCP-1.[171]

In contrast, mice fed an HFD have increased gene expression of PPARγ1 and PPARγ2 in white adipose tissue, generating obesity by hypertrophy with insulin resistance.[172,173] In subcutaneous white adipose tissue from humans with obesity (both genders), only the expression of PPARγ2 mRNA is increased in comparison to adipose tissue from lean subjects.[174]

Histological analyses reveal that epididymal adipocytes from both, wild-type mice and heterozygous PPARγ-deficient mice under HFD feeding, are significantly larger than those fed HCHDs. However, the size of adipocytes from heterozygous PPARγ-deficient mice is significantly smaller than that of adipocytes from wild-type mice fed an HFD.[173,175]

In brown adipose tissue of mice fed an HFD rich in saturated fatty acids, the gene and protein expression of PPARγ is reduced, together with an increase in lipid droplet area, [72,176] with a slight decrease in the protein expression of PPARα.[176]

5.5.2.2 High-carbohydrate Diet

In mice fed an HCHD, the gene expression of PPARγ increased in white adipose tissue, promoting the lipogenic pathway even in the fasted state.[177] HCHD feeding generates an acute increase in PPARγ mRNA expression in white adipose tissue of mice from the beginning of the intake, generating early onset and sustained adipose tissue inflammation and metabolic disarrangement that includes an increase in leptin and resistin and a decrease in adiponectin.[178] In obese female rats induced by HFD, the gene expression of PPARγ in periovarian white adipose tissue decreases in those rats fed posteriorly an HCHD in comparison to those fed a low-carbohydrate diet.[179] In particular, a high-sugar diet induces hypertrophy and hyperplasia in retroperitoneal white adipose tissue of rats, upregulating the proadipogenic PPARγ pathway.[180] Enhanced expression of PPARγ2 occurs in the stroma-vascular fraction cells of visceral white adipose tissue of rats drinking fructose-supplemented water, generating a combination of accelerated adipogenesis and final cell hypertrophy.[181] Consumption of 60%-fructose solution for 9 weeks increases the nuclear protein content of PPARγ in visceral white adipose tissue in mice, favoring hypertrophy and adipogenesis (hyperplasia) development of visceral adiposity.[182] Indeed, a longer period of fructose intake (up to 24 weeks) increases the expression of PPARγ.[183]

PPARβ/δ also regulates white adipose tissue metabolism, since high fructose intake (30% in drinking water) for 8 weeks in PPARδ-deficient mice increases glucose intolerance and leads to inflammation and macrophage infiltration through CD36 and activation of the janus kinase (JNK) pathway.[184]

5.5.2.3 Vitamin Deficiency

5.5.2.3.1 Vitamin A Low vitamin A diets favor adipose tissue formation in animals and enhance the formation of intramuscular fat. Indeed, individuals with overweight and obesity have an inadequate vitamin A status or reduced plasma levels of carotenoids, as well as an inverse association between vitamin A intake and adiposity and insulin resistance. Endogenous retinoic acid production *in vitro* regulates the expression of zinc finger protein 423 (Zfp423), a transcription factor required for PPARγ expression, suggesting that retinoic acid represses adipogenesis *via* PPARγ inhibition.[185]

5.5.2.3.2 Vitamin D Evidence shows that adipose tissue may be a direct target of vitamin D. Vitamin D deficiency contributes to the pathogenesis of obesity. Clinical studies have shown that individuals with obesity tend to have low levels of vitamin D and this may be linked to a dysregulation of white adipose tissue.[186]

Maternal vitamin D deficiency during the perinatal period increases PPARγ and vitamin D receptor in the perigonadal and subcutaneous adipose tissue of male offspring in the early adulthood.[187] There is evidence that vitamin D inhibits adipogenic transcription factors, such as PPARγ, and lipid accumulation during adipocyte differentiation reducing body fat mass. In addition, serum adiponectin levels are positively correlated with vitamin D status.[186]

5.5.3 Restoration of Adipose Tissue Dysfunction by Macronutrients and Other Compounds Through PPARs

5.5.3.1 Lipids

Different types of lipids, such as PUFAs, especially omega-3 and omega-6, SCFAs, and nitrated fatty acids activate PPARs.

5.5.3.1.1 Omega-3 and Omega-6 Polyunsaturated Fatty Acids Both omega-3 and omega-6 PUFAs can bind and/or regulate transcription factors that control the expression of genes involved in preadipocyte differentiation. Dietary omega-3 PUFAs regulate fundamental adipose functions through modulation of the activity and abundance of key transcription factors that act as nutrient sensors, such as PPAR α, γ and β/δ. As a consequence, adiposity is reduced, adipose tissue inflammation is relieved, with systemic benefits through the reduction in circulating concentrations of proinflammatory cytokines, and insulin sensitivity is incremental.[170,188] In addition, PUFAs stimulate genes involved in peroxisomal and mitochondrial fatty acid β-oxidation through the activation of PPARs.[170]

Omega-3 PUFAs modulate several PPARγ target genes in mice and 3T3-L1 adipocytes, acting as ligands of PPARγ.[158] Elevated concentrations of omega-6 and omega-3 PUFAs in adipose tissue composition correlate with reduced adipocyte size in subcutaneous adipose tissue from humans.[189] Supplementation with omega-3 in rats with metabolic syndrome induced by high-fructose and HFD resulted in an increase in the content of PPARγ protein, accompanied by a reduction in insulin resistance, an increment in adiponectin, and a normalization of hyperlipidemia.[190] In particular, arachidonic acid and its metabolites induce fat cell differentiation and accelerate maturation by elevating lipoprotein lipase expression *in vitro*, acting as ligands of PPARγ.[158,191,192] Adipogenesis promotion and improvement in hypoxia-induced adipocyte inflammation and insulin resistance occur through the enhancement of PPARγ binding to PPAR response element by omega-3 PUFAs, principally in the promoter region of vascular endothelial growth factor-A (VEGF-A).[193] Fish oil, which is rich in omegas, normalized the decrease in PPARγ protein in adipose tissue of insulin-resistant rats fed a sucrose-rich diet, ameliorating adipose tissue dysfunction.[194] A high-fat fish oil diet increases the gene and protein expression of the three isoforms of PPAR, namely α, β/δ, and γ, resulting in a decrease in insulin resistance and inflammation in comparison to animals fed a high-fat lard diet.[195]

DHA transactivates PPARγ.[196] Dietary supplementation of diet with DHA increases PPARγ gene expression in the intraperitoneal adipose tissue of grass carp (*Ctenopharyngodon idella*), decreasing the number of adipocytes but increasing its size, accompanied by an intraperitoneal fat index reduction.[197]

Supplementation with omega-3 oil, composed of EPA (13%) and DHA (9%), in rats with metabolic syndrome induced by high-fructose and HFD, results in an increase in the content of PPARγ protein content and in serum adiponectin (a PPARγ target gene), accompanied by a reduction in insulin resistance and a normalization of hyperlipidemia.[190] Supplementation with long-chain omega-3 PUFAs in adolescents with obesity results in the downregulation of the expression of PPARγ in subcutaneous adipose tissue. The metabolic consequences of supplementation with EPA and DHA in combination with dietary restriction are improved anthropometric parameters and fewer circulating triglycerides.[198]

In rodents, omega-3 PUFAs have also been shown to increase mitochondrial biogenesis and fatty acid oxidation in adipose tissue,[199] possibly through PPARα induction.[158] In fact, in adolescents with obesity, supplementation with long-chain omega-3 PUFAs results in the upregulation of PPARα in subcutaneous adipose tissue.[198]

In brown adipose tissue, supplementation with fish oil increases the protein levels of PPARγ and decreases those of PPARα in the setting of HFD consumption, generating a higher increase in the thermogenic UCP-1 protein and restoring the brown adipocyte cytoarchitecture as denoted by multilocular adipocytes.[176]

5.5.3.1.2 Short-chain Fatty Acids Dietary SCFAs, which are the main products of dietary fiber fermentation, prevented and reversed the metabolic abnormalities induced by HFD by decreasing the expression and activity of PPARγ in adipose tissue, switching from lipid synthesis to fatty oxidation *via* UCP-2–5′ adenosine monophosphate-activated protein kinase (AMPK).[122]

5.5.3.1.3 Nitrated Fatty Acids Nitrolinoleic acid, a nitroalkene derivative of linoleic acid, is a potent ligand of PPARγ and induces the expression of CD36 in macrophages, as well as adipocyte differentiation and glucose uptake in 3T3-L1 adipocytes.[200] *In vivo* administration of the PPARγ agonist nitro-oleic acid to diabetic Zucker rats reduces body weight gain, ameliorating diabetic symptoms.[201]

5.5.3.2 Proteins and Amino Acids

5.5.3.2.1 Soy Protein Soy protein has many metabolic beneficial effects, in particular it reduces body weight gain, limits the accretion of visceral adiposity, and decreases adipose tissue cell volume.[202] The expression of PPARγ in epididymal white adipose tissue is increased by soy protein only when there is not an excess of dietary fat.[72] Moreover, the gene expression of PPARγ and PPARα in epididymal white adipose tissue of obese Zucker *fa/fa* rats is increased by soy protein consumption in comparison to casein, independent of the content of soybean oil as a fat source, but is reduced when coconut oil is the fat dietary source.[203] In rats fed a sucrose-rich diet, soy protein increased the protein expression of PPARγ to its normal values, contributing to the restoration of adipose tissue functionality.[202]

Soy protein, in comparison to casein, increases the mRNA levels of PPARγ in brown adipose tissue, and the effect is accompanied by a reduction in lipid droplet area, in rats fed either a normal diet or HFD.[72] In comparison to casein, in the brown adipose tissue of rats, soy protein increases the gene expression of the thermogenic marker UCP-1, a target gene of PPARγ and PPARα, reducing fasting glucose.[204]

5.5.3.2.2 Gluten The exclusion of gluten, a protein complex present in many cereals, from the diet, results in an upregulation of PPARα, which is related to fatty acid oxidation, and in an overexpression of PPARγ related to an improvement in glucose homeostasis and proinflammatory profile, as well as to an increase in adiponectin (PPARγ target gene) and GLUT4. As a consequence, adiposity gain, inflammation, and insulin resistance are ameliorated.[205]

5.5.3.2.3 Bovine Serum Albumin In mice fed an HFD in combination with bovine serum albumin (BSA) as a protein source, the percentage of fat mass is reduced in comparison to those mice fed casein as a protein source. In subcutaneous adipose tissue, BSA induces the mRNA expression of PPARα and its levels are negatively correlated with the mass of subcutaneous adipose tissue.[206]

5.5.3.2.4 Rice Bran Dietary supplementation with rice bran enzymatic extract (RBEE) restores PPARγ and adiponectin mRNA levels in adipose tissue of obese mice induced by HFD. RBEE also increases the serum adiponectin concentration that was reduced by HFD. Moreover, insulin resistance, dyslipidemia, and morphological and functional alterations of adipose tissue induced by HFD are also attenuated by RBEE supplementation.[207] The administration of rice bran protein hydrolysate for 6 weeks to rats fed an HCHD/HFD diet increases the mRNA expression of PPARα in adipose tissue as well as adiponectin secretion, with insulin sensitivity restoration.[208]

5.5.3.2.5 L-Carnitine Finally, supplementation with L-carnitine, an amine that is synthesized from two essential amino acids, namely, lysine and methionine, in rats with metabolic syndrome induced by high-fructose and HFD results in an increase in the content of PPARγ protein and in serum adiponectin, accompanied by a reduction in insulin resistance and a normalization of hyperlipidemia.[190]

5.5.3.3 Other Compounds

5.5.3.3.1 Vitamins

Vitamin D 1,25-Dihydroxyvitamin D3 (1,25(OH)2D_3), the physiologically active form of vitamin D_3, reduces fat mass caused by vitamin D deficiency. In particular, 1,25(OH)2D_3 inhibits porcine preadipocyte differentiation through the suppression of PPARγ mRNA levels and interferes with the induction of RXR mRNA, resulting in the downregulation of adipogenesis-related genes.[209]

In 3T3-L1 preadipocyte cells, the early events in the adipogenesis process are those that are inhibited by 1,25(OH)2D_3 through the suppression of PPARγ mRNA levels. This downregulation of PPARγ mRNA also occurs in cultures of mouse epididymal fat tissues.[210]

Vitamin B_3 Niacin, or vitamin B_3, stimulates the *in vitro* gene expression of PPARγ in cultured adipocytes from hypercholesterolemic rabbits in a dose-dependent manner, promoting the HDL-induced cholesterol efflux.[209]

Vitamin A *All-trans* retinoic acid or vitamin A (at-RA) treatment induces body fat loss in mice through the induction of lipolysis in white adipose tissue due to the activation of PPARβ/δ and an enhanced capability of fatty acid oxidation by the upregulation of the mRNA levels of PPARα.[185,211] In addition, at-RA reduces the capability of lipogenesis in white adipose tissue in response to a decrease in the expression of PPARγ, thus increasing the appearance of smaller and multilocular adipocytes.[211] Different at-RA doses and administration routes in diet-induced obese mice ameliorate obesity, reduce body weight and adiposity, enhance insulin sensitivity and glucose tolerance, and decrease dyslipidemia and hepatosteatosis.[211]

In brown adipose tissue, at-RA also increases fatty acid oxidation *via* PPARα as well as thermogenesis, since retinoic acid positively regulates the expression of UCP-1 by PPARγ.[211]

5.5.3.3.2 Polyphenols The regulation of PPARγ can be carried out directly and/or indirectly *via* the antiadipogenic activities of a range of natural products that suppress adipogenesis *via* downregulation of PPARγ expression and/or activity.[212]

Some phytochemicals can suppress the inflammatory mediators secreted by obese adipose tissue, acting as PPARγ agonists.[213] Polyphenols from vegetable sources, such as coffee, rice, olives, berries, and others, are proposed to restore the adequate expression and activity of PPARs induced by HFD consumption, which favors obesity, dyslipidemia, insulin resistance, and other conditions.[75]

Resveratrol Resveratrol inhibits lipogenesis and adipocyte proliferation and differentiation.[214] Resveratrol from red wine and isohumulone found in *Humulus lupulus* hop is a non-flavonoid polyphenol that regulates obesity-induced inflammation in a PPARγ-dependent manner.[213] Freeze-dried strawberry and blueberry, which are rich sources of resveratrol, inhibit adipogenesis and lipogenesis through downregulation of PPARγ gene expression in 3T3-L1 cells and in retroperitoneal adipocytes from high-fat-sucrose-fed rats.[215] Resveratrol exerts its antiadipogenic effects by decreasing lipid accumulation in maturing adipocytes through the downregulation of PPARγ mRNA expression[216] or protein levels[217] or by decreasing PPARγ1–3 mRNA in human visceral adipocytes.[218] Cell differentiation of 3T3-L1 adipocytes[217,219] or primary pig preadipocyte proliferation and differentiation[220] are inhibited by resveratrol, together with a decrease in protein levels and/or mRNA of PPARγ. The early or terminal phase of adipogenesis in murine preadipocytes is inhibited by resveratrol, together with a reduction in PPARγ protein expression as well as a decrease in PPARγ2 expression and activity levels, respectively.[221,222] Resveratrol also reduces proliferation, differentiation, and triglyceride accumulation of human Simpson–Golabi–Behmel syndrome preadipocytes with a reduction in the PPARγ mRNA and protein levels.[223]

Basal lipolysis is reduced by resveratrol, whereas stimulated lipolysis induced by TNFα is inhibited in mature 3T3-L1 adipocytes, together with the downregulation of PPARγ mRNA and protein levels.[224]

The resveratrol metabolite *trans*-resveratrol-3-*O*-sulfate reduces the triglyceride content by decreasing PPARγ mRNA levels in maturing 3T3-L1 adipocytes.[225] Resveratrol attenuates the inhibition of adiponectin expression induced by TNFα in 3T3-L1 adipocytes by increasing PPARγ DNA-binding activity, an effect that may improve insulin resistance.[226]

Soy Isoflavones Genistein, the principal soy phytoestrogen, inhibits mitotic clonal expansion, triglyceride accumulation, PPARγ1 and PPARγ2 protein expression and increases basal and stimulated lipolysis in 3T3-L1 adipocytes.[227,228] In murine preosteoblastic cells and mouse bone marrow cells, low concentrations of genistein inhibit adipogenesis.[229]

The soy isoflavone daidzein and its metabolite equol enhance adipocyte differentiation and insulin-stimulated glucose uptake in adipocytes *via* the activation of PPARγ.[135] Daidzein stimulates adipogenic differentiation and increases adiponectin expression in 3T3-L1 adipocytes with the activation of PPARγ. In obese mice induced by a high-fat high-sucrose diet, daidzein improves insulin sensitivity through the upregulation of PPARγ, inhibiting adipocytes hypertrophy and increasing the expression of adiponectin.[230] Moreover, daidzein also reduces the gene expression of proinflammatory cytokines through the activation of PPARα and PPARγ in adipocyte and macrophage cocultures.[231]

Diet supplementation with a mix of soy isoflavones (mainly daidzein, genistein daidzin, genistin, and its metabolites) increases the mRNA levels of PPARγ and its target gene UCP-1 in brown adipose tissue of rats in a dose-dependent manner, improving fasting triglycerides and glucose values.[204]

Soy Leaf Extracts Supplementation of an HFD with extracts of soy leaf, rich in kaempferol and coumestrol, regulates the expression of PPARγ and its target gene ap2 in white adipose tissue of mice, together with a suppression of body weight gain and fat accumulation. In accordance, *in vitro* treatment of kaempferol and coumestrol in 3T3-L1 adipocytes exhibits antiadipogenic effects *via* downregulation of PPARγ.[232]

Chia Seed In rats fed a sucrose-rich diet for a long period, the addition of chia seed reverses the decrease in PPARγ protein level induced by an HCHD, reducing the weight of visceral white adipose tissue and improving adipose tissue dysfunction.[233]

Curcuma comosa Treatment with an extract or with a phytoestrogen derived from *Curcuma comosa* prevents the visceral adiposity increase caused by ovariectomization, decreasing the diameter of adipocytes with an induction of PPARα protein levels and the suppression of PPARγ target genes, such as leptin and adiponectin.[234]

Ginger The main ginger-derived component, 6-shogaol, in 3T3-L1 adipocytes is a potent agonist of PPARγ, upregulating the expression of its target genes adiponectin and aP2. In addition, 6-shogaol pretreatment significantly inhibits the TNFα-mediated downregulation of adiponectin expression.[235]

Corn Silk Mayasin, a major flavonoid of corn silk, inhibits lipid accumulation and adipocyte differentiation through the reduction of PPARγ protein levels in 3T3-L1 cells.[236] In mice fed an HFD, mayasin decreases weight gain, epididymal fat weight, and dyslipidemia, with a reduction in mRNA expression of PPARγ1 and PPARγ2 in epididymal adipose tissue.[236,237]

Naringenin The flavonoid naringenin is a grapefruit-derived compound that enhances the expression of PPARγ1 and PPARγ2 in 3T3-L1 adipocytes and promotes the regulation of adipogenesis with an increase in the accumulation of triglycerides.[228]

Lucidone The process of adipogenesis *in vitro* of 3T3-L1 preadipocytes is suppressed by lucidone, the fruit of *Lindera erythrocarpa Makino*, through a reduction in the transcription levels of PPARγ, and its various target genes, including adiponectin, lipoprotein lipase, and ap2. In accordance, lucidone given to HFD-induced obese mice reduces body weight and white visceral adipose tissue and improves the biochemical parameters associated with metabolic disorders such as obesity and Type 2 diabetes.[238]

Defatted Safflower Seed Extract The differentiation of 3T3-L1 preadipocytes is inhibited by defatted safflower seed extract, with a decrease in the gene expression of PPARγ and its target genes. In accordance, in *ob*/*ob* mice (without the gene that codes for leptin) fed an HFD, supplementation with defatted safflower seed extract decreases the amount of visceral white adipose tissue in comparison to mice fed an HFD without supplementation.[239]

Cyanidin-3-glucoside Preadipocyte 3T3-L1 cells in the presence of anthocyanin cyanidin-3-glucoside (Cy3G) derived from black soybean differentiate into smaller insulin-sensitive adipocytes. In accordance, Cy3G induces PPARγ mRNA and increases the secretion of one of its target genes, adiponectin.[240]

Combination of Polyphenols Catechin, querecetin, and the combination of both attenuate the decrease in adiponectin (a PPARγ target gene) induced by TNFα in 3T3-L1 adipocytes and prevent the downregulation of PPARγ induced by high-fructose diet in rats, improving insulin sensitivity through an increase in adiponectin.[241]

In contrast, the combination of soybean genistein, green tea EGCG, and/or grape resveratrol additively inhibits preadipocyte differentiation, both in 3T3-L1 cells and in human primary adipocytes. Protein expression of PPARγ is additively reduced with this combination, as well as its target genes, such as fatty acid binding protein-4 (FABP-4) and perilipin.[242]

5.5.3.3.3 Terpenes Terpenes are present in many dietary and herbal plants. Auraptene, a monoterpene derivate found in the peel of citrus fruit, is a dual PPARγ and α agonist, increasing the activation of both of them, thus modulating the transcription of target genes in adipocytes, in particular decreasing the gene expression of monocyte chemoattractant protein-1 (MCP-1) and increasing adiponectin gene expression as well as its secretion.[243,244]

5.5.4 Adipose Tissue Dysfunction by Macronutrients Through LXRα

The mRNA expression of LXRα increases in the epididymal white adipose tissue of rats with the consumption of an HFD.[245] In particular, rats fed an HFD composed of lard and sunflower as a fat source have higher body weight gain due to an increase in subcutaneous and visceral adipose tissue with a decrease in insulin signaling in the epididymal adipose tissue.[246] In accordance, the protein levels of LXRα increase in the perirenal adipose tissue of rats fed an HFD and also an HCHD.[247]

5.5.5 Restoration of Adipose Tissue Dysfunction by Lipids and Other Compounds Through LXRα

Dietary omega-3 PUFAs reduce adiposity and increase insulin sensitivity in rodents, regulating fundamental adipose functions by modulating the activity and abundance of key transcription factors that act as nutrient sensors, such as LXR α/β. The inhibition of LXR α/β by PUFAs results in a decrease in the expression of genes involved in fatty acid synthesis and cholesterol excretion.[170]

Niacin, also known as vitamin B_3, stimulates *in vitro* the gene expression of LXRα in cultured adipocytes from hypercholesterolemic rabbits in a dose-dependent manner, improving the serum values of total cholesterol, LDL-C and HDL-C, as well as promoting HDL-induced cholesterol efflux.[209]

The oligosaccharide chitosan reduces the mRNA levels of LXRα induced by HFD in epididymal white adipose tissue of rats and inhibits the proliferation of mesenteric and subcutaneous adipocytes.[245]

Dietary supplementation with an extract or with a phytoestrogen derived from *Curcuma comosa* decreases the gene and protein expression of LXRα induced by ovariectomization, returning it to values similar to those in the control condition. The gene expression of the LXRα target lipogenic enzymes FAS and ACC are also decreased and the suppression of lipogenic genes is reflected in a reduced adipocyte diameter.[234]

5.6 Skeletal Muscle

Skeletal muscle is mainly composed of two types of muscle fibers: type I fibers that utilize aerobic metabolism (oxidative) and are rich in mitochondria and myoglobin and type II fibers (glycolytic) that utilize anaerobic glycolysis and are low in mitochondria and myoglobin[248] (Figure 5.5A). Skeletal muscle accounts for ~40% of the total body mass and >35% of energy expenditure; hence, it has a paramount role in energy balance and in insulin sensitivity because it is the primary tissue of insulin-stimulated glucose uptake, disposal, and storage. Concerning lipid metabolism, skeletal muscle is the most predominant tissue for whole-body lipid oxidation, with up to 90% of the energy requirements at rest derived from fatty acids. Skeletal muscle also regulates cholesterol efflux and strongly influences metabolism *via* modulation of circulating and stored lipid flux.[249–251]

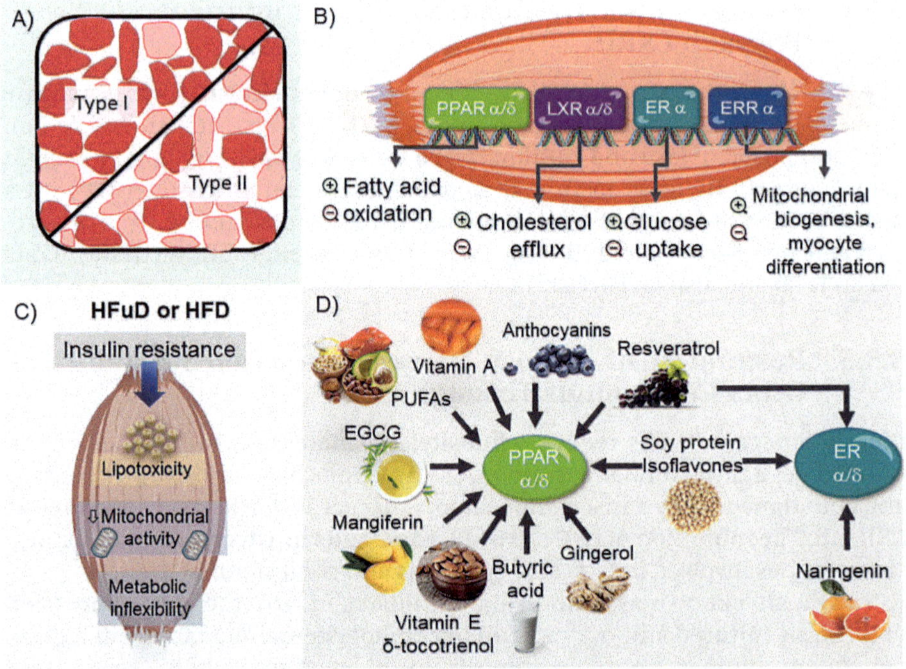

Figure 5.5 (**A**) Muscle fibers: type I (oxidative) are rich in mitochondria and myoglobin and type II (glycolytic) are low in mitochondria and myoglobin. (**B**) Role of nuclear receptors in insulin sensitivity, obesity, and diabetes. PPARs play essential roles in the regulation of fatty acid oxidation. LXRs are regulators of cholesterol metabolism. ERα plays a role in glucose uptake-mediated GLUT4 expression. ERRα regulates genes implicated in myocyte differentiation and mitochondrial biogenesis: (+) during insulin sensitivity/(−) during obesity and diabetes. (**C**) Effects of chronic consumption of high-fructose (HFuD) or high-fat diet (HFD) on skeletal muscle. Skeletal muscle insulin resistance comprises perturbations in glucose and fatty acid metabolism and a defect in mitochondrial activity. HFuD or HFD induces skeletal muscle insulin resistance by altering lipid metabolism in skeletal muscle, associated with an imbalance between fatty acid uptake and oxidation, resulting in lipotoxicity. (**D**) Nuclear receptor targeting of dietary components. Natural compounds derived from food that modulate PPARδ (activity/expression) in skeletal muscle are soy protein, PUFAs, butyric acid, RA, δ-tocotrienol, resveratrol, EGCG, robinetinidol, fisetinidol, isoflavones, anthocyanins, mangiferin, and gingerols. Natural compounds derived from food that modulate ERα (activity/expression) are naringenin, resveratrol, and isoflavones.

5.6.1 Physiological Role of PPARs, LXR, ER, and ERRs in Skeletal Muscle

Nuclear receptors involved in energy expenditure and the control of lipid and cholesterol metabolism in skeletal muscle include PPARs, LXRs, estrogen receptors (ERs), estrogen-related receptors (ERRs), thyroid hormone receptors (TRs), glucocorticoid receptor (GR), FXR, RORα1, nuclear receptor subfamily 1, group D, member 1 α,β (NR1D1α,β), and nuclear receptor subfamily 4 group A member 1 (NR4A1)[252] (Figure 5.5B). This section focuses on the interaction of nutrients with nuclear receptors, with an emphasis on the impact of nuclear receptors concerning insulin resistance, obesity, and diabetes.

PPARs play essential roles in the regulation of lipid metabolism. They sense and respond to free fatty acids and their derivatives to regulate genes involved at almost all levels of lipid metabolism, including lipid import/export, synthesis, storage, breakdown, and oxidation[253] (Figure 5.5B). Both PPARα and PPARδ play essential roles in muscle fatty acid oxidation. PPARδ is the most abundant PPAR in skeletal muscle and is predominantly expressed in type I fibers relative to type II fibers.[254,255] Several studies demonstrated that PPARδ induces genes involved in energy expenditure and lipid catabolism. Moreover, PPARδ activation leads to a predominant type I fiber phenotype, with an increase in mitochondrial density, insulin sensitivity, and resistance to obesity.[256]

Although initially identified as liver-enriched transcription factors, LXRs are also expressed in skeletal muscle. LXRα and LXRβ are regulators of cholesterol and lipid metabolism as well as carbohydrate metabolism (Figure 5.5B). LXRα is required for the control of cholesterol metabolism in the liver. In contrast, LXRβ is more widely expressed and appears to be a major regulator of glucose tolerance, insulin sensitivity, and energy utilization in muscle.[166] LXRβ is the most abundant isoform in skeletal muscle and is involved in enhancing the insulin-stimulated glucose disposal rate, decreasing triglycerides, and increasing lipid catabolism, energy expenditure, reverse cholesterol transport and plasma HDL levels.[257] Treatment of wild-type mice and C2C12 myoblasts with T0901317 (synthetic LXR agonist) induces the expression of genes associated with cholesterol synthesis and increases the efflux of intracellular cholesterol to extracellular acceptors such as HDL.[258]

Although ERs exist in two main forms, α and β, which have multiple splice variants of unknown function, ERs exhibit tissue specificity in expression and function. Both are recognized as relevant regulators of metabolic homeostasis and lipid metabolism.[259] ERα is the most abundant isoform in skeletal muscle. In skeletal muscle, the metabolic role of ERα is to improve insulin sensitivity and glucose tolerance mediated by GLUT4 expression[260] (Figure 5.5B). Barros and Gustafsson demonstrated that ERβ plays a suppressive role in GLUT4 expression in skeletal muscle, which suggests a prodiabetogenic function.[261]

Regarding ERRs, it is important to specify that, although their name accurately reflects their molecular origin, the ERRs do not bind natural estrogens, and they do not directly participate in classic estrogen signaling pathways or biological processes. The ERR subfamily includes three members: ERRα, ERRβ, and ERRγ. ERRα is uniformly expressed in both muscle fibers (I and II) and regulates genes implicated in fatty acid metabolism, myocyte differentiation, and mitochondrial biogenesis in skeletal muscle[262,263] (Figure 5.5B). ERRα cooperates with PPARγ coactivator-1α (PGC1α) in muscle cells to activate the transcription of mitochondrial genes, such as ATP-synthase 5O and cytochrome c oxidase 5b, which are necessary for maximal respiration induction by PGC1α.[255,262,264]

ERRγ is selectively expressed in oxidative type I fibers, such as the soleus and red gastrocnemius. Expression of ERRγ in glycolytic type II fibers triggers an aerobic transformation, vascularization, and mitochondrial biogenesis, accompanied by the induction of genes of different metabolic pathways such as fatty acid oxidation, the TCA cycle, and oxidative phosphorylation.[255,262] However, the role that ERRγ plays in mitochondrial function remains unclear.

5.6.2 Skeletal Muscle Dysfunction by Macronutrients Through PPARs, LXR, ER, and ERRs

Obesity-related skeletal muscle insulin resistance is a hallmark of Type 2 diabetes and a primary risk factor for promoting Type 2 diabetes development. Insulin resistance is often accompanied by lipotoxicity, which is defined as the accumulation of lipids in non-adipose tissues such as skeletal muscle and liver[265] (Figure 5.5C). Preservation of skeletal muscle mass and functionality is critical for glucose homeostasis. Reduced muscle mass and strength are commonly associated with obesity and insulin resistance. In addition, other chronic metabolic perturbations and abnormalities in insulin signal transduction have been associated with reduced activity of insulin-dependent glucose transport in skeletal muscles in patients with diabetes.[266,267] Moreover, the skeletal muscles of insulin-resistant individuals are characterized by reduced fatty acid oxidation during fasting conditions and by a "metabolic inflexibility" in the switch between oxidation of fatty acid and glucose in the transition from the fasting to the postprandial state[268,269] (Figure 5.5C).

Several authors indicated a significant downregulation of LXRα, PPARα, and PPARδ mRNA levels in skeletal muscle from individuals with insulin resistantce[270] and diabetic hamsters.[250] Supporting this, PPARδ knockout mice display glucose intolerance and insulin resistance. In human myotubes isolated from healthy subjects and those with diabetes, treatment with the PPARδ agonist did not influence insulin resistance compared to those myotubes isolated from healthy subjects.[271] Interestingly, the fiber-type switch in muscle-specific PPARδ knockdown animals was followed by the development of obesity and Type 2 diabetes, underlining the importance of PPARδ in the muscle.

Reduced whole-body ERα expression or impaired ERα function due to genetic alterations (including genetic variants) has been linked with an increased prevalence of specific features of metabolic syndrome, including insulin resistance and obesity in both male and female humans and rodents.[260,272] Furthermore, a study using ERα-selective agonists indicates that activation of this nuclear receptor shows beneficial metabolic effects in several obesity models in mice. Studies demonstrate that ERα knockout mice are obese and insulin-resistant and exhibit glucose intolerance.[252]

On the other hand, treatment of C2C12 myoblasts with a synthetic ERRα inhibitor impedes mitochondrial respiration, eliciting certain aspects of a diabetic phenotype.[273] Thus, it is implied that reduced expression of LXRα, PPARs, ERα, and ERRα might contribute to skeletal muscle lipid accumulation associated with abnormal skeletal muscle glucose and lipid metabolism, and the development of insulin resistance.

5.6.2.1 High-fructose Diet

Epidemiological studies indicate that the consumption of sugar-sweetened beverages (sucrose or high-fructose corn syrup) is associated with the development of obesity and diabetes (Figure 5.5C). The underlying mechanisms by which high-fructose feeding induces alteration of muscle metabolism are not well defined. Bioinformatics tools have indicated that possible transcription factors orchestrate this genomic regulation, particularly PPARδ and nuclear receptor subfamily 4 group A member 2 (NR4A2).[274]

5.6.2.2 High-fat Diet

Evidence indicates that an HFD induces obesity and skeletal muscle insulin resistance by altering lipid metabolism in skeletal muscle, associated with an imbalance between fatty acid uptake and oxidation, resulting in intramuscular lipid accumulation[275] (Figure 5.5C). In obese individuals, HFD consumption for 5 days fails to increase PPARα expression in skeletal muscle, in contrast to lean subjects in which an HFD upregulates genes involved in lipid oxidation, reflecting metabolic flexibility.[276] Otherwise, an HFD downregulates the expression of ERRα and PGC1α in the skeletal muscle.

5.6.3 Restoration of Insulin Resistance by Macronutrients, Vitamins, and Other Compounds Through PPARs, LXR, ER, and ERRs

5.6.3.1 Dietary Sources of Protein

Studies in humans have shown more insulin response with soy *vs.* casein protein.[71,277] The mechanisms underlying the beneficial effect of soy protein on metabolic abnormalities in the skeletal muscle of insulin-resistant rats include increased expression of PPARα[278,279] (Figure 5.5D).

5.6.3.2 Lipids

Several natural compounds derived from food that modulate PPARδ activity have been reported. Fatty acids are endogenous ligands for PPARs. Forman *et al.* demonstrated that PUFAs, such as linoleic acid, arachidonic acid, and EPA, are activators of PPARδ.[280,281] An HFD supplemented with fish oil (omega-3 PUFAs) improves oxygen consumption, stimulates mitochondrial biogenesis, and upregulates mRNA expression of PPARα in skeletal muscle.[282] Cow's milk, ice cream, butter, and yogurt have been described as activators of PPARδ in reporter assays.[283] Moreover, supplementation with butyric acid, an SCFA present in milk fat, enhanced type I muscle fibers and upregulated the expression of PPARδ, leading to an increase in fatty acid oxidation in skeletal muscle.[45,284] (Figure 5.5D).

5.6.3.3 Other Compounds

5.6.3.3.1 Vitamins Shaw *et al.* have shown that RA is a high-affinity selective ligand for PPARδ.[285] RA is the carboxylic acid form of vitamin A. Studies have demonstrated that in obese mice treated with RA, the expression of PPARδ and retinoic acid receptor target genes involved in the regulation of lipid homeostasis is induced, leading to weight loss and improved insulin responsiveness[286] (Figure 5.5D).

Tocotrienols are a unique family of four natural vitamin E compounds (α-, β-, δ-, γ-tocotrienols) found in fruits, vegetables, cereal grains, and essential oils that have distinct biologic activity from tocopherols (α-, β-, δ-, γ-tocopherols).[287] δ-Tocotrienol seems to activate PPARδ in luciferase assays, and administration of tocotrienol-enriched palm oil to mice improves glucose intolerance and enhances PPARδ target genes in their muscles[160] (Figure 5.5D).

5.6.3.3.2 Bioactive Compounds Flavonoids are plant secondary metabolites widely found in fruits, vegetables, nuts, and seeds. Based on the differences in the structure of the C ring, flavonoids can be classified into six groups: flavanones (naringenin), flavones (luteolin), flavonols (quercetin), flavanols (catechins such as resveratrol, EGCG, robinetinidol, and fisetinidol), isoflavones (genistein, daidzein, and equol), and anthocyanins.[288] Flavonoids were shown to regulate glucose and lipid metabolism through various intracellular signaling pathways[289] (Figure 5.5D).

Naringenin is found mainly in citrus fruits and is responsible for the bitter taste in grapefruit. Naringenin can directly activate glucose uptake in L6 myoblasts and improve insulin action in skeletal muscle in fructose-fed insulin-resistant rats.[290] Additionally, naringenin activates the phosphorylation of p38/ mitogen-activated protein kinase (MAPK), which is involved in reducing ROS formation in L6 myoblasts.[291] Resveratrol is an agonist of ERs[292] and restores skeletal muscle dysfunction by increasing mitochondrial biogenesis and fatty acid metabolism. Resveratrol, *via* ERα, increases caveolin 3 (CAV-3) expression and enhances GLUT4 translocation to the plasma membrane to promote glucose uptake in skeletal muscle, exerting its protective effects against HFD-induced insulin resistance.[293] EGCG is the most abundant catechin in green tea and induces fatty acid oxidation as well as an upregulation of PPARα activity in the skeletal muscle of high-fat-fed mice.[294] Robinetinidol and fisetinidol derived from *Acacia mearnsii* decreased body weight, glycemia, and fat mass, and increased the skeletal muscle expression of PPARα and PPARδ in KKAy mice (a model of obese Type 2 diabetes).[295]

Isoflavones, such as daidzein and genistein, have a significant estrogenic effect and are therefore termed phytoestrogens. Isoflavones possess high affinity for PPARs and ERβ.[279] Genistein decreases body weight in mice fed an HFD compared to that of mice without genistein in their diet. The addition of genistein to the HFD also increases the mRNA abundance of PPARδ, PGC1α, and AMPK phosphorylation. As a result, there is an increase in the metabolic rate and a decrease in respiratory exchange ratio (RER) during the light and dark cycle.[296] Anthocyanins confer red, blue, purple, and black pigmentation in plants. In HFD-fed rats, the intake of blueberry increases skeletal muscle PPARα activity and affects PPAR transcripts involved in fatty acid oxidation. The specific mechanisms of PPAR activation by anthocyanins are unknown; however, studies suggest that anthocyanins interact with kinase signaling that then affects PPAR nuclear translocation and DNA binding.[297]

Mangiferin is a xanthone present at significant levels in higher plants and different parts of the mango fruit. Mangiferin has been shown to have a hypolipidemic effect on hyperlipidemia induced by HFD feeding in hamsters by the upregulation of mRNA expression of PPARα in skeletal muscle.[298] Another dietary compound that possesses therapeutic activity is gingerols, the main pungent component of ginger. Misawa *et al.* have shown that ginger extract prevents HFD-induced obesity in mice *via* PPARδ. Furthermore, ginger extracts increased the number of type I muscle fibers; thus, the authors suggest that dietary 6-shogaol and 6-gingerol may be responsible for the regulatory effects of dietary ginger on PPARδ signaling.[299]

5.7 Kidney

The kidneys have long been known to play an important role in the metabolism of carbohydrates, proteins, lipids, and other nutrients.[300–303] Proximal tubular cells generate glucose through gluconeogenesis, clear insulin from the circulation, perform critical steps of the urea cycle, and regulate the metabolism of fat-soluble vitamins such as A and D.[303] In addition, the kidneys produce or activate hormones that are involved in erythrogenesis, Ca^{2+} metabolism, and the regulation of blood pressure and blood flow.[304]

5.7.1 Physiological Role of PPARs, LXR, FXR, and VRL in the Kidney

5.7.1.1 PPARα in the Kidney

In the kidney, PPARα is abundantly expressed in epithelial cells of the proximal tubule, but little is expressed in glomeruli and it is not expressed in the loop of Henle, the distal tubules, or collecting ducts.[305–308] This distribution indicates a pivotal role for PPARα in orchestrating the energy metabolism of proximal tubule cells.[308] PPARα can be activated by exogenous compounds, such as fibrates, and also by endogenous ligands, including fatty acids and prostaglandins.[4] The PPARα agonist clofibrate significantly induced the expression of β-oxidation enzymes in the renal cortex, including long-chain acyl CoA dehydrogenase (LCAD), medium-chain acyl CoA dehydrogenase (MCAD), and ACOX. This data suggests that renal PPARα plays a major role in triggering fatty acid utilization and the adaptive response to dietary lipids in the kidney.[309] Specifically, downregulation of PPARα can decrease fatty acid β-oxidation, which can increase the lipid content of proximal tubular cells and initiate the accumulation of lipid, and induce lipotoxicity.[308,310] Additionally, Kang *et al.*[311] identified by gene ontology analysis that the levels of genes related to fatty acid metabolism and their key transcriptional regulator PPARα were markedly lower in chronic kidney disease (CKD) samples compared to samples obtained from people with normal kidney function (controls).

In addition, PPARα has been shown to play an important role in maintaining a sustained balance of energy production/utilization in the kidney through regulating the expression of genes involved in fatty acid β-oxidation.[312] The anti-inflammatory effects of PPARα activation in the kidney have been shown in previous studies.[313,314] Wilson *et al.* demonstrated the anti-inflammatory mechanisms of PPARα activation involve a decrease in renal MCP-1, renal expression of intercellular adhesion molecule 1 (ICAM-1), Cyclooxygenase 2 (COX-2), and plasma plasma interleukine (IL-6) during angiotensin II (Ang II) hypertension in mice.[315] PPARs are not only transcription factors but they also have properties that are not associated with the expression of genes. They are capable of binding different proteins, thereby inactivating them. It is predominantly the binding of PPARα[316,317] to p65 NF-κB that is responsible for the anti-inflammatory properties[318] of p65 NF-κB, which reduces the proinflammatory response. Thus, PPARα activation is essential for renal fatty acid metabolism, energy homeostasis, and anti-inflammatory regulation.

In addition, the anti-inflammatory effects of PPARα agonists have been shown to decrease blood pressure in various models of hypertension.[315] In fact, several mechanisms have been proposed to explain the antihypertensive effects of PPARα agonists such as increased excretion of Na^+ through reduced Na^+-K^+ ATPase activity in proximal tubules,[319] increased cytochrome P450 4A (CYP4A) expression,[320] and increased renal tubular 20-hydroxyeicosatetraenoic acid (20-HETE) production,[321] which exerts a natriuretic effect.[322,323] Recently, Wilson *et al.* demonstrated that the PPARα agonist decreases mean arterial pressure and plasma IL-6 during an acute model of DOCA-salt hypertension [17], suggesting that there is crosstalk between PPARα and IL-6 in regulating blood pressure.

Since the proximal tubule is a major player in the maintenance of salt and water hemostasis, as it accounts for the absorption of approximately 70% of Na^+ and the water-filtered load,[324] the activation of PPARα suggests that it is involved in lowering blood pressure (Figure 5.6).

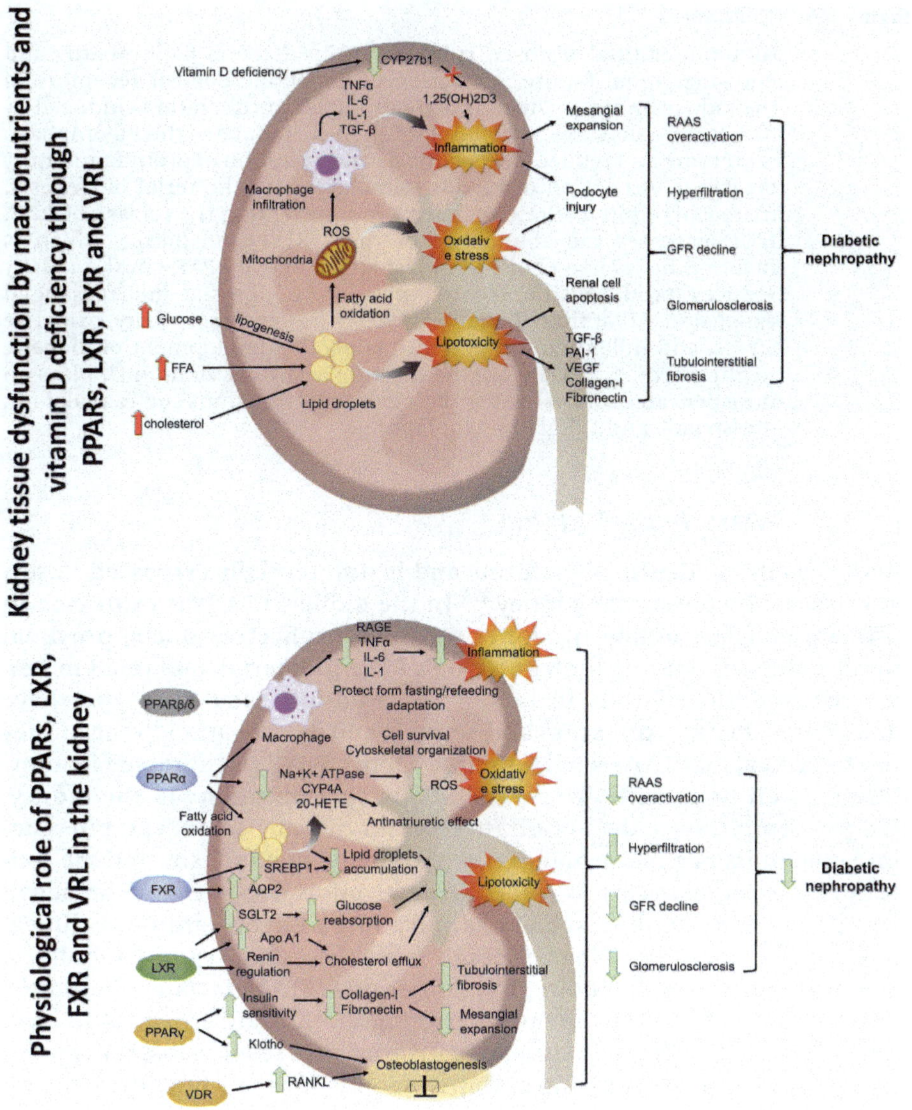

Figure 5.6 Obesity/diabetes induces systemic alterations that impair kidney integrity and function. Available glucose, free fatty acids, and cholesterol enter the kidney, leading to lipid accumulation. Excess intracellular lipids induce cellular damage directly by lipotoxicity or by the generation of ROS by the mitochondria. Oxidative stress induces an inflammatory response mediated by increased macrophage recruitment, leading to proinflammatory cytokine secretion. Lipotoxic damage, oxidative stress, and inflammation induce mesangial expansion, podocyte injury, renal cell apoptosis, and profibrotic cytokine secretion as well as collagen and fibronectin. All these renal insults lead to RAAS overactivation, hyperfiltration, GFR decline, glomerulosclerosis, and

(*continued*)

Figure 5.6 (*Continued*)

tubulointerstitial fibrosis, culminating in diabetic nephropathy and end-stage renal failure. Activation of metabolic nuclear receptors in the kidney by dietary ligands can prevent all the alterations induced by obesity and diabetes. Activation of PPARδ modulates the macrophage secretome and reduces the synthesis and secretion of proinflammatory cytokines. Activation of PPARα stimulates mitochondrial biogenesis, increasing lipid oxidative capacity and preventing the release of ROS. LXR increases the abundance of the ABCA1 transporter, which is involved in reverse cholesterol transport. Active PPARγ in the kidney reduces profibrogenic factors, preventing fibrosis and mesangial expansion. All of these beneficial effects prevent lipotoxicity, oxidative stress, and inflammation and ultimately the development of diabetic nephropathy. 1,25(OH)2D_3, the hormonal form of vitamin D, plays an indispensable role in the regulation of renal functions such as calcium, phosphate, and sulfate metabolism.

5.7.1.2 PPARγ in the Kidney

PPARγ has three identified isoforms and is differentially expressed in several tissues, including the kidney.[325] In the kidney, PPARγ is expressed in different cells, including the inner medullary collecting ducts, proximal tubules, thick ascending limb of Henle's loop, glomeruli, and renal microvascular endothelial cells in rats.[14,307,326] Multiple renal cell types have endogenous PPARγ expression and activity, and its activation in the kidney may be critical for PPARγ renal function.[14] Since PPARγ is expressed in many different cell types with diverse metabolic specializations in the kidney, this gives emphasis to the important roles for this nuclear receptor in renal lipid metabolism, glucose management, mineral metabolism, and the control of systemic blood pressure. PPARγ activation improves insulin sensitivity and reduces cardiovascular complications and renal injury in clinical practice.[327] Under physiological conditions, the kidney can metabolize a variety of substrates, including free fatty acids (FFAs), lactate, glutamine, 3-hydroxybutyrate, citrate, pyruvate, α-ketoglutarate, glycerol, proline, and other amino acids. Proximal tubule reabsorption is responsible for approximately 70% of these substrates, the metabolic fate of which depends on the extracellular medium, hormonal influences, and metabolic conditions (acid–base). Once reabsorbed, FFAs are metabolized within the mitochondria of proximal tubule cells by β-oxidation (as their glycolytic capacity is poor), constituting the largest source of ATP. An increase in the availability of intracellular fatty acids for mitochondria results in competition with other oxidizable substrates, triggering a decrease in the use of glutamine, with a consequent reduction in ammoniagenesis.[328] Accordingly, mice deficient in PPARγ present alterations in renal lipid metabolism, and the extent of renal damage induced by an HFD is lower in PPARγ heterozygous knockout mice than that in wild-type mice concomitant with a decrease in lipid accumulation and lipogenesis and an attenuation of lipid-mediated kidney damage.[329] The kidney contributes to glucose homeostasis not only through the processes of utilization (*i.e.*, glucose filtration, reabsorption, and consumption), but it is also increasingly recognized as having a significant role in gluconeogenesis (~20% of all glucose production) and uniquely contributes to plasma glucose regulation by controlling glucose reabsorption from renal tubules following glomerular filtration.[330] Whereas the poorly vascularized, and consequently relatively hypoxic, medulla is a site of considerable glycolysis, the cortex is the renal site of gluconeogenesis. Moreover, the proximal tubule is the only region of the kidney with the appropriate enzymes necessary for gluconeogenesis.[330]

CKD is directly related to the development of abnormalities in bone mineral metabolism. The kidney is known to regulate levels of vitamin D_3 by producing its most active form, 1,25 dihydroxyvitamin D_3 (calcitriol), which participates in calcium and phosphate metabolism.

The key players in this hormonal bone–parathyroid–kidney axis include fibroblast growth factor 23 (FGF 23), Klotho, parathyroid hormone (PTH), and calcitriol, and recent work has also implicated the involvement of PPARγ.[331] Klotho is expressed the highest in the kidney, and interestingly, the phenotype of the Klotho hypomorph or knockout mice has striking similarities to CKD patients, such as ectopic soft-tissue calcification, hyperphosphatemia, and high plasma FGF 23,[332] suggesting that CKD might be a state of Klotho deficiency. Consistent with its activation by PPARγ, oral administration of the PPARγ agonist troglitazone augments renal Klotho mRNA expression in Otsuka Long-Evans Tokushima fatty rats and protects against the endothelial dysfunction induced in this model of atherogenesis.[333] Unfortunately, thiazolinedione treatment is associated with an increased risk of hip fractures and is linked to the formation of excessive calcium phosphate precipitates in the urinary bladder.[327,334] This effect is likely caused by excess PPARγ activity, which increases adipogenesis and downregulates the β-catenin pathway and osteoblastogenesis. Overall, this data points to PPARγ as an important player in maintaining mineral metabolism, both at renal and systemic levels (Figure 5.6).

5.7.1.3 PPARβ/δ in the Kidney

Although PPARβ/δ seems to be abundant in the kidney[258,305] and ubiquitously expressed along the nephron,[305] the role of this PPAR isotype in the kidney remains poorly understood. PPARβ/δ may play an important role in renal metabolic adaptation to fasting and refeeding.[335] One study[336] raised the possibility that PPARβ/δ may contribute to renal metabolic adaptation to fasting and refeeding. After an overnight fast, PPARβ/δ mRNA in the kidney was dramatically downregulated and was rapidly restored to control levels upon refeeding. Moreover, this tight nutritional regulation is independent of the presence of PPARα, because similar regulation of PPARβ/δ is observed in PPARα knockout mice. The regulation of PPARβ/δ expression may be related to metabolic fates, for example, gluconeogenesis and lipogenesis. As gluconeogenesis is increased during fasting, PPARβ/δ would be a negative regulator of this pathway. These findings suggest its involvement in metabolic kidney diseases such as diabetic nephropathy.[337] Additionally, overexpression of PPARβ/δ can protect cultured medullary interstitial cells from hypertonicity-induced cell death, suggesting that PPARβ/δ is an important survival factor for medullary interstitial cells in the hypertonic condition in renal medulla.[338] Additionally, Letavernier *et al.* demonstrated that PPARβ/δ can provide strong protection against ischemia-induced renal injury as a result of its combined action on cell survival and cytoskeletal reorganization.[339] Moreover, PPARβ/δ downregulates the expression of the receptor for advanced glycation end-products (RAGE) and proinflammatory cytokines (TNFα, IL-1, and IL-6) in the kidney of streptozotocin-induced diabetic mice.[340] Collectively, PPARβ/δ agonists may be considered a novel means of conferring renal protection in diabetic nephropathy and other diseases (Figure 5.6).

5.7.1.4 LXRs in the Kidney

LXRs are members of a nuclear receptor family consisting of two isoforms, LXRα and LXRβ. They play a major role in energy metabolism, including glucose and lipid metabolism.[341–343] In the kidney, LXRs are specifically expressed in renin-producing JG cells, in which LXRα is specifically enriched, and interact with the renin promoter.[344] The hormone renin is the rate-limiting step in the renin–angiotensin–aldosterone system (RAAS), which is the most critical regulator of blood pressure and salt volume homeostasis in physiological and pathophysiological conditions. LXRs were shown to regulate renin expression *in vivo*, suggesting a cross-talk between the RAAS and lipid metabolism.[344] Natural ligands for LXRs are oxysterols, but strong synthetic agonists such as T0901317 (T09) and GW3965 have been developed.[14,345] Additionally, a study conducted by Wu *et al.* demonstrated that glomerular LXRα expression was markedly induced by TZD.[337] A similar effect was observed with the use of the LXR agonist T-0901317, which markedly increased ApoA1-mediated cholesterol efflux in cultured mesangial cells, suggesting that LXRα may participate, at least in part, in cholesterol transport in renal mesangial cells.[337] In addition to lipid metabolism, LXR activation regulates renal glucose transport by decreasing sodium-dependent glucose cotransporters 2 (SGLT2) transport function in human renal proximal cells in both normal and high-glucose conditions.[343] In addition to energy metabolism, LXR activation has been reported to regulate kidney function. LXRs have been shown to regulate membrane transport proteins, including cystic fibrosis transmembrane conductance,[346] epithelial sodium channel,[347] organic anion transporter 1,[348] sodium–phosphate cotransporter,[349] and aquaporins 1 (AQP1) and 2 (AQP2)[46,350] (Figure 5.6).

5.7.1.5 FXR in the Kidney

FXR performs multiple physiological functions in the kidney. The kidney has been known to secrete and reabsorb BAs that escape the enterohepatic circulation.[351] A proposed mechanism involves organic solute transporter (OST), a BA and sterol transporter, as suggested by its basolateral localization in the proximal renal tubule cell[352] and its FXR-dependent upregulation.[353–356] Remarkably, renal expression of FXR, the major BA sensor, is predominantly tubular[357] and is less abundant in the glomerulus, podocytes, and mesangial cells.[358] Another effect attributed to FXR is modulation of urine volume and osmolality.[359] FXR upregulates renal AQP2, a vasopressin-regulated water channel protein critical for body water homeostasis,[359] and protects the collecting duct cells from the hypertonic stress that normally prevails in the renal medulla and would otherwise lead to their apoptosis.[360] Regarding lipid metabolism.[361] FXR agonists downregulate renal overexpression of sterol regulatory element-binding protein 1 (SREBP-1), which leads to lipid accumulation. Additionally, the activation of FXR could prevent the induction of profibrotic growth factors, proinflammatory cytokines, and oxidative stress-related enzymes in the kidney[361] (Figure 5.6).

5.7.1.6 VDR in the Kidney

VDR is a ligand-activated transcription factor that belongs to the nuclear receptor superfamily. 1,25(OH)2D_3, the hormonal metabolite of vitamin D, is the natural ligand of the VDR. After binding to 1,25(OH)2D_3, VDR enters the nucleus and forms a heterodimer with RXR, which interacts with response elements in target gene promoters to regulate gene transcription. Immunohistochemical studies indicate that VDR is expressed in proximal and distal tubular epithelial cells, glomerular parietal epithelial cells, and collecting duct cells; VDR is also expressed by cultured podocytes and mesangial cells.[362–364]

Vitamin D_3 is either acquired from dietary sources or is generated *via* solar ultraviolet irradiation of 7-dehydrocholesterol in the skin. Vitamin D_3 is then processed into the active hormone 1,25(OH)2D_3 *via* two consecutive hydroxylation reactions.[365,366] The first such reaction takes place in the liver and is catalyzed by 25-hydroxylase, and the second is catalyzed by 1α-hydroxylase, which is expressed predominantly in the kidney. The primary signal for activating CYP27B1 transcription in the kidney is PTH. In response to low serum calcium, the parathyroid gland secretes PTH, which (among its many functions) stimulates the transcription of CYP27B1, resulting in the production of 1,25-(OH)2D_3. Very low blood phosphate also increases CYP27B1 expression.[366] In turn, 1,25-(OH)2D_3 directly stimulates intestinal calcium and phosphate absorption and, together with PTH, stimulates the synthesis and secretion of a cytokine known as receptor activator of nuclear factor-κB ligand (RANKL), which has a key role in osteoclastogenesis and the activation of giant osteoclasts to induce bone resorption.[366,367] Additionally, 1,25-(OH)2D_3 and PTH are required in the distal renal tubule to increase calcium reabsorption.[366] Briefly, 1,25(OH)2D_3, the hormonal form of vitamin D, plays an indispensable role in the regulation of renal functions such as calcium, phosphate, and sulfate metabolism.[363] 1,25(OH)2D_3 also negatively regulates the renin–angiotensin system,[368] which is considered to play a key role in the development of diabetic nephropathy.[369] Vitamin D has been shown to have a renoprotective role[363] (Figure 5.6).

5.7.2 Kidney Tissue Dysfunction by Macronutrients and Vitamin D Deficiency Through PPARs, LXR, FXR, and VRL

5.7.2.1 Carbohydrates

It is now recognized that the kidney has a significant role in glucose homeostasis under both physiological and pathological conditions.[370] The kidney contributes to glucose homeostasis not only through the processes of utilization (*i.e.*, glucose filtration, reabsorption, and consumption), but it is also increasingly recognized as having a significant role in gluconeogenesis (~20% of all glucose production). In addition, the kidney uniquely contributes to plasma glucose regulation by controlling glucose reabsorption from renal tubules following glomerular filtration.[327,330]

All three PPAR isoforms seem to play important roles in the development of diabetic nephropathy in Type 2 diabetes. Several studies have demonstrated that PPARγ activation by troglitazone and rosiglitazone prevented glucose-induced TGF-β expression in the kidney, a central event in the pathogenesis of diabetic glomerulosclerosis.[371] *In vitro* studies have shown that PPARγ activation in cultured renal mesangial cells modulated cell proliferation and differentiation,[372] and inhibited high-glucose- or TGF-β-stimulated synthesis of type I collagen.[373] PPARγ mRNA is reduced by 77% in glomeruli of diabetic mice.[373,374]

Hyperglycemic conditions lead to a dysfunction of SGLT-mediated glucose transport in proximal tubular cells and promote epithelial–mesenchymal transition (EMT), leading to renal fibrosis. PPARγ agonists reverse this hyperglycemia-induced EMT and restore the two functional SGLT proteins situated in different segments of the epithelial cells of the proximal tubule (SGLT2 in the S1 segment and SGLT1 in the S3 segment).[327,370]

Additionally, chronic hyperglycemia leads to various types of protein homeostasis collapse. Advanced glycation end products (AGEs) are produced by glycation, which is a non-enzymatic reaction between glucose and proteins.[375] AGEs accumulate in the body when humans are exposed to high levels of glucose, such as in diabetes. AGEs levels increase as CKD progresses, since the kidney plays an important role in AGEs clearance;[376] renal proximal tubule cells absorb AGEs and catabolize them.[377,378] In a study conducted by Matsui *et al.*,[379] they demonstrated that nifedipine inhibited AGEs and their receptor (RAGE) interaction-mediated proximal tubular cell injury *via* PPARγ.

In vitro high glucose was sufficient to downregulate PPARα, FXR, LXRα, and LXRβ expression, associated with severe dyslipidemia and inflammation in diabetic kidneys[380,381] (Figure 5.6). On the other hand, both mRNA and protein of PPARα are elevated in glomeruli, cortical tubules, and renal arterial vessels of *db/db* mice.[382]

Fructose is a monosaccharide found in fruits, vegetables, and honey. However, not only fructose but also high-fructose corn syrup is one of the most widely used food ingredients in nearly all soft drinks, canned jams, breakfast cereals, and baked goods.[383] Fructose may contribute to kidney injury through several mechanisms. In a study conducted by Magliano *et al.*, the anti-inflammatory effect of GW50156, a PPARβ/δ agonist in the kidney of a high-fructose-fed model, was demonstrated.[384] Vitamin D receptor is decreased in both intestine and kidney of genetically diabetic *db/db* mice, which may link to hyperplasia of kidney[385] (Figure 5.6).

5.7.2.2 Lipids

The kidney is usually not regarded as an important organ in lipid metabolism because the glomerular filtration barrier prevents large molecules such as lipoprotein particles from being filtered into the pre-urine. Nevertheless, the proximal tubule epithelium is equipped with a high-capacity receptor system for the uptake of a broad spectrum of ligands, including various apolipoproteins, albumin, and other proteins binding lipophilic substances.

Lipid accumulation in non-adipose tissue is increasingly recognized to contribute to organ injury though a process termed lipotoxicity. Several studies have shown an association between kidney dysfunction and renal lipid accumulation, including models of obesity and diabetes.[386] The role of palmitate in the progression of kidney disease has been demonstrated in several reports. In cultured proximal tubular cells, palmitate increases the expression of MCP-1 and leads to intracellular accumulation of diacylglycerol and subsequent activation of the protein kinase C protein family. In a study conducted by Kume *et al.*,[329] mice deficient in PPARγ presented alterations in renal lipid metabolism, and the extent of renal damage induced by an HFD is lower in heterozygous knockout mice than that in wild-type mice concomitant with a decrease in lipid accumulation and lipogenesis, as well as an attenuation of lipid-mediated kidney damage.[329] Furthermore, in podocytes treated with saturated palmitic acid, decreased PPARγ1 expression occurs and is correlated with a proinflammatory and proapoptotic state, as well as with changes in the gene expression of enzymes involved in fatty acid synthesis, including a decrease in ACOX and FAS.[387]

In a study conducted by Shin *et al.*,[388] PPARα deficiency in HFD-induced obesity in genetically hypertensive SHR rats was associated with an increase in lipotoxicity in kidneys. Consequently, renal lipotoxicity likely contributed to the aggravation of blood pressure and pathological changes by intrarenal RAS activation and oxidative stress. Similarly, Chung *et al.*[389] demonstrated that an HFD induced renal cell apoptosis and oxidative stress in spontaneously hypertensive rates (SHRs) and that these alterations were related to a decreased PPARα expression in the kidney (Figure 5.6).

In a series of studies using rodent models of Type 1 diabetes,[381] HFD-induced obesity[390] with insulin resistance and Type 2 diabetes mellitus, FXR expression and its target genes were downregulated in the kidney and correlated with the level of inflammation and fibrosis.[357,358,391–394]

5.7.2.3 Proteins

Dietary protein intake can modulate renal function and its role in renal disease.[395,396] In a study conducted by Tovar-Palacio *et al.*[397] the authors demonstrated that dietary protein from both vegetable and animal sources can modify renal lipogenesis, triglyceride accumulation, and glomerulosclerosis. These results have practical importance since high-protein diets are frequently recommended for obese subjects who are attempting to lose weight.[395,396] Furthermore, excess fatty acids accompanied by triglyceride accumulation in the parenchymal cells of multiple tissues, including the kidney, may result in chronic cellular dysfunction and injury.[398]

5.7.2.4 Vitamin D Deficiency

Several studies have suggested that kidney diseases are closely tied to inadequate vitamin D levels. The prevalence of vitamin D deficiency is very high in patients with kidney disease, mainly due to decreased CYP27b1 activity for 1,25(OH)2D_3 synthesis, impaired reabsorption of 25(OH)D in the proximal tubular cells, and increased levels of FGF 23 that suppress the biosynthesis of 1,25(OH)2D_3. As the only nuclear receptor that mediates the biological activity of 1,25(OH)2D_3, VDR activity is compromised in vitamin D deficiency, which is believed to contribute to the pathogenesis of kidney disease[144] (Figure 5.6).

5.7.3 Restoration of Kidney Tissue Dysfunction by Macronutrients and Other Components Through PPARs, LXRs, FXR, and VDR

5.7.3.1 Lipids

Palmitate-induced renal tubular cell damage was attenuated by PUFAs such as oleate and EPA.[399] Additionally, El-Ashmawy *et al.* showed that fish oil treatment upregulated PPARγ gene expression in renal tissue.[400] These results were in agreement with Han *et al.*,[401] who found that treatment with EPA and DHA significantly increased the expression of PPARγ and decreased the expression of integrin β_1 in rat glomerular mesangial cells treated with lipopolysaccharide. Therefore, fish oil containing omega-3 PUFAs, including EPA and DHA, have been shown to prevent the progression of nephropathy and retard the progression of kidney disease *via* PPARγ upregulation.

5.7.3.2 Proteins

In a study conducted by Yeh *et al.*,[402] kidney PPARγ protein expression in diabetic nephropathy (DN) rats was lower than that in the control group, and soy protein and β-conglycinin increased PPARγ protein expression and subsequently decreased TNFα levels. These results suggest that soy protein and β-conglycinin decrease inflammation and increase insulin sensitivity *via* activation of PPARγ, retarding the progression of DN.

5.7.3.3 Other Components

5.7.3.3.1 Vitamin D Treatment with VDR agonists has been shown to reduce albuminuria and alleviate glomerular injury in experimental models of kidney diseases.[144] A large body of evidence has demonstrated that VDR agonists can effectively attenuate the progression of DN. Podocyte injury is a key event in the development of DN, which results in proteinuria and often leads to progressive renal injury. A study conducted by Verouti *et al.* showed that calcitriol and paricalcitol can maintain and reactivate the expression of specialized components of podocytes, including podocalyxin, hence providing protection against loss of the permselective renal barrier.[403] Wang *et al.* reported that 1,25(OH)2D_3 suppresses high glucose-induced podocyte apoptosis by blocking p38- and ERK-mediated pro-apoptotic pathways, and podocyte-specific expression of human VDR in transgenic mice protects against diabetic renal injury in experimental models.[404] VDR can also inhibit podocyte apoptosis by targeting the activation of the NF-κB pathway.[405] Other studies further showed that treating DN rats with calcitriol can normalize the expression of podocyte markers, including nephrin and podocin,[406] and that the phosphoinositide 3-kinases/phospho-protein Kinase B (PI3K/p-Akt) signaling pathway participates in calcitriol amelioration of podocyte injury in DN rats.[407]

5.7.3.3.2 Polyphenols A recent study found that Huangkui capsule, an extract from *Abelmoschus manihot* (L.) Medic., can ameliorate DN by increasing PPARα/PPARγ signaling, leading to lowered endoplasmic reticulum stress in rats.

Anthocyanin ameliorates DN in *db/db* mice by decreasing renal lipid concentrations, with improvement in glomerular matrix expansion and inflammation, which are related to increased phosphorylation of AMPK and activation of PPARα and PPARγ and inhibited activity of ACC and SREBP-1.[193] In HK-2 cells, anthocyanins inhibit high-glucose-induced cholesterol accumulation and inflammation by activating the LXRα pathway.[26]

5.7.3.3.3 Oxysterols Ouabagenin belongs to a novel class of naturally occurring LXR ligands that do not cause the unwanted side effects observed with conventional LXR ligands such as lipid accumulation in the liver. Ouabagenin induces downregulation of the epithelial sodium channel (ENaC) in an LXRβ-specific manner. ENaC is required for sodium reabsorption in the kidney, which is essential for the maintenance of sodium and water homeostasis and hence blood pressure stability.[408]

5.8 Conclusion and Perspectives

The understanding of the mechanism for the direct activation of gene transcription through specific dietary components has improved significantly due to the discovery of metabolic nuclear receptors. Understanding the interaction between dietary components and nuclear receptors in metabolic tissues has been labeled nutrigenomics. Nutrigenomics is a fertile area of research to advance our understanding of how to further prevent and treat ailments promoted by dietary factors related to obesity and diabetes.

In the intestine, further research is needed to establish how nutrients can modulate PPARs, LXRs, and FXR pathways. Additionally, more research is required at the level of nuclear receptors to identify how a microbiota that provides beneficial effects in the human body can be obtained. In the liver, several questions regarding the discrepant transcriptional effects of PPARs and LXRs depending on dietary ligand binding are still unanswered. The induction of hyperplasia in white adipose tissue and fatty acid oxidation and thermogenesis in brown adipose tissue by diverse dietary components opens an opportunity for the treatment of obesity and diabetes.

Current guidelines for obesity and diabetes prevention in many countries are focused on the reduction in saturated fat and animal protein intake. However, these dietary guidelines do not take into consideration the important health effects posed by bioactive molecules that can modulate the metabolic response of the different body organs to these saturated fats and animal protein. Advances in the understanding of the delicate interaction between dietary components and the activity of metabolic nuclear receptors will provide healthcare professionals with unparalleled tools to develop new and more advantageous dietary guidelines to prevent or treat obesity and diabetes.

Activation of nuclear receptors by particular food components such as oxidized fatty acids, partially hydrogenated fatty acids, or lipid molecules arising from fructose-mediated lipogenesis induces severe metabolic derangements in skeletal muscle, liver, kidney, and adipose tissues through unconventional activation of metabolic nuclear receptors. Conversely, polyphenols, plant sterols, and long-chain PUFAs exert clear beneficial effects in these metabolic organs by differential activation of the same nuclear receptors. Thus, independent of its energy density, the particular composition of our dietary selections can favor or prevent the development of metabolic diseases. Consequently, policies and advice for the general population must be focused on increasing the intake of fresh vegetables, fruits, and whole grain cereals and simultaneously encouraging the exclusion of ultraprocessed foods from their dietary habits. This approach could increase the consumption of polyphenols, sterols, and prebiotic fiber and simultaneously reduce the intake of foods rich in fat and refined sugars. These general recommendations should not be viewed as dietary prescription for older and sick people: rather these recommendations should be observed by healthy young people in order to prevent the development of chronic diseases in older age.

Abbreviations

1,25(OH)2D_3	1,25-dihydroxyvitamin D_3
20-HETE	20-hydroxyeicosatatraenoic acid
ABCA1	ATP-binding cassette transporter
AA	arachidonic acid
ABCG5/ABCG8	ATP-binding cassette sub-family G member 5 and 8
ACAT2	acyl-coenzyme A:cholesterol acyltransferase 2
ACC	acetyl-CoA carboxylase
ACOX	acyl-CoA oxidase
ACS	acyl CoA synthase
ACSLs	long-chain-fatty-acid-CoA ligases
ADPN	adiponutrin
AF1	activation function 1
AGEs	advanced glycation end-products
AGPATs	acylglycerol-3-phosphate acyltransferases
AMPK	5′ adenosine monophosphate-activated protein kinase
Ang II	angiotensin II
aP2	fatty acid-binding protein
ApoA1	apolipoprotein A1
AQP1	aquaporin 1
AQP2	aquaporin 2
AQP7	aquaporin 7
ASBT	apical sodium-dependent bile acid transporter
ATP	adenosine triphosphate
at-RA	*all-trans* retinoic acid
BAs	bile acids
BAT	brown adipose tissue
BSA	bovine serum albumin
BSEP	bile salt export pump
CA	cholic acid
CAV-3	caveolin 3
CD36	cluster of differentiation 36
CDCA	chenodeoxycholic acid
ChREBP	carbohydrate-responsive element binding protein
CKD	chronic kidney disease
CLPR3	ATP-dependent Clp protease proteolytic subunit-related protein 3 (chloroplastic)

COX-2	cyclooxygenase 2
CREB	cAMP response element-binding protein
Cy3G	cyanidin-3-glucoside
CYP4A	cytochrome P450 4A
CYP7A1	bile acid-synthesizing enzyme 7α hydroxylase
CYP27	bile acid-synthesizing enzyme 27α hydroxylase
DA	deoxycholic acid
DHA	docosahexaenoic acid
DN	diabetic nephropathy
DNA	deoxyribonucleic acid
EGCG	epigallocatechin
ELOVs	fatty acid elongases
EMT	epithelial–mesenchymal transition
ENaC	epithelial sodium channel
EPA	eicosapentaenoic acid
ERRα	estrogen-related receptor α
ERRβ	estrogen-related receptor β
ERRγ	estrogen-related receptor γ
ERRs	estrogen-related receptors α, β, γ
ERs	estrogen receptors
FABPs	fatty acid-binding proteins
FADS	fatty acid desaturase
FAS	fatty acid synthase
FFA	free fatty acid
FGF 15/19	fibroblast growth factor 15/19
FGF 23	fibroblast growth factor 23
FGFR4	fibroblast growth factor receptor 4
FXR	farnesoid X receptor
G-6-Pase	glucose 6 phosphatase
GFR	glomerular filtration rate
GK	glycerol kinase
GLP1	glucagon-like peptide-1
GLUT2	glucose transporter 2
GLUT4	glucose transporter 4
GR	glucocorticoid receptor
HCHD	high-carbohydrate diet
HDL	high-density lipoprotein
HDL-C	high-density lipoprotein cholesterol
HFD	high-fat diet
HFuD	high-fructose diet
HNF4α	hepatocyte nuclear factor 4 α
HNF4γ	hepatocyte nuclear factor 4 γ
IBABP	ileal bile acid-binding protein
ICAM-1	intercellular adhesion molecule 1
IL-1	interleukin 1
IL-6	interleukin 6

IL-10	interleukin 10
IL-22	interleukin 22
iNOS	inducible nitric oxide synthase
JG	juxtaglomerular
JNK	janus kinase
LA	lithocholic acid
LBD	ligand-binding domain
LCAD	long-chain acyl CoA dehydrogenase
LDH	lactate dehydrogenase
LDL	low-density lipoprotein
LDL-C	low-density lipoprotein cholesterol
LDLr	low-density lipoprotein receptor
Lip	lipoprotein lipase
LXRE	liver X response element
LXRs	liver X receptors
LXRα	liver X receptor α
LXRβ	liver X receptor β
MAPK	mitogen-activated protein kinase
MCAD	medium-chain acyl CoA dehydrogenase
MCP-1	monocyte chemoattractant protein-1
MOGAT	acylglycerol *O*-acyltransferase
MTTP	microsomal triglyceride transfer protein
NAFLD	non-alcoholic fatty liver disease
NFA	nitrated fatty acid
NF-κB	nuclear factor kappa-light-chain-enhancer of activated B cells
NPC1L1	Niemann–Pick C1-like 1
NR1D1α,β	nuclear receptor subfamily 1, group D, member 1 α,β
NR4A1	nuclear receptor subfamily 4 group A member 1
NR4A2	nuclear receptor subfamily 4 group A member 2
NTCP	bile salt transporters sodium/taurocholate cotransporting polypeptide
OST	organic solute and steroid transporter
OSTα	organic solute and steroid transporter α
OSTβ	organic solute and steroid transporter β
PAI-1	plasminogen activator inhibitor 1
PDK1	pyruvate dehydrogenase kinase 1
PEPCK	phosphoenolpyruvate carboxykinase
PGC1α	PPARγ coactivator-1α
PI3K/p-Akt	phosphoinositide 3-kinase/phospho-protein kinase B
PNPLA	patatin-like phospholipase domain-containing protein
PPARE	peroxisome proliferator-activated response element
PPARs	peroxisome proliferator-activated receptors
PPARα	peroxisome proliferator-activated receptor alpha
PPARβ/δ	peroxisome proliferator-activated receptor beta/delta
PPARγ	peroxisome proliferator-activated receptor gamma

PTH	parathyroid hormone
PUFAs	polyunsaturated fatty acids
PXR	pregnane X receptor
RA	retinoic acid
RAAS	renin–angiotensin–aldosterone system
RAGE	receptor advanced glycation end-products
RANKL	receptor activator of nuclear factor-κB ligand
RAS	renin–angiotensin system
RBEE	rice bran enzymatic extract
RER	respiratory exchange ratio
RORα1	retinoic acid receptor-related orphan receptor α1
RORγ	retinoic acid receptor-related orphan receptor γ
ROS	reactive oxygen species
RXR	retinoid X receptor
RXRα	retinoid X receptor α
SCFAs	short-chain fatty acids
SGLT2	sodium-dependent glucose cotransporters 2
SHR	spontaneously hypertensive rats
SR-B1	scavenger receptor class B type 1
SREBP-1c	sterol regulatory element-binding protein 1c
TA	taurocholic acid
T-B-MCA	tauro-B-muricholic acid
TCA cycle	tricarboxylic acid cycle
TGF-β	transforming growth factor beta 1
TGR5	G-protein-coupled bile acid receptor, Gpbar1
TLR4	Toll-like receptor 4
TNFα	tumor necrosis factor-α
Treg	regulatory T cells
TRs	thyroid hormone receptors
TZD	thiazolidinedione
UCP-1	uncoupling protein-1
UCP-2	uncoupling protein-2
VDR	vitamin D receptor
VEGF	vascular endothelial growth factor
VLDL	very-low-density lipoprotein
VLDLr	very-low-density lipoprotein receptor
Zfp423	zinc finger protein 423

References

1. D. Toft, G. Shyamala, and J. Gorski, *Proc. Natl. Acad. Sci. U. S. A.*, 1967, **57**, 1740–1743.
2. E. V. Jensen and E. R. DeSombre, *Annu. Rev. Biochem.*, 1972, **41**, 203–230.
3. L. Nagy and J. W. Schwabe, *Trends Biochem, Sci.*, 2004, **29**(6), 317–324.
4. T. Kushiro, E. Nambara and P. McCourt, *Nature*, 2003, **422**(6928), 122.

5. G. A. Francis, E. Fayard, F. Picard and J. Auwerx, *Annu. Rev. Physiol.*, 2003, **65**, 261–311.
6. V. K. Dhiman, M. J. Bolt and K. P. White, *Nat. Rev. Genet.*, 2018, **19**(3), 160–174.
7. R. Sever and C. K. Glass, *Cold Spring Harb. Perspect. Biol.*, 2013, **5**(3), a016709.
8. C. Degirolamo, C. Sabba and A. Moschetta, *J. Lipid Res.*, 2015, **56**(7), 1262–1270.
9. H. Gronemeyer, J. A. Gustafsson and V. Laudet, *Nat. Rev. Drug Discov.*, 2004, **3**(11), 950–964.
10. D. J. Mangelsdorf and R. M. Evans, *Cell*, 1995, **83**(6), 841–850.
11. D. J. Mangelsdorf, C. Thummel, M. Beato, P. Herrlich, G. Schutz, K. Umesono, *et al.*, *Cell*, 1995, **83**(6), 835–839.
12. J. Sonoda, L. Pei and R. M. Evans, *FEBS Lett.*, 2008, **582**(1), 2–9.
13. A. L. Bookout, Y. Jeong, M. Downes, R. T. Yu, R. M. Evans and D. J. Mangelsdorf, *Cell*, 2006, **126**(4), 789–799.
14. C. Tovar-Palacio, N. Torres, A. Diaz-Villasenor and A. R. Tovar, *Genes Nutr.*, 2012, **7**(4), 483–498.
15. A. K. Sikalidis, *J. Am. Coll. Nutr.*, 2019, **38**(1), 84–95.
16. S. Sauer and A. Plauth, *Appl. Microbiol. Biotechnol.*, 2017, **101**(3), 951–961.
17. S. Gothai, P. Ganesan, S. Y. Park, S. Fakurazi, D. K. Choi and P. Arulselvan, *Nutrients*, 2016, **8**(8), 461.
18. M. Lu, Y. Cao, J. Xiao, M. Song and C T. Ho, *Food Funct.*, 2018, **9**(9), 4569–4581.
19. P. R. Kiela and F. K. Ghishan, *Best Pract. Res. Clin. Gastroenterol.*, 2016, **30**(2), 145–159.
20. F. M. Gribble and F. Reimann, *Nat. Rev. Endocrinol.*, 2019, **15**(4), 226–237.
21. J. C. Clemente, L. K. Ursell, L. W. Parfrey and R. Knight, *Cell*, 2012, **148**(6), 1258–1270.
22. W. Wahli, *J. Intern. Med.*, 2008, **263**(6), 613–619.
23. C. J. Sinal, M. Tohkin, M. Miyata, J. M. Ward, G. Lambert and F. J. Gonzalez, *Cell*, 2000, **102**(6), 731–744.
24. S. A. Kliewer and D. J. Mangelsdorf, *Dig. Dis.*, 2015, **33**(3), 327–331.
25. C. Jiang, C. Xie, F. Li, L. Zhang, R. G. Nichols, K. W. Krausz, *et al.*, *J. Clin. Invest.*, 2015, **125**(1), 386–402.
26. M. S. Trabelsi, M. Daoudi, J. Prawitt, S. Ducastel, V. Touche, S. I. Sayin, *et al.*, *Nat, Commun.*, 2015, **6**, 7629.
27. P. Li, L. Zhu, X. Yang, W. Li, X. Sun, B. Yi, *et al.*, *J. Cell Physiol.*, 2019, **234**(8), 12839–12846.
28. B. Bonamassa and A. Moschetta, *Trends Endocrinol. Metab.*, 2013, **24**(3), 120–128.
29. G. Lo Sasso, S. Murzilli, L. Salvatore, I. D'Errico, M. Petruzzelli, P. Conca, *et al.*, *Cell Metab.*, 2010, **12**(2), 187–193.
30. T. L. Errico, K. A. Mendez-Lara, D. Santos, N. Cabrerizo, L. Baila-Rueda, J. Metso, *et al.*, *Transl. Res.*, 2017, **186**, 19–35 e5.

31. E. Grenier, G. Mailhot, D. Dion, Z. Ravid, S. Spahis, M. Bendayan, *et al.*, *Biochem. Cell Biol.*, 2013, **91**(6), 476–486.
32. Z. Ravid, M. Bendayan, E. Delvin, A. T. Sane, M. Elchebly, J. Lafond, *et al.*, *Am. J. Physiol.: Gastrointest. Liver Physiol.*, 2008, **295**(5), G873–G885.
33. A. Alvaro, R. Rosales, L. Masana and J. C. Vallve, *J. Nutr. Biochem.*, 2010, **21**(6), 518–525.
34. O. Briand, V. Touche, S. Colin, G. Brufau, A. Davalos, M. Schonewille, *et al.*, *Gastroenterology*, 2016, **150**(3), 650–658.
35. L. Cruz-Garcia and A. Schlegel, *J. Lipid Res.*, 2014, 55(9), 1944–1958.
36. T. Benitez-Santana, S. E. Hugo and A. Schlegel, *Front. Physiol.*, 2017, **8**, 280.
37. A. Mohammadi and E. A. Oshaghi, *J. Diabetes Metab. Disord.*, 2014, **13**(1), 20.
38. M. Gonzalez-Granillo, K. R. Steffensen, O. Granados, N. Torres, M. Korach-Andre, V. Ortiz, *et al.*, *Diabetologia*, 2012, 55(9), 2469–2478.
39. P. J. Turnbaugh, M. Hamady, T. Yatsunenko, B. L. Cantarel, A. Duncan, R. E. Ley, *et al.*, *Nature*, 2009, **457**(7228), 480–484.
40. P. Louis, K. P. Scott, S. H. Duncan and H. J. Flint, *J. Appl. Microbiol.*, 2007, **102**(5), 1197–1208.
41. M. X. Byndloss, E. E. Olsan, F. Rivera-Chavez, C. R. Tiffany, S. A. Cevallos, K. L. Lokken, *et al.*, *Science*, 2017, **357**(6351), 570–575.
42. M. Kinoshita, Y. Suzuki and Y. Saito, *Biochem. Biophys. Res. Commun.*, 2002, **293**(2), 827–831.
43. J. Tomas, C. Mulet, A. Saffarian, J. B. Cavin, R. Ducroc, B. Regnault, *et al.*, *Proc. Natl. Acad. Sci. U. S. A.*, 2016, **113**(40), E5934–E5943.
44. G. Jakobsdottir, J. Xu, G. Molin, S. Ahrne, and M. Nyman, *PLoS One*, 2013, **8**(11), e80476.
45. Z. Gao, J. Yin, J. Zhang, R. E. Ward, R. J. Martin, M. Lefevre, *et al.*, *Diabetes*, 2009, **58**(7), 1509–1517.
46. Y. Lu, C. Fan, P. Li, Y. Lu, X. Chang and K. Qi, *Sci. Rep.*, 2016, **6**, 37589.
47. M. Fumery, S. Speca, A. Langlois, A. M. Davila, C. Dubuquoy, M. Grauso, *et al.*, *EMBO Mol. Med.*, 2017, **9**(11), 1471–1481.
48. A. U. Hasan, A. Rahman and H. Kobori, *Int. J. Mol. Sci.*, 2019, **20**(2), 387.
49. E. Zigmond, B. Bernshtein, G. Friedlander, C. R. Walker, S. Yona, K. W. Kim, *et al.*, *Immunity*, 2014, 40(5), 720–733.
50. I. Manoharan, A. Suryawanshi, Y. Hong, P. Ranganathan, A. Shanmugam, S. Ahmad, *et al.*, *J. Immunol.*, 2016, **196**(11), 4739–4749.
51. R. Kubeck, C. Bonet-Ripoll, C. Hoffmann, A. Walker, V. M. Muller, V. L. Schuppel, *et al.*, *Mol. Metab.*, 2016, **5**(12), 1162–1174.
52. A. Parseus, N. Sommer, F. Sommer, R. Caesar, A. Molinaro, M. Stahlman, *et al.*, *Gut*, 2017, **66**(3), 429–437.
53. D. F. Romagnolo, M. G. Donovan, T. C. Doetschman and O. I. Selmin, *Nutrients*, 2019, **11**(1), 171.
54. F. Li, C. Jiang, K. W. Krausz, Y. Li, I. Albert, H. Hao, *et al.*, *Nat. Commun.*, 2013, **4**, 2384.

55. N. Kazgan, M. R. Metukuri, A. Purushotham, J. Lu, A. Rao, S. Lee, *et al.*, *Gastroenterology*, 2014, **146**(4), 1006–1016.
56. S. Fang, J. M. Suh, S. M. Reilly, E. Yu, O. Osborn, D. Lackey, *et al.*, *Nat. Med.*, 2015, **21**(2), 159–165.
57. M. L. Chen, K. Takeda, and M. S. Sundrud, *Mucosal Immunol.*, 2019, **12**(4), 851–861.
58. O. Garcia-Irigoyen and A. Moschetta, *Cell Metab.*, 2017, **25**(4), 763–764.
59. R. M. Gadaleta, B. Oldenburg, E. C. Willemsen, M. Spit, S. Murzilli, L. Salvatore, *et al.*, *Biochim. Biophys. Acta*, 2011, **1812**(8), 851–858.
60. P. Vavassori, A. Mencarelli, B. Renga, E. Distrutti and S. Fiorucci, *J. Immunol.*, 2009, **183**(10), 6251–6261.
61. Y. D. Wang, W. D. Chen, M. Wang, D. Yu, B. M. Forman and W. Huang, *Hepatology*, 2008, **48**(5), 1632–1643.
62. R. M. Gadaleta, K. J. van Erpecum, B. Oldenburg, E. C. Willemsen, W. Renooij, S. Murzilli, *et al.*, *Gut*, 2011, **60**(4), 463–472.
63. V. Massafra, N. Ijssennagger, M. Plantinga, A. Milona, J. M. Ramos Pittol, M. Boes, *et al.*, *Biochim. Biophys. Acta*, 2016, **1862**(2), 166–173.
64. J. H. Cummings and G. T. Macfarlane, *J. Appl. Bacteriol.*, 1991, **70**(6), 443–459.
65. D. L. Topping and P. M. Clifton, *Physiol. Rev.*, 2001, **81**(3), 1031–1064.
66. S. Silvi, C. J. Rumney, A. Cresci and I. R. Rowland, *J. Appl. Microbiol.*, 1999, 86(3), 521–530.
67. A. Cresci, C. Orpianesi, S. Silvi, V. Mastrandrea and P. Dolara, *J. Appl. Microbiol.*, 1999, **86**(2), 245–250.
68. G. S. Raza, H. Putaala, A. A. Hibberd, E. Alhoniemi, K. Tiihonen, K. A. Makela, *et al.*, *Sci. Rep.*, 2017, **7**(1), 5294.
69. L. Madsen, L. S. Myrmel, E. Fjaere, B. Liaset and K. Kristiansen, *Front. Physiol.*, 2017, **8**, 1047.
70. D. J. Millward, D. K. Layman, D. Tome and G. Schaafsma, *Am. J. Clin. Nutr.*, 2008, **87**(5), 1576S–1581S.
71. L. Noriega-Lopez, A. R. Tovar, M. Gonzalez-Granillo, R. Hernandez-Pando, B. Escalante, P. Santillan-Doherty, *et al.*, *J. Biol. Chem.*, 2007, **282**(28), 20657–20666.
72. I. Torre-Villalvazo, A. R. Tovar, V. E. Ramos-Barragan, M. A. Cerbon-Cervantes and N. Torres, *J. Nutr.*, 2008, **138**(3), 462–468.
73. C. A. Edwards, J. Havlik, W. Cong, W. Mullen, T. Preston, D. J. Morrison, *et al.*, *Nutr. Bull.*, 2017, **42**(4), 356–360.
74. T. Ozdal, D. A. Sela, J. Xiao, D. Boyacioglu, F. Chen and E. Capanoglu, *Nutrients*, 2016, **8**(2), 78.
75. J. A. Dominguez-Avila, G. A. Gonzalez-Aguilar, E. Alvarez-Parrilla and L. A. de la Rosa, *Int. J. Mol. Sci.*, 2016, **17**(7), 1002.
76. F. F. Anhe, R. T. Nachbar, T. V. Varin, J. Trottier, S. Dudonne, M. Le Barz, *et al.*, *Gut*, 2019, **68**, 453–464.
77. L. Sheng, P. K. Jena, H. X. Liu, Y. Hu, N. Nagar, D. N. Bronner, *et al.*, *FASEB J.*, 2018, **32**, 6371–6384.

78. Y. Tian, J. Cai, W. Gui, R. G. Nichols, I. Koo, J. Zhang, *et al.*, *Drug Metab. Dispos.*, 2019, 47(2), 86–93.
79. M. Alves-Bezerra and D. E. Cohen, *Compr. Physiol.*, 2018, **8**(1), 1–8.
80. L. P. Bechmann, R. A. Hannivoort, G. Gerken, G. S. Hotamisligil, M. Trauner and A. Canbay, *J. Hepatol.*, 2012, **56**(4), 952–964.
81. N. Tanaka, T. Aoyama, S. Kimura and F. J. Gonzalez, *Pharmacol. Ther.*, 2017, **179**, 142–157.
82. S. Kersten, J. Seydoux, J. M. Peters, F. J. Gonzalez, B. Desvergne and W. Wahli, *J. Clin. Invest.*, 1999, **103**(11), 1489–1498.
83. M. Rakhshandehroo, B. Knoch, M. Muller and S. Kersten, *PPAR Res.*, 2010, 2010, 612089.
84. J. Li, G. Ning and S. A. Duncan, *Genes Dev.*, 2000, **14**(4), 464–474.
85. K. A. Rye and P. J. Barter, *Circ Res.*, 2014, **114**(1), 143–156.
86. M. F. Linton, H. Tao, E. F. Linton and P. G. Yancey, *Trends Endocrinol. Metab.*, 2017, **28**(6), 461–472.
87. M. V. Chakravarthy, I. J. Lodhi, L. Yin, R. R. Malapaka, H. E. Xu, J. Turk, *et al.*, *Cell*, 2009, **138**(3), 476–488.
88. M. V. Chakravarthy, Z. Pan, Y. Zhu, K. Tordjman, J. G. Schneider, T. Coleman, *et al. Cell Metab.*, 2005, **1**(5), 309–322.
89. M. E. Frigolet and R. Gutierrez-Aguilar, *Adv. Nutr.*, 2017, **8**(1), 173S–181S.
90. R. Hertz, N. Ben-Haim, A. D. Petrescu, B. Kalderon, I. Berman, N. Eldad, *et al.*, *J. Biol. Chem.*, 2003, **278**(25), 22578–22585.
91. N. Y. Kalaany and D. J. Mangelsdorf, *Annu. Rev. Physiol.*, 2006, **68**, 159–191.
92. S. B. Joseph, B. A. Laffitte, P. H. Patel, M. A. Watson, K. E. Matsukuma, R. Walczak, *et al.*, *J. Biol. Chem.*, 2002, **277**(13), 11019–11025.
93. H. El Hadi, A. Di Vincenzo, R. Vettor and M. Rossato, *Int. J. Mol. Sci.*, 2019, **20**(9), 2215.
94. S. Marchisello, A. Di Pino, R. Scicali, F. Urbano, S. Piro, F. Purrello, *et al.*, *Int. J. Mol. Sci.*, 2019, **20**(8), 1948.
95. M. Ma, R. Duan, H. Zhong, T. Liang and L. Guo, *J. Immunol. Res.*, 2019, **2019**, 3954890.
96. G. Kolios, V. Valatas and E. Kouroumalis, *World J. Gastroenterol.*, 2006, **12**(46), 7413–7420.
97. S. Sanz, J. B. Pucilowska, S. Liu, C. M. Rodriguez-Ortigosa, P. K. Lund, D. A. Brenner, *et al.*, *Gut*, 2005, **54**(1), 134–141.
98. C. A. Monteiro, G. Cannon, R. B. Levy, J. C. Moubarac, M. L. Louzada, F. Rauber, *et al.*, *Public Health Nutr.*, 2019, **22**(5), 936–941.
99. C. A. Monteiro, G. Cannon, J. C. Moubarac, R. B. Levy, M. L. C. Louzada and P. C. Jaime, *Public Health Nutr.*, 2018, **21**(1), 5–17.
100. F. Juul, E. Martinez-Steele, N. Parekh, C. A. Monteiro and V. W. Chang, *Br. J. Nutr.*, 2018, **120**(1), 90–100.
101. F. M. Silva, L. Giatti, R. C. de Figueiredo, M. Molina, L. de Oliveira Cardoso, B. B. Duncan, *et al.*, *Public Health Nutr.*, 2018, **21**(12), 2271–2279.
102. C. S. Costa, B. Del-Ponte, M. C. F. Assuncao and I. S. Santos, *Public Health Nutr.*, 2018, **21**(1), 148–159.

103. G. Longo-Silva, J. A. C. Silveira, R. C. E. Menezes and M. H. A. Toloni, *J. Pediatr. (Rio J.)*, 2017, **93**(5), 508–516.
104. G. Torres-Villalobos, N. Hamdan-Perez, A. R. Tovar, G. Ordaz-Nava, B. Martinez-Benitez, I. Torre-Villalvazo, *et al.*, *Ann. Hepatol.*, 2015, **14**(4), 540–546.
105. X. J. Liu, N. N. Duan, C. Liu, C. Niu, X. P. Liu and J. Wu, *Lab. Invest.*, 2018, **98**(9), 1184–1199.
106. N. Mitro, P. A. Mak, L. Vargas, C. Godio, E. Hampton, V. Molteni, *et al.*, *Nature*, 2007, **445**(7124), 219–223.
107. R. C. Sales, P. C. Medeiros, F. Spreafico, P. C. de Velasco, F. K. A. Goncalves, R. Martin-Hernandez, *et al.*, *Int. J. Mol. Sci.*, 2018, **20**(1), 8.
108. P. M. Chao, M. F. Yang, Y. N. Tseng, K. M. Chang, K. S. Lu, C. J. Huang, *J. Nutr. Sci. Vitaminol. (Tokyo)*, 2005, **51**(5), 361–368.
109. R. Micha and D. Mozaffarian, *Prostaglandins Leukot. Essent. Fatty Acids*, 2008, **79**(3–5), 147–152.
110. P. G. Illesca, J. V. Lavandera, C. D. Gerstner, M. A. Gonzalez and C. A. Bernal, *Nutr. Hosp.*, 2015, **32**(2), 738–748.
111. R. M. Machado, J. T. Stefano, C. P. Oliveira, E. S. Mello, F. D. Ferreira, V. S. Nunes, *et al.*, *J. Nutr.*, 2010, **140**(6), 1127–1132.
112. V. Perissi and M. G. Rosenfeld, *Nat. Rev. Mol. Cell Biol.*, 2005, **6**(7), 542–554.
113. A. V. Contreras, C. Rangel-Escareno, N. Torres, G. Aleman-Escondrillas, V. Ortiz, L. G. Noriega, *et al.*, *Genes Nutr.*, 2015, **10**(2), 452.
114. T. Mello, M. Materozzi and A. Galli, *PPAR Res.*, 2016, **2016**, 7403230.
115. Y. Sumida, E. Niki, Y. Naito and T. Yoshikawa, *Free Radic. Res.*, 2013, **47**(11), 869–880.
116. P. Pettinelli and L. A. Videla, *J. Clin. Endocrinol. Metab.*, 2011, **96**(5), 1424–1430.
117. K. Sikder, S. K. Shukla, N. Patel, H. Singh and K. Rafiq, *Cell Physiol. Biochem.*, 2018, **48**(3), 1317–1331.
118. N. Torres, I. Torre-Villalvazo and A. R. Tovar, *Rev. Invest. Clin.*, 2019, **71**(3), 157–167.
119. A. R. Tovar, I. Torre-Villalvazo, M. Ochoa, A. L. Elias, V. Ortiz, C. A. Aguilar-Salinas, *et al.*, *J. Lipid Res.*, 2005, **46**(9), 1823–1832.
120. M. J. Ronis, Y. Chen, J. Badeaux and T. M. Badger, *J. Nutr.*, 2009, **139**(8), 1431–1438.
121. C. A. Aguilar Salinas, *Rev. Invest. Clin.*, 2003, **55**(1), 84–86.
122. G. den Besten, A. Bleeker, A. Gerding, K. van Eunen, R. Havinga, T. H. van Dijk, *et al.*, *Diabetes*, 2015, **64**(7), 2398–2408.
123. S. A. Kliewer, S. S. Sundseth, S. A. Jones, P. J. Brown, G. B. Wisely, C. S. Koble, *et al.*, *Proc. Natl. Acad. Sci. U. S. A.*, 1997, **94**(9), 4318–4323.
124. H. E. Xu, M. H. Lambert, V. G. Montana, D. J. Parks, S. G. Blanchard, P. J. Brown, *et al.*, *Mol. Cell.* 1999, **3**(3), 397–403.
125. L. Hodson, L. Bhatia, E. Scorletti, D. E. Smith, N. C. Jackson, F. Shojaee-Moradie, *et al.*, *Eur. J. Clin. Nutr.*, 2017, **71**(8), 973–979.

126. A. Z. Lalia, M. L. Johnson, M. D. Jensen, K. C. Hames, J. D. Port, I. R. Lanza, *Diabetes Care*, 2015, **38**(7), 1228–1237.
127. E. Scorletti, L. Bhatia, K. G. McCormick, G. F. Clough, K. Nash, L. Hodson, *et al.*, *Hepatology*, 2014, **60**(4), 1211–1221.
128. H. M. Parker, J. S. Cohn, H. T. O'Connor, M. L. Garg, I. D. Caterson, J. George, *et al.*, *Nutrients,* 2019, **11**(2), 475.
129. T. Shang, L. Liu, J. Zhou, M. Zhang, Q. Hu, M. Fang, *et al.*, *Lipids Health Dis.*, 2017, **16**(1), 65.
130. C. S. Ku, Y. Yang, Y. Park and J. Lee, *J. Med. Food.*, 2013, **16**(2), 103–111.
131. Y. Jia, J. Y. Kim, H. J. Jun, S. J. Kim, J. H. Lee, M. H. Hoang, *et al.*, *Biochim. Biophys. Acta,* 2013, **1831**(4), 698–708.
132. F. F. Anhe, R. T. Nachbar, T. V. Varin, V. Vilela, S. Dudonne, G. Pilon, *et al.*, *Mol. Metab.* 2017, **6**(12), 1563–1573.
133. P. Vitaglione, G. Mazzone, V. Lembo, G. D'Argenio, A. Rossi, M. Guido, *et al.*, *J. Nutr. Sci.*, 2019, **8**, e15.
134. R. A. Chavez-Santoscoy, J. A. Gutierrez-Uribe, O. Granados, I. Torre-Villalvazo, S. O. Serna-Saldivar, N. Torres, *et al.*, *Br. J. Nutr.*, 2014, **112**(6), 886–899.
135. K. W. Cho, Y. O. Kim, J. E. Andrade, J. R. Burgess and Y. C. Kim, *Eur. J. Nutr.*, 2011, **50**(2), 81–88.
136. C. J. Chang, T. F. Tzeng, S. S. Liou, Y. S. Chang, I. M. Liu, *Planta Med.*, 2011, 77(17), 1876–1882.
137. Y. Zhang, Y. Cui, X. L. Wang, X. Shang, Z. G. Qi, J. Xue, *et al.*, *Cytokine*, 2015, **75**(1), 127–135.
138. A. Montagner, A. Polizzi, E. Fouche, S. Ducheix, Y. Lippi, F. Lasserre, *et al.*, *Gut,* 2016, **65**(7), 1202–1214.
139. C. Rodriguez-Rodriguez, N. Torres, J. A. Gutierrez-Uribe, L. G. Noriega, I. Torre-Villalvazo, A. M. Leal-Diaz, *et al.*, *Food Funct.*, 2015, **6**(3), 805–815.
140. P. Lopez, M. Sanchez, C. Perez-Cruz, L. A. Velazquez-Villegas, T. Syeda, M. Aguilar-Lopez, *et al.*, *Mol. Nutr. Food Res.*, 2018, **62**(16), e1800313.
141. A. M. Leal-Diaz, L. G. Noriega, I. Torre-Villalvazo, N. Torres, G. Aleman-Escondrillas, P. Lopez-Romero, *et al.*, *Sci. Rep.*, 2016, **6**, 34242.
142. J. Plat, J. A. Nichols, R. P. Mensink, *J. Lipid Res.,* 2005, **46**(11), 2468–2476.
143. J. Plat, T. Hendrikx, V. Bieghs, M. L. Jeurissen, S. M. Walenbergh, van Gorp PJ, *et al.*, *PLoS One*, 2014, **9**(10), e110758.
144. S. Feng, L. Gan, C. S. Yang, A. B. Liu, W. Lu, P. Shao, *et al.*, *J. Agric. Food Chem.*, 2018, **66**(13), 3417–3425.
145. C. Yan, Y. Zhang, X. Zhang, J. Aa, G. Wang and Y. Xie, *Biomed. Pharmacother.*, 2018, **105**, 274–281.
146. F. Xiong and Y. S. Guan, *World J. Gastroenterol.*, 2017, **23**(19), 3388–3395.
147. A. M. Schoepfer, A. Engel, K. Fattinger, U. A. Marbet, D. Criblez, J. Reichen, *et al.*, *J. Hepatol.,* 2007, **47**(4), 521–526.
148. G. C. Chen, V. S. Ramanathan, D. Law, P. Funchain, G. C. Chen, S. French, *et al.*, *World J. Hepatol.*, 2010, **2**(11), 410–415.
149. D. P. Hayes, *Eur. J. Clin. Nutr.*, 2007, **61**(2), 147–159.
150. M. M. Ibrahim, *Obes. Rev.*, 2010, **11**(1), 11–18.

151. G. H. Goossens, *Obes. Facts*, 2017, **10**(3), 207–215.
152. S. S. Choe, J. Y. Huh, I. J. Hwang, J. I. Kim, J. B. Kim, *Front. Endocrinol. (Lausanne)*, 2016, 7, 30.
153. I. Shimomura, R. E. Hammer, J. A. Richardson, S. Ikemoto, Y. Bashmakov, J. L. Goldstein, *et al.*, *Genes Dev.*, 1998, **12**(20), 3182–3194.
154. R. H. Unger, P. E. Scherer, *Trends Endocrinol. Metab.*, 2010, **21**(6), 345–352.
155. V. R. Narala, P. A. Subramani, V. R. Narasimha, F. B. Shaik and K. Panati, *Int. Immunopharmacol.*, 2014, **23**(1), 283–287.
156. M. A. Lazar, *Biochimie*, 2005, **87**(1), 9–13.
157. A. Hammarstedt, C. X. Andersson, V. Rotter Sopasakis and U. Smith, *Prostaglandins Leukot. Essent. Fatty Acids*, 2005, **73**(1), 65–75.
158. K. Albracht-Schulte, N. S. Kalupahana, L. Ramalingam, S. Wang, S. M. Rahman, J. Robert-McComb, *et al.*, *J. Nutr. Biochem.*, 2018, **58**, 1–16.
159. R. Marion-Letellier, G Savoye and S. Ghosh, *Eur. J. Pharmacol.*, 2016, **785**, 44–49.
160. T. Goto, Y. I. Kim, N. Takahashi and T. Kawada, *Mol. Nutr. Food Res.*, 2013, **57**(1), 20–33.
161. S. H. Hong, M. Ahmadian, R. T. Yu, A. R. Atkins, M. Downes and R. M. Evans, *Diabetologia*, 2014, **57**(5), 860–867.
162. F. Villarroya, R. Iglesias and M. Giralt, *PPAR Res.*, 2007, **2007**, 74364.
163. A. S. Grewal, M. Beniwal, D. Pandita, B. S. Sekhon and V. Lather, *Med. Chem.*, 2016, **12**(1), 3–21.
164. L. Cruz-Garcia, J. Sanchez-Gurmaches, J. Gutierrez and I. Navarro, *Comp. Biochem. Physiol. A Mol. Integr. Physiol.*, 2012, **163**(1), 120–126.
165. L. K. Juvet, S. M. Andresen, G. U. Schuster, K. T. Dalen, K. A. Tobin, K. Hollung, *et al.*, *Mol. Endocrinol.*, 2003, **17**(2), 172–182.
166. M. Korach-Andre, A. Archer, C. Gabbi, R. P. Barros, M. Pedrelli, K. R. Steffensen, *et al.*, *Am. J. Physiol.: Endocrinol. Metab.*, 2011, **301**(1), E210–E222.
167. B. M. Stenson, M. Ryden, N. Venteclef, I. Dahlman, A. M. Pettersson, A. Mairal, *et al.*, *J. Biol. Chem.*, 2011, **286**(1), 370–379.
168. J. B. Seo, H. M. Moon, W. S. Kim, Y. S. Lee, H. W. Jeong, E. J. Yoo, *et al.*, *Mol. Cell Biol.*, 2004, **24**(8), 3430–3444.
169. E. Kim, E. J. Kim, S. W. Seo, C. G. Hur, R. A. McGregor and M. S. Choi, *Crit. Rev. Food Sci. Nutr.*, 2014, **54**(7), 910–923.
170. H. Al-Hasani and H. G. Joost, *Best Pract. Res. Clin. Endocrinol. Metab.*, 2005, **19**(4), 589–603.
171. A. R. Tovar, A. Diaz-Villasenor, N. Cruz-Salazar, G. Ordaz, O. Granados, B. Palacios-Gonzalez, *et al.*, *Arch. Med. Res.*, 2011, **42**(6), 540–553.
172. A. Vidal-Puig, M. Jimenez-Linan, B. B. Lowell, A. Hamann, E. Hu, B. Spiegelman, *et al.*, *J. Clin. Invest.*, 1996, **97**(11), 2553–2561.
173. N. Kubota, Y. Terauchi, H. Miki, H. Tamemoto, T, Yamauchi, K. Komeda, *et al.*, *Mol. Cell*, 1999, **4**(4), 597–609.
174. A. J. Vidal-Puig, R. V. Considine, M. Jimenez-Linan, A. Werman, W. J. Pories, J. F. Caro, *et al.*, *J. Clin. Invest.*, 1997, **99**(10), 2416–2422.

175. T. Kadowaki, K. Hara, N. Kubota, K. Tobe, Y. Terauchi, T. Yamauchi, *et al.*, *J. Diabetes Complications,* 2002, **16**(1), 41–45.
176. T. C. Bargut, A. C. Silva-e-Silva, V. Souza-Mello, C. A. Mandarim-de-Lacerda and M. B. Aguila, *Eur. J. Nutr.*, 2016, **55**(1), 159–169.
177. A. V. Ferreira, Z. Menezes-Garcia, J. B. Viana, E. G. Mario and L. M. Botion, *Nutrition,* 2014, **30**(10), 1138–1143.
178. M. C. Oliveira, Z. Menezes-Garcia, M. C. Henriques, F. M. Soriani, V. Pinho, A. M. Faria, *et al.*, *Obesity* (*Silver Spring*), 2013, **21**(9), E396–E406.
179. R. L. McNeel and H. J. Mersmann, *Obes. Res.,* 2005, **13**(10), 1651–1660.
180. K. B. de Queiroz, R. S. Coimbra, A. R. Ferreira, C. M. Carneiro, N. C. Paiva, D. C. Costa, *et al.*, *Mol. Nutr. Food Res.*, 2014, **58**(12), 2331–2341.
181. M. G. Zubiria, J. P. Farina, G. Moreno, J. J. Gagliardino, E. Spinedi and A. Giovambattista, *FEBS J.,* 2013, **280**(22), 5864–5874.
182. B. N. Bursac, A. D. Vasiljevic, N. M. Nestorovic, N. A. Velickovic, D. D. Vojnovic Milutinovic, G. M. Matic, *et al.*, *J. Nutr. Biochem.,* 2014, **25**(4), 446–455.
183. M. B. Pektas, H. B. Koca, G. Sadi and F. Akar, *Biomed. Res. Int.,* 2016, **2016**, 8014252.
184. E. Barroso, R. Rodriguez-Rodriguez, M. R. Chacon, E. Maymo-Masip, L. Ferrer, L. Salvado, *et al.*, *Biochim. Biophys. Acta*, 2015, **1852**(5), 1049–1058.
185. S. Flajollet, B. Staels and P. Lefebvre, *Horm. Mol. Biol. Clin. Investig.*, 2013, **14**(3), 75–86.
186. C. Ding, D. Gao, J. Wilding, P. Trayhurn, C. Bing, *Br. J. Nutr.*, 2012, **108**(11), 1915–1923.
187. A. M. Belenchia, K. L. Jones, M. Will, D. Q. Beversdorf, V. Vieira-Potter, C. S. Rosenfeld, *et al.*, *Eur. J. Nutr.*, 2018, **57**(2), 723–730.
188. Y. Wang and F. Huang, *Biomed. Res. Int.,* 2015, **2015**, 581469.
189. M. Garaulet, J. J. Hernandez-Morante, J. Lujan, F. J. Tebar and S. Zamora, *Int. J. Obes.* (*Lond.*), 2006, **30**(6), 899–905.
190. E. A. Zayed, A. A. AinShoka, K. A. El Shazly and H. A. Abd El Latif, *J. Biochem. Mol. Toxicol.,* 2018, **32**(11), e22218.
191. B. M. Forman, P. Tontonoz, J. Chen, R. P. Brun, B. M. Spiegelman and R. M. Evans, *Cell,* 1995, **83**(5), 803–812.
192. M. K. Mater, D. Pan, W. G. Bergen and D. B. Jump, *J. Lipid Res.*, 1998, **39**(7), 1327–1334.
193. A. U. Hasan, K. Ohmori, K. Konishi, J. Igarashi, T. Hashimoto, K. Kamitori, *et al.*, *Mol. Cell Endocrinol.*, 2015, **406**, 10–18.
194. D. Selenscig, M. D. R. Ferreira, A. Chicco and Y. B. Lombardo, *Food Funct.,* 2018, **9**(4), 2496–2507.
195. T. C. Bargut, C. A. Mandarim-de-Lacerda and M. B. Aguila, *J. Nutr. Biochem.*, 2015, **26**(9), 960–969.
196. T. Itoh and K. Yamamoto, *Naunyn Schmiedebergs Arch. Pharmacol.,* 2008, **377**(4–6), 541–547.
197. A. Jin, C. X. Lei, J. J. Tian, J. Sun and H. Ji, *Fish Physiol. Biochem.*, 2018, **44**(1), 197–207.

198. C. M. Mejia-Barradas, B. E. Del-Rio-Navarro, A. Dominguez-Lopez, R. Campos-Rodriguez, M. Martinez-Godinez, S. Rojas-Hernandez, *et al.*, *Endocrine,* 2014, **45**(1), 98–105.
199. P. Flachs, O. Horakova, P. Brauner, M. Rossmeisl, P. Pecina, N. Franssen-van Hal, *et al.*, *Diabetologia,* 2005, **48**(11), 2365–2375.
200. F. J. Schopfer, Y. Lin, P. R. Baker, T. Cui, M. Garcia-Barrio, J. Zhang, *et al.*, *Proc. Natl. Acad. Sci. U. S. A.*, 2005, **102**(7), 2340–2345.
201. H. Wang, H. Liu, Z. Jia, G. Guan and T. Yang, *PPAR Res.*, 2010, **2010**, 601562.
202. M. E. Oliva, D. Selenscig, M. E. D'Alessandro, A. Chicco and Y. B. Lombardo, *Br. J. Nutr.*, 2011, **105**(8), 1188–1198.
203. A. Diaz-Villasenor, O. Granados, B. Gonzalez-Palacios, C. Tovar-Palacio, I. Torre-Villalvazo, V. Olivares-Garcia, *et al.*, *J. Nutr. Biochem.*, 2013, **24**(11), 1798–1809.
204. Y. Takahashi and T. Ide, *J. Nutr. Biochem.*, 2008, **19**(10), 682–693.
205. F. L. Soares, R. de Oliveira Matoso, L. G. Teixeira, Z. Menezes, S. S. Pereira, A. C. Alves, *et al.*, *J. Nutr. Biochem.*, 2013, **24**(6), 1105–1111.
206. B. L. McManus, R. Korpela, J. R. Speakman, J. F., Cryan, P. D. Cotter and K. N. Nilaweera, *Br. J. Nutr.*, 2015, **114**(4), 654–662.
207. M. L. Justo, C. Claro, M. Zeyda, T. M. Stulnig, M. D. Herrera and R. Rodriguez-Rodriguez, *Eur. J. Nutr.*, 2016, **55**(6), 2011–2019.
208. K. Boonloh, V. Kukongviriyapan, B. Kongyingyoes, U. Kukongviriyapan, S. Thawornchinsombut and P. Pannangpetch, *Nutrients,* 2015, 7(8), 6313–6329.
209. S. P. Zhao, J. Yang, J. Li, S. Z. Dong and Z. H. Wu, *Int. J. Cardiol.*, 2008, **124**(2), 172–178.
210. J. Kong and Y. C. Li, *Am. J. Physiol.: Endocrinol. Metab.*, 2006, **290**(5), E916–E924.
211. M. L. Bonet, J. Ribot and A. Palou, *Biochim. Biophys. Acta*, 2012, **1821**(1), 177–189.
212. S. Feng, L. Reuss and Y. Wang, *Molecules,* 2016, **21**(10), 1278.
213. S. Hirai, N. Takahashi, T. Goto, S. Lin, T. Uemura, R. Yu, *et al.*, *Mediators Inflamm.*, 2010, **2010**, 367838.
214. L. Aguirre, A. Fernandez-Quintela, N. Arias, M. P. Portillo, *Molecules,* 2014, **19**(11), 18632–18655.
215. P. Aranaz, A. Romo-Hualde, M. Zabala, D. Navarro-Herrera, M. Ruiz de Galarreta, A. G. Gil, *et al.*, *Food Funct.*, 2017, **8**(11), 3999–4013.
216. S. Rayalam, J. Y. Yang, S. Ambati, M. A. Della-Fera and C. A. Baile, *Phytother. Res.*, 2008, **22**(10), 1367–1371.
217. N. E. Kang, A. W. Ha, J. Y. Kim and W. K. Kim, *Nutr. Res. Pract.*, 2012, **6**(6), 499–504.
218. S. Costa Cdos, F. Rohden, T. O. Hammes, R. Margis, J. W. Bortolotto, A. V. Padoin, *et al.*, *Obes. Surg.*, 2011, **21**(3), 356–361.
219. S. Chen, Z. Li, W. Li, Z. Shan, W. Zhu, *Can. J. Physiol. Pharmacol.*, 2011, **89**(11), 793–799.

220. L. Bai, W. J. Pang, Y. J. Yang, G. S. Yang, *Mol. Cell Biochem.*, 2008, **307**(1–2), 129–140.
221. J. Y. Kwon, S. G. Seo, S. Yue, J. X. Cheng, K. W. Lee, K. H. Kim, *Nutr. Res.*, 2012, **32**(8), 607–616.
222. M. C. Mitterberger and W. Zwerschke, *J. Gerontol. A Biol. Sci. Med. Sci.*, 2013, **68**(11), 1356–1376.
223. P. Fischer-Posovszky, V. Kukulus, D. Tews, T. Unterkircher, K. M. Debatin, S. Fulda, *et al.*, *Am. J. Clin. Nutr.*, 2010, **92**(1), 5–15.
224. C. C. Chang, K. Y. Lin, K. Y. Peng, Y. J. Day and L. M. Hung, *Endocr. J.* 2016, **63**(2), 169–178.
225. A. Lasa, I. Churruca, I. Eseberri, C. Andres-Lacueva and M. P. Portillo, *Mol. Nutr. Food Res.*, 2012, **56**(10), 1559–1568.
226. H. Y. Zhang, Z. X. Du and X. Meng, *Clin. Exp. Med.*, 2013, **13**(3), 193–199.
227. A. Mosqueda-Solis, A. Lasa, S. Gomez-Zorita, I. Eseberri, C. Pico and M. P. Portillo, *Food Funct.*, 2017, **8**(10), 3576–3586.
228. A. W. Harmon and J. B. Harp, *Am. J. Physiol.: Cell Physiol.*, 2001, **280**(4), C807–C813.
229. Z. C. Dang, V. Audinot, S. E. Papapoulos, J. A. Boutin and C. W. Lowik, *J. Biol. Chem.*, 2003, **278**(2), 962–967.
230. Y. Sakamoto, A. Naka, N. Ohara, K. Kondo and K. Iida, *Mol. Nutr. Food Res.*, 2014, **58**(4), 718–726.
231. Y. Sakamoto, J. Kanatsu, M. Toh, A. Naka, K. Kondo and K. Iida, *PLoS One*, 2016, **11**(2), e0149676.
232. H. Li, J. H. Kang, J. M. Han, M. H. Cho, Y. J. Chung, K. H. Park, *et al.*, *J. Med. Food*, 2015, **18**(8), 899–908.
233. M. R. Ferreira, S. M. Alvarez, P. Illesca, M. S. Gimenez and Y. B. Lombardo, *Eur. J. Nutr.*, 2018, **57**(1), 83–94.
234. N. Sutjarit, J. Sueajai, N. Boonmuen, N. Sornkaew, A. Suksamrarn, P. Tuchinda, *et al.*, *J. Ethnopharmacol.*, 2018, **215**, 167–175.
235. Y. Isa, Y. Miyakawa, M. Yanagisawa, T. Goto, M. S. Kang, T. Kawada, *et al.*, *Biochem. Biophys. Res. Commun.*, 2008, **373**(3), 429–434.
236. C. W. Lee, J. Y. Seo, S. L. Kim, J. Lee, J. W. Choi and Y. I. Park, *Biomed. Pharmacother.*, 2017, **93**, 267–275.
237. E. Y. Lee, S. L. Kim, H. J. Kang, M. H. Kim, A. W. Ha and W. K. Kim, *Nutr. Res. Pract.*, 2016, **10**(6), 575–582.
238. Y. H. Hsieh and S. Y. Wang, *Phytomedicine*, 2013, **20**(5), 394–400.
239. E. Y. Hwang, M. H. Yu, Y. S. Jung, S. P. Lee, J. H. Shon and S. O. Lee, *Nutr. Res.*,2016, **36**(9), 995–1003.
240. T. Matsukawa, T. Inaguma, J. Han, M. O. Villareal and H. Isoda, *J. Nutr. Biochem.*, 2015, **26**(8), 860–867.
241. M. A. Vazquez Prieto, A. Bettaieb, C. Rodriguez Lanzi, V. C. Soto, D. J. Perdicaro, C. R. Galmarini, *et al.*, *Mol. Nutr. Food Res.*, 2015, **59**(4), 622–633.
242. B. Ahmed, S. Liu, and H. Si, *J. Med. Food*, 2017, **20**(2), 162–170.
243. K. Kuroyanagi, M. S. Kang, T. Goto, S. Hirai, K. Ohyama, T. Kusudo, *et al.*, *Biochem. Biophys. Res. Commun.*, 2008, **366**(1), 219–225.

244. N. Takahashi, M. Senda, S. Lin, T. Goto, M. Yano, T. Sasaki, *et al.*, *Mol. Nutr. Food Res.*, 2011, **55**(12), 1791–1797.
245. L. Huang, J. Chen, P. Cao, H. Pan, C. Ding, T. Xiao, *et al.*, *Mar. Drugs*, 2015, **13**(5), 2732–2756.
246. H, Sun, T. Jiang, S. Wang, B. He, Y. Zhang, D. Piao, *et al.*, *Diabetes Res. Clin. Pract.*, 2013, **102**(3), 183–192.
247. M. Vida, P. Rivera, A. L. Gavito, J. Suarez, F. J. Pavon, S. Arrabal, *et al.*, *PLoS One*, 2014, **9**(2), e90016.
248. W. Fan and R. Evans, *Curr. Opin. Cell Biol.*, 2015, **33**, 49–54.
249. X. Yang, P. Bi and S. Kuang, *Adipocyte*, 2014, **3**(4), 280–289.
250. G. S. Li, X. H. Liu, H. Zhu, L. Huang, Y. L. Liu and C. M. Ma, *Exp. Ther. Med.*, 2016, **11**(6), 2259–2269.
251. A. J. McAinch, J. S. Lee, C. R. Bruce, R. J. Tunstall, J. A. Hawley and D. Cameron-Smith, *Obes. Res.*, 2003, **11**(12), 1471–1479.
252. A. G. Smith and G. E. Muscat, *Int. J. Biochem. Cell Biol.*, 2005, **37**(10), 2047–2063.
253. R. M. Evans, G. D. Barish and Y. X. Wang, *Nat. Med.*, 2004, **10**(4), 355–361.
254. D. M. Muoio, P. S. MacLean, D. B. Lang, S. Li, J. A. Houmard, J. M. Way, *et al.*, *J. Biol. Chem.*, 2002, **277**(29), 26089–26097.
255. W. Fan, A. R. Atkins, R. T. Yu, M. Downes and R. M. Evans, *J. Mol. Endocrinol.*, 2013, **51**(3), T87–T100.
256. K. D. Wagner, and N. Wagner, *Pharmacol. Ther.*, 2010, **125**(3), 423–435.
257. N. P. Hessvik, M. V. Boekschoten, M. A. Baltzersen, S. Kersten, X. Xu, H. Andersen, *et al.*, *Am. J. Physiol.: Endocrinol. Metab.*, 2010, **298**(3), E602–E613.
258. G. E. Muscat, B. L. Wagner, J. Hou, R. K. Tangirala, E. D. Bischoff, P. Rohde, *et al.*, *J. Biol. Chem.*, 2002, **277**(43), 40722–40728.
259. V. Ribas, B. G. Drew, Z. Zhou, J. Phun, N. Y. Kalajian, T. Soleymani, *et al.*, *Sci. Transl. Med.*, 2016, **8**(334), 334ra54.
260. A. L. Z. Z. Hevener, B. G. Drew and V. Ribas, in *Sex and Gender Factors Affecting Metabolic Homeostasis, Diabetes and Obesity*, ed. F. Mauvais-Jarvis, Springer, Cham, 2017, pp. 257–284.
261. R. P. Barros and J. A. Gustafsson, *Cell Metab.*, 2011, **14**(3), 289–299.
262. E. Audet-Walsh and V. Giguere, *Acta Pharmacol. Sin.*, 2015, **36**(1), 51–61.
263. J. M. Huss, W. G. Garbacz and W. Xie, *Biochim. Biophys. Acta*, 2015, **1852**(9), 1912–1927.
264. J. Perez-Schindler and A. Philp, *Clin. Sci.* (*Lond.*), 2015, **129**(7), 589–599.
265. J. E. Galgani, C. Moro and E. Ravussin, *Am. J. Physiol.: Endocrinol. Metab.*, 2008, **295**(5), E1009–E1017.
266. R. Burrows, P. Correa-Burrows, M. Reyes, E. Blanco, C. Albala and S. Gahagan, *Pediatr. Diabetes*, 2017, **18**(8), 895–902.
267. N. Buchmann, J. Nikolov, D. Spira, I. Demuth, E. Steinhagen-Thiessen, R. Eckardt, *et al.*, *J. Gerontol., Ser. A*, 2016, **71**(2), 265–272.
268. D. M. Muoio, *Cell*, 2014, **159**(6), 1253–1262.
269. B. Ukropcova, O. Sereda, L. de Jonge, I. Bogacka, T. Nguyen, H. Xie, *et al.*, *Diabetes*, 2007, **56**(3), 720–727.

270. A. Jans, L. M. Sparks, A. M. van Hees, I. M. Gjelstad, A. C. Tierney, U. Riserus, *et al.*, *Obesity* (*Silver Spring*), 2011, **19**(11), 2158–2166.
271. C. Debard, D. Cozzone, N. Ricard, J. Vouillarmet, E. Disse, B. Husson, *et al.*, *J. Mol. Med.* (*Berl.*), 2006, **84**(9), 747–752.
272. K. Ikeda, K. Horie-Inoue and S. Inoue, *J. Steroid Biochem. Mol. Biol.*, 2019, **191**, 105375.
273. V. K. Mootha, C. Handschin, D. Arlow, X. Xie, J. St Pierre, S. Sihag, *et al.*, *Proc. Natl. Acad. Sci. U. S. A.*, 2004, **101**(17), 6570–5.
274. K. Seyssel, E. Meugnier, K. A. Le, C. Durand, E. Disse, E. Blond, *et al.*, *Mol. Nutr. Food Res.*, 2016, **60**(12), 2691–2699.
275. S. L. Roseno, P. R. Davis, L. M. Bollinger, J. J. Powell, C. A. Witczak and J. J. Brault, *Nutr. Metab.* (*Lond.*), 2015, **12**, 39.
276. K. E. Boyle, J. P. Canham, L. A. Consitt, D. Zheng, T. R. Koves, T. P. Gavin, *et al.*, *J. Clin. Endocrinol. Metab.*, 2011, **96**(3), 775–781.
277. R. Gougeon, *Can. J. Diabetes*, 2013, **37**(2), 115–120.
278. M. E. Oliva, A. Chicco and Y. B. Lombardo, *Eur. J. Nutr.*, 2015, **54**(3), 407–419.
279. J. D. Wagner, L. Zhang, M. K. Shadoan, K. Kavanagh, H. Chen, K. Tresnasari, *et al.*, *Metabolism,* 2008, **57**(7 Suppl 1), S24–S31.
280. B. M. Forman, J. Chen and R. M. Evans, *Proc. Natl. Acad. Sci. U. S. A.*, 1997, **94**(9), 4312–4317.
281. M. Moreno, A. Lombardi, E. Silvestri, R. Senese, F. Cioffi, F. Goglia, *et al.*, *PPAR Res.,* 2010, 2010, 435689.
282. A. R. Martins, A. R. Crisma, L. N. Masi, C. L. Amaral, G. N. Marzuca-Nassr, L. H. M. Bomfim, *et al.*, *J. Nutr. Biochem.*, 2018, **55**, 76–88.
283. W. Suhara, H. Koide, T. Okuzawa, D. Hayashi, T. Hashimoto and H. Kojo, *J. Dairy Sci.,* 2009, **92**(9), 4180–4187.
284. P. W. Parodi, *Med. Hypotheses*, 2016, **89**, 1–7.
285. N. Shaw, M. Elholm and N. Noy, *J. Biol. Chem.*, 2003, **278**(43), 41589–41592.
286. J. Amengual, J. Ribot, M. L. Bonet and A. Palou, *Obesity* (*Silver Spring*), 2008, **16**(3), 585–591.
287. C. K. Sen, S. Khanna and S. Roy, *Mol. Aspects Med.*, 2007, **28**(5–6), 692–728.
288. Y. Avior, D. Bomze, O. Ramon and Y. Nahmias, *Food Funct.*, 2013, **4**(6), 831–844.
289. P. V. Babu, D. Liu and E. R. Gilbert, *J. Nutr. Biochem.,* 2013, **24**(11), 1777–1789.
290. D. J. Den Hartogh and E. Tsiani, *Biomolecules*, 2019, **9**(3), 99.
291. M. Pellegrini, P. Bulzomi, P. Galluzzo, M. Lecis, S. Leone, V. Pallottini, *et al.*, *Genes Nutr.*, 2014, **9**(5), 425.
292. S. S. Kulkarni and C. Canto, *Biochim. Biophys. Acta*, 2015, **1852**(6), 1114–1123.
293. Z. Tan, L. J. Zhou, P. W. Mu, S. P. Liu, S. J. Chen, X. D. Fu, *et al.*, *J. Nutr. Biochem.*, 2012, **23**(12), 1716–1724.

294. S. Sae-Tan, K. A. Grove, M. J. Kennett and J. D. Lambert, *Food Funct.*, 2011, **2**(2), 111–116.
295. N. Ikarashi, T. Toda, T. Okaniwa, K. Ito, W. Ochiai and K. Sugiyama, *Evid. Based Complement Alternat. Med.*, 2011, **2011**, 952031.
296. B. Palacios-Gonzalez, A. Zarain-Herzberg, I. Flores-Galicia, L. G. Noriega, G. Aleman-Escondrillas, T. Zarinan, *et al.*, *Biochim. Biophys. Acta*, 2014, **1841**(1), 132–140.
297. E. M. Seymour, I. I. Tanone , D. E. Urcuyo-Llanes, S. K. Lewis, A. Kirakosyan, M. G. Kondoleon, *et al.*, *J. Med. Food*, 2011, **14**(12), 1511–1518.
298. P. Mirmiran, Z. Bahadoran and F. Azizi, *World J. Diabetes*, 2014, **5**(3), 267–281.
299. K. Misawa, K. Hashizume, M. Yamamoto, Y. Minegishi, T. Hase and A. Shimotoyodome, *J. Nutr. Biochem.*, 2015, **26**(10), 1058–1067.
300. J. Joven, E. Vilella, S. Ahmad, M. C. Cheung and J. D. Brunzell, *Kidney Int.*, 1993, **43**(2), 410–418.
301. R. H. Hutchings, R. Hickman and B. H. Scribner, *Pediatrics,* 1966, **37**(1), 68–73.
302. C. G. Zubrod, S. L. Eversole and G. W. Dana, *N. Engl. J. Med.*, 1951, **245**(14), 518–528.
303. I. H. de Boer and K. M. Utzschneider, *Nephrol. Dial. Transplant.*, 2017, **32**(4), 588–590.
304. W. F. Boron and E. L. Boulpaep, *Medical Physiology,* 3rd edn, Elsevier, Philadelphia, PA, 2017, xii.
305. Y. Guan, Y. Zhang, L. Davis and M. D. Breyer, *Am. J. Physiol.,* 1997, **273**(6), F1013–F1022.
306. X. Z. Ruan, J. F. Moorhead, R. Fernando, D. C. Wheeler, S. H. Powis and Z. Varghese, *J. Am. Soc. Nephrol.*, 2003, **14**(3), 593–600.
307. T. Yang, D. E. Michele, J. Park, A. M. Smart, Z. Lin, F. C. Brosius, 3rd, *et al.*, *Am. J. Physiol.*, 1999, **277**(6), F966–F973.
308. T. M. Jao, M. Nangaku, C. H. Wu, M. Sugahara, H. Saito, H. Maekawa, *et al.*, *Kidney Int.*, 2019, **95**(3), 577–589.
309. Y. Guan and M. D. Breyer, *Kidney Int.,* 2001, **60**(1), 14–30.
310. E. S. Koh, J. H. Lim, M. Y. Kim, S. Chung, S. J. Shin, B. S. Choi, *et al.*, *J. Transl. Med.*, 2015, **13**, 203.
311. H. M. Kang, S. H. Ahn, P. Choi, Y. A. Ko, S. H. Han, F. Chinga, *et al.*, *Nat. Med.,* 2015, **21**(1), 37–46.
312. D. Portilla, G. Dai, J. M. Peters, F. J. Gonzalez, M. D. Crew and A. D. Proia, *Am. J. Physiol.: Renal Physiol.,* 2000, **278**(4), F667–F675.
313. P. Delerive, J. C. Fruchart and B. Staels, *J. Endocrinol.*, 2001, **169**(3), 453–459.
314. M. E. Poynter and R. A. Daynes, *J. Biol. Chem.*, 1998, **273**(49), 32833–32841.
315. J. L. Wilson, R. Duan, A. El-Marakby, A. Alhashim and D. L. Lee, *PPAR Res.*, 2012, **2012**, 645969.
316. P. Delerive, K. De Bosscher, S. Besnard, W. Vanden Berghe, J. M. Peters, F. J. Gonzalez, *et al.*, *J. Biol. Chem.*, 1999, **274**(45), 32048–32054.

317. J. Zuniga, M. Cancino, F. Medina, P. Varela, R. Vargas, G. Tapia, *et al.*, *PLoS One*, 2011, **6**(12), e28502.
318. S. W. Chung, B. Y. Kang, S. H. Kim, Y. K. Pak, D. Cho, G. Trinchieri, *et al.*, *J. Biol. Chem.*, 2000, **275**(42), 32681–32687.
319. M. A. Newaz, K. Ranganna and A. O. Oyekan, *BMC Pharmacol.*, 2004, **4**, 1.
320. J. M. Williams, X. Zhao, M. H. Wang, J. D. Imig and D. M. Pollock, *Hypertension*, 2005, **46**(2), 366–371.
321. T. Vera, M. Taylor, Q. Bohman, A. Flasch, R. J. Roman and D. E. Stec, *Hypertension*, 2005, **45**(4), 730–735.
322. Y. Zhou, P. Luo, H. H. Chang, H. Huang, T. Yang, Z. Dong, *et al.*, *Kidney Int.*, 2008, **74**(8), 1040–1048.
323. Y. Zhou, H. Huang, H. H. Chang, J. Du, J. F. Wu, C. Y. Wang, *et al.*, *J. Pharmacol. Exp. Ther.*, 2006, **317**(1), 11–18.
324. E. Feraille, M. L. Carranza, S. Gonin, P. Beguin, C. Pedemonte, M. Rousselot, *et al.*, *Mol. Biol. Cell*, 1999, **10**(9), 2847–2859.
325. J. Xiao, J. C. Leung, L. Y. Chan, H. Guo, K. N. Lai, *Nephrol. Dial. Transplant.*, 2009, **24**(7), 2067–2077.
326. E. Kiss-Toth and T. Roszer, *PPAR Res.*, 2008, **2008**, 183108.
327. P. Corrales, A. Izquierdo-Lahuerta and G. Medina-Gomez, *Int. J. Mol. Sci.*, 2018, **19**(7), 2063.
328. I. A. Bobulescu, Y. Lotan, J. Zhang, T. R. Rosenthal, J. T. Rogers, B. Adams-Huet, *et al.*, *PLoS One*, 2014, **9**(8), e101285.
329. S. Kume, T. Uzu, S. Araki, T. Sugimoto, K. Isshiki, M. Chin-Kanasaki, *et al.*, *J. Am. Soc. Nephrol.*, 2007, **18**(10), 2715–2723.
330. A. Mather and C. Pollock, *KidneyInt. Suppl.*, 2011, 120, S1–S6.
331. M. C. Hu, M. Kuro-o and O. W. Moe, *Contrib. Nephrol.*, 2013, **180**, 47–63.
332. M. C. Hu, M. Shi, J. Zhang, H. Quinones, C. Griffith, M. Kuro-o, *et al.*, *J. Am. Soc. Nephrol.*, 2011, **22**(1), 124–136.
333. T. Yamagishi, Y. Saito, T. Nakamura, S. Takeda, H. Kanai, H. Sumino, *et al.*, *Hypertens. Res.*, 2001, **24**(6), 705–709.
334. M. A. Dominick, M. R. White, T. P. Sanderson, T. Van Vleet, S. M. Cohen, L. E. Arnold, *et al.*, *Toxicol. Pathol.*, 2006, **34**(7), 903–920.
335. P. Escher, O. Braissant, S. Basu-Modak, L. Michalik, W. Wahli and B. Desvergne, *Endocrinology*, 2001, **142**(10), 4195–4202.
336. J. Rieusset, F. Touri, L. Michalik, P. Escher, B. Desvergne, E. Niesor, *et al.*, *Mol. Endocrinol.*, 2002, **16**(11), 2628–2644.
337. J. Wu, Z. W. Zhang and Y. F. Guan, *Sheng Li Ke Xue Jin Zhan*, 2004, **35**(1), 69–72.
338. C. M. Hao, F. Yull, T. Blackwell, M. Komhoff, L. S. Davis and M. D. Breyer, *J. Clin. Invest.*, 2000, **106**(8), 973–982.
339. E. Letavernier, J. Perez, E. Joye, A. Bellocq, B. Fouqueray, J. P. Haymann, *et al.*, *J. Am. Soc. Nephrol.*, 2005, **16**(8), 2395–2402.
340. Y. J. Liang, S. A. Chen and J. H. Jian, *Eur. J. Pharm. Sci.*, 2011, **43**(1–2), 65–70.
341. I. Kuipers, P. van der Harst, F. Kuipers, L. van Genne, M. Goris, J. Y. Lehtonen, *et al.*, *Lab. Invest.*, 2010, **90**(4), 630–636.

342. P. Tontonoz and D. J. Mangelsdorf, *Mol. Endocrinol.*, 2003, **17**(6), 985–993.
343. P. Chonlaket, T. Wongwan and S. Soodvilai, *Exp. Physiol.*, 2018, **103**(2), 250–260.
344. F. Morello, R. A. de Boer, K. R. Steffensen, M. Gnecchi, J. W. Chisholm, F. Boomsma, *et al.*, *J. Clin. Invest.*, 2005, **115**(7), 1913–1922.
345. B. A. Janowski, P. J. Willy, T. R. Devi, J. R. Falck and D. J. Mangelsdorf, *Nature*, 1996, **383**(6602), 728–731.
346. P. Raksaseri, V. Chatsudthipong, C. Muanprasat and S. Soodvilai, *Am. J. Physiol.: Renal Physiol.*, 2013, **305**(4), F583–F591.
347. S. Soodvilai, Z. Jia, S. Fongsupa, V. Chatsudthipong and T. Yang, *Am. J. Physiol.: Renal Physiol.*, 2012, **303**(12), F1610–F1616.
348. S. Kittayaruksakul, S. Soodvilai, N. Asavapanumas, C. Muanprasat and V.Chatsudthipong, *Am. J. Physiol.: Renal Physiol.*, 2012, **302**(5), F552–F560.
349. Y. A. Caldas, H. Giral, M. A. Cortazar, E. Sutherland, K. Okamura, J. Blaine, *et al.*, *Kidney Int.*, 2011, **80**(5), 535–544.
350. C. Gabbi, X. Kong, H. Suzuki, H. J. Kim, M. Gao, X. Jia, *et al.*, *Proc. Natl. Acad. Sci. U. S. A.*, 2012, **109**(8), 3030–3034.
351. M. V. St-Pierre, G. A. Kullak-Ublick, B. Hagenbuch and P. J. Meier, *J. Exp. Biol.*, 2001, **204**(Pt 10), 1673–1686.
352. N. Ballatori, W. V. Christian, J. Y. Lee, P. A. Dawson, C. J. Soroka, J. L. Boyer, *et al.*, *Hepatology*, 2005, **42**(6), 1270–1279.
353. H. Lee, Y. Zhang, F. Y. Lee, S. F. Nelson, F. J. Gonzalez and P. A. Edwards, *J. Lipid Res.*, 2006, **47**(1), 201–214.
354. J. L. Boyer, M. Trauner, A. Mennone, C. J. Soroka, S. Y. Cai, T. Moustafa, *et al.*, *Am. J. Physiol.: Gastrointest. Liver Physiol.*, 2006, **290**(6), G1124–G1130.
355. T. Frankenberg, A. Rao, F. Chen, J. Haywood, B. L. Shneider and P. A. Dawson, *Am. J. Physiol.: Gastrointest. Liver Physiol.*, 2006, **290**(5), G912–G922.
356. J. F. Landrier, J. J. Eloranta, S. R. Vavricka and G. A. Kullak-Ublick, *Am. J. Physiol.: Gastrointest. Liver* Physiol., 2006, **290**(3), G476–G485.
357. A. Marquardt, M. M. Al-Dabet, S. Ghosh, S. Kohli, J. Manoharan, A. ElWakiel, *et al.*, *J. Am. Soc. Nephrol.*, 2017, **28**(11), 3182–3189.
358. T. Jiang, X. X. Wang, P. Scherzer, P. Wilson, J. Tallman, H. Takahashi, *et al.*, *Diabetes*, 2007, **56**(10), 2485–2493.
359. X. Zhang, S. Huang, M. Gao, J. Liu, X. Jia, Q. Han, *et al.*, *Proc. Natl. Acad. Sci. U. S. A.*, 2014, **111**(6), 2277–2282.
360. S. Xu, S. Huang, Z. Luan, T. Chen, Y. Wei, M. Xing, *et al.*, *Proc. Natl. Acad. Sci. U. S. A.*, 2018, **115**(21), 5600–5605.
361. T. Jiang, Z. Wang, G. Proctor, S. Moskowitz, S. E. Liebman, T. Rogers, *et al.*, *J. Biol. Chem.*, 2005, **280**(37), 32317–32325.
362. R. Kumar, J. Schaefer, J. P. Grande and P. C. Roche, *Am. J. Physiol.*, 1994, **266**(3 Pt 2), F477–F485.

363. Y. F. Lin, N. Zhang, H. S. Guo, D. S. Kong, T. Jiang, W. Liang, *et al.*, *Apoptosis,* 2007, **12**(12), 2163–2173.
364. Y. Zhang, X. Zhang, L. Chen, J. Wu, D. Su, W. J. Lu, *et al.*, *Am. J. Physiol.: Renal Physiol.*, 2006, **290**(5), F1065–F1073.
365. M. F. Holick, *N. Engl. J. Med.,* 2007, **357**(3), 266–281.
366. L. A. Plum and H. F. DeLuca, *Nat. Rev. Drug Discov.*, 2010, **9**(12), 941–955.
367. J. E. Aubin and E. Bonnelye, *Medscape Womens Health,* 2000, **5**(2), 5.
368. Y. C. Li, J. Kong, M. Wei, Z. F. Chen, S. Q. Liu and L. P. Cao, *J. Clin. Invest.*, 2002, **110**(2), 229–238.
369. R. M. Carey and H. M. Siragy, *Trends Endocrinol. Metab.,* 2003, **14**(6), 274–281.
370. O. Marsenic, *Am. J. Kidney Dis.*, 2009, **53**(5), 875–883.
371. C. Weigert, K. Brodbeck, A. Bierhaus, H. U. Haring and E. D. Schleicher, *Biochem. Biophys. Res. Commun.,* 2003, **304**(2), 301–307.
372. S. Zafiriou, S. R. Stanners, S. Saad, T. S. Polhill, P. Poronnik and C. A. Pollock, *J. Am. Soc. Nephrol.*, 2005, **16**(3), 638–645.
373. F. Zheng, A. Fornoni, S. J. Elliot, Y. Guan, M. D. Breyer, L. J. Striker, *et al.*, *Am. J. Physiol.: Renal Physiol.*, 2002, **282**(4), F639–F648.
374. X. Z. Ruan, Z. Varghese, S. H. Powis and J. F. Moorhead, *Kidney Int.,* 2005, **68**(6), 2444–2461.
375. S. Hasegawa, T. M. Jao and R. Inagi, *Nutrients*, 2017, **9**(4), 358.
376. S. Y. Goh and M. E. Cooper, *J. Clin. Endocrinol. Metab.*, 2008, **93**(4), 1143–1152.
377. A. Gugliucci and M. Bendayan, *Diabetologia,* 1996, **39**(2), 149–160.
378. T. Miyata, Y. Ueda, K. Horie, M. Nangaku, S Tanaka, C. van Ypersele de Strihou, *et al.*, *Kidney Int.,* 1998, **53**(2), 416–422.
379. T. Matsui, S. Yamagishi, M. Takeuchi, S. Ueda, K. Fukami and S. Okuda, *Biochem. Biophys. Res. Commun.*, 2010, **398**(2), 326–330.
380. M. Herman-Edelstein, P. Scherzer, A. Tobar, M. Levi and U. Gafter, *J. Lipid Res.*, 2014, **55**(3), 561–572.
381. G. Proctor, T. Jiang, M. Iwahashi, Z. Wang, J. Li and M. Levi, *Diabetes,* 2006, **55**(9), 2502–2509.
382. R. Mishra, S. N. Emancipator, C. Miller, T. Kern and M. S. Simonson, *Am. J. Physiol.: Renal Physiol.,* 2004, **286**(5), F913–F921.
383. D. M. Zhang, R. Q. Jiao and L. D. Kong, *Nutrients,* 2017, **9**(4), 335.
384. D'A. C. Magliano, I. Bringhenti and V. Souza-Mello, *Nuclear Receptor Res.*, 2016, 3, 101206.
385. H. Ishida, N. S. Cunningham, H. L. Henry and A. W. Norman, *Endocrinology,* 1988, **122**(6), 2436–2443.
386. I. A. Bobulescu, *Curr. Opin. Nephrol. Hypertens.*, 2010, **19**(4), 393–402.
387. C. Martinez-Garcia, A. Izquierdo-Lahuerta, Y. Vivas, I. Velasco, T. K. Yeo, S. Chen, *et al.*, *PLoS One,* 2015, **10**(11), e0142291.
388. S. J. Shin, J. H. Lim, S. Chung, D. Y. Youn, H. W. Chung, H. W. Kim, *et al.*, *Hypertens. Res.*, 2009, **32**(10), 835–845.
389. K. W. Chung, E. K. Lee, M. K. Lee, G. T. Oh, B. P. Yu and H. Y. Chung, *J. Am. Soc. Nephrol.,* 2018, **29**(4), 1223–1237.

390. X. X. Wang, T. Jiang, Y. Shen, L. Adorini, M. Pruzanski, F. J. Gonzalez, *et al.*, *Am. J. Physiol.: Renal Physiol.*, 2009, **297**(6), F1587–F1596.
391. M. Adil, A. D. Kandhare, P. Ghosh, S. Venkata, K. S. Raygude and S. L. Bodhankar, *Renal Fail.*, 2016, **38**(6), 1007–1020.
392. T. Gui and Z. Gai, *Genom. Data*, 2015, **6**, 31–32.
393. S. J. Glastras, M. G. Wong, H. Chen, J. Zhang, A. Zaky, C. A. Pollock, *et al.*, *Nutr. Metab.* (*Lond.*), 2015, **12**, 40.
394. Z. Gai, C. Hiller, S. H. Chin, L. Hofstetter, B. Stieger, D. Konrad, *et al.*, *Biochim. Biophys. Acta,* 2014, **1842**(9), 1870–1878.
395. D. K. Layman and D. A.Walker, *J. Nutr.*, 2006, **136**(1 Suppl), 319S–323S.
396. A. J. King and A. S. Levey, *J. Am. Soc. Nephrol.*, 1993, **3**(11), 1723–1737.
397. C. Tovar-Palacio, A. R. Tovar, N. Torres, C, Cruz, R. Hernandez-Pando, G. Salas-Garrido, *et al.*, *Am. J. Physiol.: Renal Physiol.*, 2011, **300**(1), F263–F271.
398. S. Li, K. Nagothu, G. Ranganathan, S. M. Ali, B. Shank, N. Gokden, *et al.*, *Am. J. Physiol.: Renal Physiol.*, 2012, **303**(3), F437–F448.
399. M. Soumura, S. Kume, K. Isshiki, N. Takeda, S. Araki, Y. Tanaka, *et al.*, *Biochem. Biophys. Res. Commun.*, 2010, **402**(2), 265–271.
400. N. E. El-Ashmawy, N. F. Khedr, H. A. El-Bahrawy and S. A. Helal, *Biomed. Pharmacother.*, 2018, **99**, 504–510.
401. W. Han, H. Zhao, B. Jiao and F. Liu, *Biosci. Trends,* 2014, **8**(2), 120–125.
402. Y. Y. Yang, S. L. Hsieh, P. C. Lee, Y. C. Yeh, K. C. Lee, Y. C. Hsieh, *et al.*, *J. Hepatol.*, 2014, **61**(5), 1004–1013.
403. S. N. Verouti, E. C. Tsilibary, E. Fragopoulou, C. Iatrou, C. A. Demopoulos, A. S. Charonis, *et al.*, *Nephron Exp. Nephrol.*, 2012, **122**(1–2), 36–50.
404. Y. Wang, D. K. Deb, Z. Zhang, T. Sun, W. Liu, D. Yoon, *et al.*, *J. Am. Soc. Nephrol.*, 2012, **23**(12), 1977–1986.
405. H. Yao, Y. J. Qiao, Y. L. Zhao, X. F. Tao, L. N. Xu, L. H. Yin, *et al.*, *World J. Gastroenterol.*, 2016, **22**(30), 6890–6905.
406. X. Zhang, Z. Song, Y. Guo and M. Zhou, *Mol. Cell Biochem.*, 2015, **399**(1–2), 155–165.
407. Z. Song, Y. Guo, M. Zhou and X. Zhang, *Metabolism*, 2014, **63**(10), 1324–1333.
408. S. Tamura, M. Okada, S. Kato, Y. Shinoda, N. Shioda, K. Fukunaga, *et al.*, *Sci. Rep.*, 2018, **8**(1), 2305.

CHAPTER 6

Nutritional Regulation of Mitochondrial Health and Its Implication in Treating Obesity and Diabetes: Lessons Learned From the Resveratrol–Sirt1 Axis

Z. TAO,[a,b] L. SHI[c] AND Z. CHENG[a,c] *

[a]Department of Human Nutrition, Foods, and Exercise, Virginia Tech, Blacksburg VA 24061, USA
[b]Cutaneous Biology Research Center, Massachusetts General Hospital, Harvard Medical School, Charlestown MA 02129, USA
[c]Food Science and Human Nutrition Department, University of Florida, Gainesville FL 32611, USA
*Email: z.cheng@ufl.edu

6.1 Introduction

Resveratrol (RSV, or 3,5,4′-trihydroxystilbene) is a nutraceutical primarily found in grapes, berries, peanuts, and some other dietary plants, as well as associated processed foods (Figure 6.1A). It has attracted vast attention since 1992, when it was postulated to explain the "French paradox" effect, *i.e.*, red wine consumption reducing the incidence of coronary artery disease. The content

Nutritional Signaling Pathway Activities in Obesity and Diabetes
Food Chemistry, Function and Analysis No. 24
Edited by Zhiyong Cheng

Published by the Royal Society of Chemistry, www.rsc.org

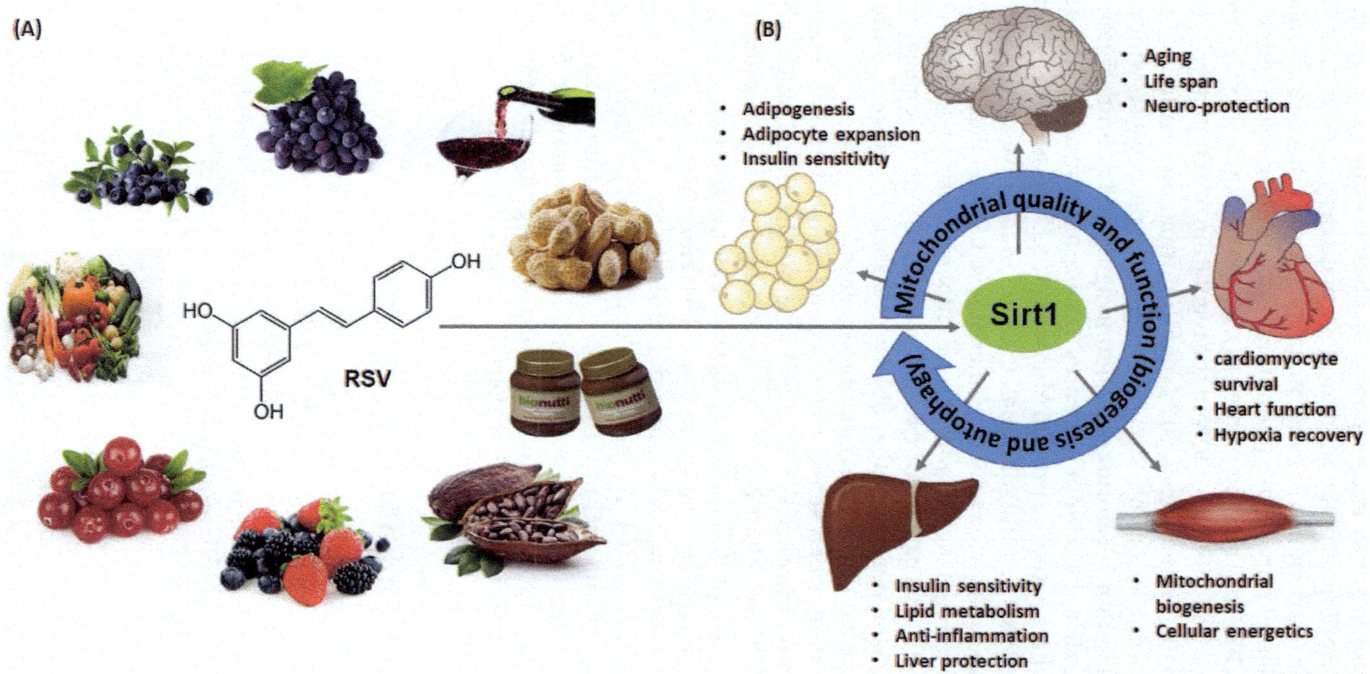

Figure 6.1 Food sources of resveratrol (RSV) and its interaction with Sirt1 in the regulation of systemic metabolism. (A) Foods like grapes, berries, peanuts, and some other dietary plants, as well as associated processed foods are relatively rich in the nutraceutical RSV. (B) RSV interacts with energy sensor Sirt1, which has been shown to regulate metabolism across tissues such as brain, heart, skeletal muscle, liver, and adipose tissues. Detailed discussion will follow in the following sections.

of RSV in these foods may vary during storage or processing procedures.[1] The beneficial effects of RSV have been reported on obesity, immune defense, cardiovascular disease, Type 2 diabetes, neurological disease, and cancer.[2–6] RSV can mimic caloric restriction to improve metabolism and it also acts as an antioxidant, anti-inflammatory, anti-cell proliferation, or anti-invasive agent.[5,6] At the cellular level, RSV may regulate autophagy, apoptosis, and cell cycle arrest.[6]

Among the targets identified for RSV in the regulation of autophagy, Sirtuin1 (Sirt1) (a mammalian homolog of Sir2 (silent information regulator 2)) is a nicotinamide adenine dinucleotide (NAD^+)-dependent deacetylase, which was previously found to mimic caloric restriction in aging and neurodegenerative disease,[7] as well as liver[8] and kidney diseases[9] (Figure 6.1B). In addition, adenosine monophosphate-activated protein kinase (AMPK),[8,9] forkhead box O transcription factors (FoxOs),[10] and mammalian target of rapamycin (mTOR)[11] are targets of RSV in response to various cellular states. RSV-mediated autophagy has been shown to regulate cell apoptosis,[12] inflammatory nuclear factor-κB (NF-κB) signaling,[13] and oxidative stress.[14] As such, RSV has been examined, as an autophagy regulator, to combat obesity,[15] diabetes,[16] cancer,[17] immune dysfunction,[18] aging, and neurodegenerative disease.[7]

Sirt1 is known as an energy sensor that mediates metabolic homeostasis *via* regulating mitochondria, the key energy producers in the cell. Ablation of Sirt1 compromising mitochondrial health has been implicated in metabolic diseases, including obesity, diabetes, liver diseases, kidney diseases, aging, and neurodegenerative disease[2,7–9] (Figure 6.1B). Therefore, Sirt1 activators have been developed to promote mitochondrial function and metabolism, with the goal of preventing or treating metabolic derangements.[19] In this chapter, we will review the role of Sirt1 and the RSV–Sirt1 axis in mitochondrial and metabolic regulation across tissues. This includes mitochondrial biogenesis, mitochondrial dynamics, and mitochondrial autophagy (quality control). In addition, we discuss the potential of the RSV–Sirt1 axis in the treatment of metabolic disorders (*i.e.*, obesity and diabetes).

6.2 Sirt1 in Metabolic Regulation

The Sirts, a family of NAD-dependent deacetylases, include Sirt1–Sirt7 and are involved in the regulation of various cellular metabolisms and metabolic diseases,[20–23] in which Sirt1 is most abundantly studied. Sirt1 plays its role in metabolism through its deacetylating effects on histones and non-histone proteins that participate in different organ metabolisms and diseases, including adipose tissue,[24,25] muscle,[26–28] liver,[29–31] heart,[32–34] and brain[35–39] (Figure 6.1B).

6.2.1 Sirt1 in Adipose Tissue

Obesity may arise from increased energy intake and/or reduced energy expenditure.[40] White adipose tissue (WAT) is the major energy reservoir for storing redundant energy in the form of triglyceride.[41] Morphologically and functionally distinct from WAT, brown adipose tissue (BAT) is composed of brown adipocytes with multilocular morphology, dense mitochondria, and the ability to activate thermogenesis upon cold treatment.[42] Recent studies indicated that activation of BAT promotes lipid metabolism in rodents[43] and humans,[44] proposing its anti-obesity potential.

In WAT, Sirt1 is downregulated in obesity.[24] In high-fat diet-induced and genetic obesity mice, the full length of Sirt1 was cleaved by caspase 1 into an inactive form in adipose tissue.[45] Furthermore, gene expression in mice fed a high-fat diet was greatly recapitulated in adipose tissue Sirt1 knockout mice.[45] Specific activation or gain function of Sirt1 protected rodents from diet-induced obesity and insulin resistance.[46,47] Administration of the potent Sirt1 activator SRT1720 protected mice from diet-induced obesity by elevating fatty acid oxidation by deacetylation of peroxisome proliferator-activated receptor-γ coactivator 1α (PGC-1α), FoxO1, and stimulation of AMPK signaling.[46] However, on normal chow diet transgenic mice overexpression of Sirt1 showed normal insulin sensitivity compared with wild-type mice.[47] It was only on a high-fat diet challenge or with a *db*/*db* background that Sirt1 overexpression improved glucose tolerance.[47] Sirt1 knockout female mice showed higher body weight and fat percentage than wild-type mice, but shared similar glucose and insulin tolerance test profile on normal chow diet.[45] After high-fat diet challenge, the glucose and insulin tolerance test profile of Sirt1 knockout mice deteriorated, with lower adiponectin in plasma.[45] Consistently, Sirt1 knockout males have higher body weight and fat percentage on normal chow diet.[48] However, Sirt1 knockout mice showed low-grade inflammation but glucose intolerance and insulin resistance, which was similarly exhibited during short-term (5-week) high-fat diet challenge. In contrast to short-term high-fat diet, on chronic (15-week) high-fat diet Sirt1 knockout mice showed improved glucose intolerance, insulin resistance and inflammation.[48]

At cellular level, Sirt1 suppressed preadipocyte proliferation[49] and preadipocyte differentiation[24] and promoted browning of white adipocyte[25] (Figure 6.1B). During 3T3L1 preadipocyte proliferation, Sirt1 knockdown resulted in hyperplastic and inflamed adipocytes that were dysfunctional through hyperacetylation and activation of c-Myc, which is a transcriptional factor responsible for cell proliferation.[49] During 3T3L1 preadipocyte differentiation, Sirt1 docked with peroxisomal proliferator-activated receptor-γ's (PPARγ's) cofactors nuclear receptor co-repressor (NCoR) and silencing mediator of retinoid and thyroid hormone receptors (SMRT) to suppress PPARγ.[24] Overexpression of Sirt1 suppressed PPARγ expression and adipogenesis, while silencing Sirt1 increased PPARγ expression and adipogenesis.[24] In WAT, Sirt1 promoted the release of free fatty acid from adipocyte by repressing PPARγ during energy restriction, indicating Sirt1's inhibitory role in WAT development.[24] In addition, Sirt1 promoted browning of subcutaneous WAT following acute cold exposure.[25] After 4 °C overnight cold treatment, Sirt1 overexpression induced browning of subcutaneous WAT with increased numbers of uncoupling protein 1 (UCP1)-immunoreactive paucilocular adipocytes.[25] Gain or loss of function of Sirt1 in adipocytes had effects on recruitment of macrophage into WAT.[50] Adipose tissue-specific knockout of Sirt1 in mice had increased WAT expression of macrophage and inflammatory factor tumor necrosis factor-alpha (TNF-α) and NF-κB nuclear translocation and gene expression by reducing H3K9 deacetylation.[50] In brown adipocyte, UCP1 expression is controlled by Sirt1 and is reduced under inflammation, which could be rescued by Sirt1 activation through resveratrol treatment.[51] *In vivo*, gain of function of Sirt1 enhanced insulin sensitivity through induction of glucose uptake in BAT and improved body temperature maintenance upon cold challenge.[52] Mechanistically, Sirt1 showed no effect on differentiation of BAT, but promoted metabolic transcriptional response to β_3-adrenergic stimuli in differentiated adipocytes.[52] Thus, Sirt1 ameliorated adipocyte hyperplasticity, inflammation through inhibiting preadipocyte proliferation and differentiation, and promoted adipocyte browning. However, in adipose tissue Sirt1 may physiologically play distinct roles due to variations of dietary challenge, gender, and genetic background.

6.2.2 Sirt1 in Muscle

In skeletal muscle, mitochondrial biogenesis increases in response to metabolic stress such as fasting and exercise.[53] During these physical and physiological challenges, Sirt1 has been widely investigated in muscular mitochondrial biogenesis both in rodents and humans.[54,55] Under these conditions of metabolic stress, Sirt1 is known to regulate a number of protein targets, including PGC-1α, AMPK, and p53, which are potential signaling nodes participating in mitochondrial biogenesis.[56]

PGC1-α is the best-studied molecule linking Sirt1 to mitochondrial biogenesis in muscle. Most *in vitro* studies have demonstrated that Sirt1 positively regulates PGC1-α transcriptionally or post-transcriptionally to induce mitochondrial biogenesis.[57,58] Transcriptionally, Sirt1 bound to the proximal promoter region of the PGC1-α gene in C2C12 myotubes, which promoted a positive autoregulatory PGC1-α expression loop.[57] Post-transcriptionally, Sirt1 deacetylated PGC1-α to activate its activity and downstream mitochondrial biogenesis[58] (Figure 6.1B). Also, during fasting and low glucose concentrations, Sirt1 was responsible for deacetylating PGC-1α, accompanied by activation of fatty acid oxidation and increased expression of mitochondrial genes, as shown by gene expression of pyruvate dehydrogenase kinase 4 (PDK4), estrogen-related receptor α (ERR-α), and cytochrome c in C2C12 muscle cells.[28] However, the relationship between Sirt1 and PGC1-α seems to be less straightforward *in vivo*. For example, in skeletal muscle subjected to voluntary wheel running and electric stimulation, PGC1-α level was not closely associated with Sirt1 expression and enzymatic deacetylase activity.[59] Further, the lack of Sirt1 deacetylase activity in skeletal muscle did not impair wheel running-induced PGC-1α expression and mitochondrial biogenesis,[60] suggesting Sirt1 may not be required for PGC-1α expression and mitochondrial biogenesis. Another study even proposed that Sirt1 overexpression in rodent muscle downregulated PGC-1α, mitochondrial transcription factor A and cyclooygenase IV.[61]

6.2.3 Sirt1 in Liver

Sirt1 plays an important role in the regulation of lipid metabolism in liver, the dysfunction of which may result in the development of alcoholic liver disease[62] and non-alcoholic fatty liver disease (NAFLD), including hepatic steatosis, inflammation, insulin resistance, and liver damage[31,63,64] (Figure 6.1B).

In alcoholic liver disease, liver samples from patients with alcohol hepatitis had reduced levels of Sirt1 than healthy counterparts.[63] Hepatic deletion of Sirt1 promoted lipid accumulation and inflammatory cytokines, whereas overexpression of Sirt1 reversed ethanol induced fat accumulation and metabolic dysfunction in liver.[63] Phamathetical administration of rosiglitazone, a PPARγ, ameliorated alcoholic fatty liver in mice through enhanced hepatic adiponectin-Sirt1-AMPK signaling.[62]

NAFLD is characterized by hepatic endoplasmic reticulum (ER) stress,[31] insulin resistance[65] and inflammation.[66] Also, the expression of Sirt1 is reduced by genetic and high-fat diet-induced NAFLD.[31,67] The reduction of Sirt1 may be due to phosphorylation by C-Jun N-terminal kinase 1 (JNK1) and casein kinase 2, which promoted its subsequent degradation by proteasome and inhibited its nuclear localization.[68,69] Lack of Sirt1 activity contributed to liver steatosis in response to high-fat diet feeding, which may result from increased lipogenesis and reduced lipid export.[66] In diet-induced insulin-resistant low-density lipoprotein receptor-deficient mice and genetically obese mice, overexpression of Sirt1 in liver attenuated hepatic steatosis and insulin resistance through inhibition of ER stress, enhanced insulin receptor and mTOR complex 1 (mTORC1) signaling.[31] In liver, Sirt1 promoted insulin sensitivity through transcriptional promotion of the gene encoding Rictor, and its activation on phosphorylation of Akt and FoxO1,[70] resulting in decreased transcription of the gluconeogenic genes glucose-6-phosphatase (G6pase) and phosphoenolpyruvate carboxykinase (Pepck).[71] Liver-specific knockdown of Sirt1 contributed to hepatic glucose overproduction and hyperglycemia.[71] However, discrepancies were noted in the study that Sirt1 knockdown in Type 2 diabetes mellitus (T2DM) mice model increased insulin sensitivity and reduced glucose production through decreased Pepck, fructose-1,6-bisphosphatase (FBPase) and G6pase due to increased acetylation of signal transducer and activator of transcription 3 (STAT3), FoxO1, and PGC-1α.[72] The controversial effects of Sirt1 knockout may be explained by the fact that Sirt1 is involved in multiple signaling pathways and plays distinct roles in different disease conditions.

6.2.4 Sirt1 in Heart

Heart failure results from aging, mechanical load, and drug-caused cardiotoxicity.[73–75] Sirt1 played protective roles in preventing heart from aging,[32,76] mechanical and physiological pressure,[77,78] and cardiotoxicity[34] (Figure 6.1B). However, detrimental roles of overactivation of Sirt1 in heart have been documented.[32,79]

Compared with healthy cardiomyocytes, Sirt1 was downregulated in cardiomyocytes from human patients with advanced heart failure.[80] Both *in vitro* and *in vivo,* Sirt1 inhibition or deficiency increased ER stress-induced cardiac injury.[81] Activation of Sirt1 protects against cardiac injury and ameliorated ER stress-induced cardiomyocyte apoptosis by attenuating protein kinase RNA-like endoplasmic reticulum kinase/elF2α pathway activation.[81] *In vitro*, Sirt1 reduced acetylation of FoxO1 and downstream Rab7 (a small GTP-binding protein that mediates late autophagosome-lysosome fusion), which is essential for maintaining cardiomyocyte function during starvation.[82] In neonatal cardiomyocyte, Sirt1 coordinated with PPARα to suppress hypertrophy *in vitro*, as shown by the evidence that activation of Sirt1 by resveratrol or overexpression of Sirt1 inhibited phenylephrine (PE)-induced hypertrophy and inhibited PE-induced downregulation of fatty acid oxidation genes.[78]

Additionally, knockdown of Sirt1 led to dilated cardiomyopathy and reduced cardiomyocyte size, which was accompanied by reduced mitochondrial genes (Tfam, cytochrome b, ND1).[83] This phenotype was due to increased acetylation level of the myocyte enhancer factor 2 (Mef2) transcription factor, which is essential for normal heart development and mitochondrial integrity.[83] In dystrophin-deficient mdx mice, Sirt1 suppresses cardiac hypertrophy and fibrosis through its deacetylation of lysine residues for ubiquitination of p300, a pro-hypertrophic coactivator, to promote its subsequent degradation by ubiquitin-proteasome.[84] As to protective effects against drug-induced cardiotoxicity, resveratrol ameliorated doxorubicin (DOX)-induced cardiomyocyte apoptosis by activating Sirt1 and its deacetylation effects on p53 *in vivo.*[85] *In vivo*, low or moderate (2.5-fold or 7.5-fold) overexpression of Sirt1 in the heart attenuated age-dependent increases in cardiac hypertrophy, apoptosis/fibrosis, and cardiac dysfunction. However, a high level of Sirt1 overexpression showed opposite effects.[32]

Nevertheless, Sirt1 was upregulated by pressure overload in the heart, and knockdown of Sirt1 attenuated pressure overload-induced cardiac hypertrophy and failure.[74] Sirt1 was shown to coordinate with PPARα to suppress genes in mitochondrial function that are regulated by ERRs, leading to cardiac dysfunction. Similarly, cardiac-specific overexpression of Sirt1 impaired cardiac function associated with reduced mitochondrial function genes (like cytochrome b, cytochrome c, and Cox5a) and respiration.[79]

6.2.5 Sirt1 in Brain

In brain, Sirt1 is highly expressed in metabolically active sites, including hypothalamic arcuate, dorsomedial, ventromedial, and paraventricular nuclei and the area postrema and the nucleus of the solitary tract in the hindbrain.[86] Like adipose tissue, muscle, and heart, Sirt1 expression increased in the brain during nutrient deprivation.[86–88] Physiologically, Sirt1 regulates neural development,[36] energy balance and glucose metabolism in neurons,[89] and neural diseases[37,39,90–92] (Figure 6.1B). At cellular level, Sirt1 mediates neuronal cell differentiation,[93] cell expansion,[94] and cell death.[95]

Hypothalamic Sirt1 regulated energy balance, as shown by the evidence that pharmacological or genetic silencing of Sirt1 in hypothalamus decreased food intake and body weight gain.[35] Neural Sirt1 negatively regulated systemic and hypothalamic insulin signaling by directly deacetylating and repressing insulin receptor substrate 1 (IRS-1) function.[89] Specifically, Sirt1 knockout in neurons promoted insulin sensitivity and had lower fasting insulin levels and improved glucose tolerance and insulin tolerance at the systemic and hypothalamic level, as shown by increased PI3K signaling, pAkt, and pFoxO1.[89]

AD onset and progression are characterized by accumulation of aggregated β-amyloid.[38] Overexpression or activation of Sirt1 mimicked the effects of caloric restriction and attenuated β-amyloid content in the brain and amyloid neuropathology through repressing serine/threonine Rho kinase (ROCK1) and promoting its downstream non-amyloidogenic α-secretase's cleavage of the amyloid precursor protein.[39] In an inherited neurodegenerative disorder caused by glutamine repeat expansion in huntingtin protein (HTT), mutant HTT interfered with Sirt1 to repress TORC1–cAMP response element-binding protein (CREB) interaction to suppress brain-derived neurotrophic factor (BDNF) transcription. Sirt1 overexpression would deacetylate TORC1 to induce the effects of TORC1-CREB on BDNF to rescue the defects in this type of neurodegenerative disorder.[37] In addition, Sirt1 could deacetylate methyl-CpG-binding protein 2 (MeCP2), an epigenetic regulator, and activate its DNA-binding activity to BDNF promoter to induce its expression in brain.[96]

At cellular level, nuclear translocalization of Sirt1 was indispensable for neural precursor cell differentiation upon differentiation stimuli, through its binding with nuclear receptor corepressor (N-CoR) and synergistically increased the number of Tuj1$^+$ neurons.[93] Sirt1 was reported to be negatively involved in oligodendrocyte regeneration during adulthood. Lack of active Sirt1 in neural stem cells increased the production of new oligodendrocyte progenitor cells (OPCs) and their expansion into oligodendrocytes.[94] In neuronal cells, Sirt1 overexpression promoted neurite outgrowth and enhanced tolerance to cell death or degeneration induced by amyloid-β1–42 oligomers, probably by its inhibition of mTOR signaling.[95] Other than its cell survival effects, Sirt1 directly modulated synaptic plasticity and memory formation by microRNA. Results showed that activation of Sirt1 promoted synaptic function by repressing the expression of miR-134 by a repressor complex containing the transcription factor YY1, and consequent translational induction of CREB and BDNF.[97]

6.3 The RSV–Sirt1 Axis in Mitochondrial Regulation

6.3.1 Mitochondrial Biogenesis

RSV was shown to induce mitochondrial biogenesis *via* Sirt1–PGC1-α cascade. In human vascular endothelial cells, depletion of mitochondrial integrity was directly associated with an increase in endothelial activation/dysfunction. Resveratrol and high-flow LSS treatment induced the expression of mitochondrial biogenesis regulator genes, such as SIRT1, PGC-1α, and mitochondrial transcription factor A (TFAM), and the mitochondrial contents.[98] In human umbilical vein endothelialcells, administration of RSV increased mitochondrial mass and mitochondrial DNA content, which is concomitant with increased Sirt1 enzymatic activity and induced mitochondrial biogenesis factors such as PGC1-α, TFAM, and nuclear respiratory factor-1 (NRF-1).[99] Fluoride decreased mitochondrial biogenesis and induced neurotoxicity; RSV was reported to significantly increase mitochondrial biogenesis marker PGC-1α, NRF1, and TFAM expression concomitant with increased mtDNA contents, mitochondrial membrane potential, and mRNA levels of complex subunits CO1 and ATP8, which alleviated the fluoride-induced mitochondrial biogenesis dysfunction in SH-SY5Y cells.[100]

In pancreas tissues of T2DM mice, RSV administration reversed the streptozocin-induced downregulation of PGC-1α, Sirt1, and FoxO3a.[101] And RSV prevented T2DM by regulating the expression of mitochondrial genes associated with biogenesis, lipid metabolism, and β-cells *via* Sirt1.[101] Diabetic cardiomyopathy (DCM) is a major threat for diabetic patients.[102] Sirt1 has a regulatory effect on mitochondrial dynamics, which is associated with DCM pathological protection. RSV ameliorated cardiac injuries in DCM through activation of Sirt1 and PGC-1α-mediated mitochondrial biogenesis.[103] In retinal disease, both *in vitro* and *in vivo* experiments demonstrated that RSV activated AMPK/PGC-1α signaling pathway and induced mitochondrial biogenesis and dynamic, which could be the possible mechanism underlying RSV's mitochondrial protection against retinal disease.[104] In the seizure-induced neuronal damage mouse model, RSV increased expression of PGC-1α, NRF-1, and Tfam, NRF-1-binding activity, cyclooxygenase 1 level, and mtDNA amount, which reduced activated caspase-3 activity and attenuated neuronal cell damage in the hippocampus following status epilepticus.[105] Also, studies have proposed that RSV inhibited reactive oxygen species (ROS) generation in cardiomyocytes *via* Sirt1 and mitochondrial biogenesis signaling pathways.[106]

However, studies have raised the debate that RSV exhibited no effects or opposite effects on mitochondrial biogenesis, which is dependent on its concentration. The GRX is a model of activated hepatic stellate cells used to study liver fibrosis development and resolution. GRX cells treated with all RSV concentrations (0.1–50 μM) presented higher autophagy/mitophagy features, but only treatments with 1 and 10 μM of RSV induced mitochondrial biogenesis.[107]

It has been reported that feeding mice with RSV activated AMPK and Sirt1 in skeletal muscle, leading to deacetylation and activation of PGC-1α and increased mitochondrial biogenesis.[46,108,109] However, a study indicated that feeding rats or mice a diet containing 4 g RSV kg^{-1} diet had no effect on mitochondrial protein levels in muscle.[110] Even worse, in C_2C_{12} myotubes knockdown or dominant-negative construction of Sirt1 increased PGC-1α acetylation, but with higher PGC-1α coactivator activity and mitochondrial proteins, which is recapitulated in the animal model.[110] This study demonstrated that Sirt1 regulated mitochondrial biogenesis by inhibiting PGC-1α coactivator activity, resulting in a decrease in mitochondria.[110]

6.3.2 Mitochondrial Autophagy (Mitophagy)

Selective autophagic degradation of mitochondria is known as mitophagy.[111] Mitophagy as a cellular remodeling process plays an important role in mitochondrial health.[112,113] In contrast to mitochondrial biogenesis, mitophagy clears aging or dysfunctional mitochondria to maintain mitochondrial homeostasis. RSV has been shown to activate mitophagy to remove damaged mitochondria and protect from ROS overproduction,[114] cytotoxicity, and cell damage.[107] In C_2C_{12} myoblast cells, RSV promoted mitophagy and eliminated dysfunctional mitochondria containing high superoxide levels, which was completely abolished by silencing autophagy gene *Atg5* or mitophagy gene PTEN-induced putative kinase 1 (*Pink1*), suggesting that the effects of RSV on autophagy and mitophagy were required for the clearance of damaged mitochondria.[114] Continuous exposure of peritoneal mesothelial cells (PMCs) to high-glucose-containing peritoneal dialysis (PD) solutions impaired mitophagy, which induced ROS production, peritoneal inflammatory injury, and impairment of local peritoneal host defense. This phenotype was ameliorated by RSV treatment and the effects of RSV on mitophagy/autophagy are dependent on AMPK activation.[115] GRX is an activated hepatic stellate cells model used to study liver fibrosis development and resolution. RSV treatment induced mitophagy in GRX and protected cells from cytotoxicity and cell damage.[107]

In Duchenne muscular dystrophy, autophagy and mitophagy-related genes such as Pink1, Parkin, Bnip3, and autophagic flux were downregulated in the muscle of mdx mice compared with healthy controls, and RSV significantly reversed these phenomena concomitant with significant ROS reduction.[114] Abnormal accumulations of mitochondria-containing autophagosomes were found in the mdx mouse heart.[116] RSV was found to ameliorate the cardiac pathology of mdx mice.[84] RSV administration ameliorated the cardiomyopathy and promoted mitophagy to clean up damaged mitochondria with significant reductions in the amount of mtDNA deletion and ROS levels. Mechanistically, RSV induced translocation of FoxO3 into nuclei and increased the expression of FoxO3 downstream autophagy-related genes, Bnip3 and Pink1[116]. In piglets challenged with diquat, RSV enhanced the protein expression of PINK1 and Parkin in intestinal mitochondria, therefore resulting in induced mitophagy and alleviating mitochondrial damage.[117]

6.4 The RSV–Sirt1 Axis Regulates Metabolism *via* Autophagy/Mitophagy

Autophagy and mitophagy have been implicated in various pathways (*e.g.*, inflammation, aging, and adipogenesis) that affect metabolic health across tissues (*e.g.*, heart, liver, and kidney). Autophagy/mitophagy mediates aging and adipogenesis,[118] which may contribute to obesity and metabolic disorders. Autophagy is also induced by pathogenic stimuli such as oxidative stress, infection, and protein aggregate accumulation.[119] Thus, targeting dysregulated autophagy may lead to new therapeutics.[7,15–18] There are three forms of autophagy – chaperone-mediated autophagy, microautophagy, and macroautophagy[119] – and RSV has been implicated mainly in the regulation of macroautophagy (hereafter referred to as autophagy).

6.4.1 Immunometabolism and Inflammation

The autophagy-signaling pathway is committed to the operations of immunity and inflammation. They equilibrate the advantageous and disadvantageous aspects of immunity and inflammation, and therefore would protect organisms against infectious, autoimmune, and inflammatory diseases.[18] The NOD-like receptor protein-3 (NLRP3) inflammasome is a caspase-1-containing multiprotein complex, which regulates inflammatory cytokine interleukin-1 (IL-1) secretion and promotes development of inflammatory disease,[115] which was activated by autophagy and subsequent IL-1β secretion.[115] Atherosclerosis, an inflammatory disease characterized by overexpression of the adhesion molecules, begins with inflammatory changes in the endothelium in which inflammation centrally participates in all stages of pathogenesis.[120] Autophagy has implicated its protective role during the development of atherosclerosis.[121] Otherwise, autophagy is involved in pathophysiological responses following inflammatory diseases, such as traumatic brain injury, which is an extremely complicated type of neurological trauma involving a series of inflammatory response. In mechanism, this injury would profoundly activate autophagy subsequent neuroinflammatory signaling that induces dramatic neuronal damage and behavioral impairment, in which Toll-like receptor 4 (TLR4) acts as a vital administrator of this cascade.[13] RSV has been proposed to regulate or collaborate multiple inflammation and immunity-related molecular targets,[122] among which autophagy's role will be reviewed.

In murine macrophage cell line J774A, RSV induced autophagy by activating p38 to inhibit NLRP3 inflammasome-derived IL-1 secretion (Figure 6.2A).[123] In peritoneal inflammatory injury of PMCs, NLRP3 activation leads to IL-1β secretion, which was enhanced by knocking down beclin-1 and Atg5 or autophagy inhibition with 3-methyladenine (3-MA).[115] Furthermore, beclin-1 and Atg5 ablation sensitized cells to IL-Iβ release. On the contrary, RSV treatment attenuated these effects by inducing mitophagy and autophagy *via* AMPK.[115] RSV also attenuated endothelial inflammation by reducing tumor necrosis factor α (TNFα) and increasing autophagy. Mechanistically, the promotive effects of RSV depended on cAMP, serine/threonine protein kinase (PRKA), AMPK, and Sirt1.[120] In an animal model of traumatic brain injury, RSV administration significantly diminished brain edema, motor deficit, and neuronal loss, and ameliorated spatial cognitive function. Interestingly, RSV appeared to suppress

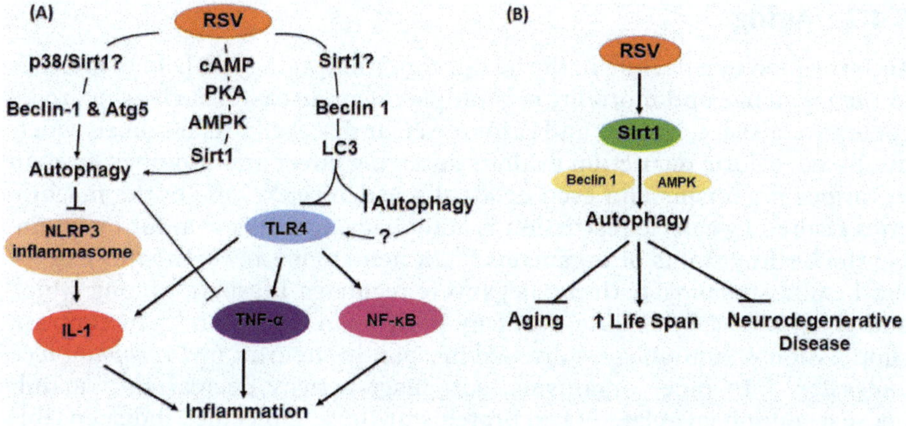

Figure 6.2 Resveratrol (RSV) regulates autophagy in inflammation and aging. (A) RSV has been shown to play a tissue- or cell type-dependent role in autophagy (*e.g.*, paradoxically inducing or suppressing autophagy) to dampen inflammation. RSV induces autophagy *via* p38/beclin 1/Atg5 cascade to inhibit NOD-like receptor protein 3 (NLRP3: NACHT, LRR, and PYD domains containing protein 3) inflammasome and interleukin-1 (IL-1) secretion in macrophage. In peritoneal mesothelial cells and human umbilical vein endothelial cells, the RSV-induced autophagy pathway includes cyclic adenosine monophosphate (cAMP), protein kinase A (PKA), AMP-activated protein kinase (AMPK), and Sirt1. By contrast, RSV suppresses autophagy and Toll-like receptor 4 (TLR4) as well as its downstream inflammatory factors IL-1, nuclear factor kappa-light-chain-enhancer of activated B cells (NF-κB) and tumor necrosis factor-α (TNF-α) secretion in hippocampal neurons. The inhibition of inflammation was shown to prevent the pathogenesis of atherosclerosis and traumatic brain injury. Whether Sirt1 plays the key role in RSV-mediated autophagy in macrophage and hippocampal neurons remains to be defined (indicated by question marks). (B) Sirt1 is the major target of RSV in aging and related metabolic diseases. Sirt1 may promote autophagy *via* beclin-1 to slow down aging and prolong lifespan. RSV promotes autophagy *via* Sirt1-AMPK cascade to ameliorate neurodegenerative disease.

autophagy in hippocampal neurons by reducing protein expression of beclin-1 and LC3, which was associated with downregulation of TLR4, NF-κB, IL-1β, and TNFα (Figure 6.2A).[13] These findings suggest that RSV may play a tissue- or cell type-dependent role in autophagy regulation.

The role of RSV in mediating inflammation *via* autophagy was confirmed by autophagy inhibitor co-treatment. The inhibitory effect of RSV on NLRP3 inflammasome-derived IL-1 secretion in macrophages was abolished by an autophagy inhibitor 3-MA *in vitro* and *in vivo*.[123] In atherosclerosis development, RSV-induced autophagy was mitigated by autophagy inhibitors 3-MA and chloroquine (CQ), or by silencing Atg5 and Beclin-1, which reversed RSV-induced downregulation of inflammatory factors.[120]

6.4.2 Aging

Accompanied by a decline in the health condition, aging will lead to increase in the incidence and mortality of multiple organ diseases, such as neurodegenerative disease, cancer, and cardiovascular disease.[124] As has been widely discussed, caloric restriction exhibits antiaging effect and prolongs lifespan in various organisms and even delays disease onset.[125] One of the machineries favored by caloric restriction is autophagy, which is vital for maintaining the healthy status of organisms.[126] In nematode *Caenorhabditis elegans* the insulin-signaling pathway negatively regulates lifespan, among which autophagy showed absolute necessity in lifespan extension.[127] In contrast, suppression of autophagy shortened lifespan in the fruit fly *Drosophila melanogaster.*[128] In older organisms, autophagy activity decreased,[129] mainly because autophagy-related (Atg) proteins or those molecules indispensable for autophagy induction, such as Sirt1, have decreased their protein expressions in aged cells or tissues and that reduced autophagy has been associated with accelerated aging.[7] There is an increasing interest in stimulation of autophagy to promote antiaging effects, including genetic manipulation or pharmacological activation of autophagy to mimic nutrient starvation or caloric restriction.[130,131]

RSV is widely used as an antiaging agent. In nematodes with autophagy-proficiency, RSV induces autophagy and extends lifespan through indirect activation of Sirt1, whose effect can be abolished by knockdown or pharmacological inhibition of Sirt1.[132] In cultured granulosa cell complex and oocytes from aged cows' early antral follicles, RSV increased Sirt1 protein expression and promoted autophagy, as demonstrated by increased Beclin-1 and LC3-II levels and decreased p62 (a protein that is selectively degraded by autophagy). Through this, RSV improved cell growth and upregulated mitochondrial biogenesis and degradation (Figure 6.2B).[133] The promoted effect of RSV on autophagy was abolished by co-treatment with Sirt1 inhibitor EX527(Selisistat, $C_{13}H_{13}ClN_2O$), suggesting RSV's role in autophagy is Sirt1-dependent.[133] Dependent on the Sirt1-AMPK signaling pathway, RSV exhibited a neuroprotective role in rat with spinal cord injury, which is a devastating condition of aging-related neurodegenerative disease.[134] Furthermore, activation of Sirt1's deacetylase activity by RSV was required for autophagy induction in human cells, *Saccharomyces cerevisiae*, and *C. elegans.*[135] At the level of acetylproteome, RSV provoked decrease in the acetylation of 560 lysine-containing motifs from 375 specific proteins, among which 170 proteins overlap with recently identified entire human autophagy protein network.[135]

The role of RSV in aging *via* autophagy has been confirmed by specific knockdown of essential autophagy proteins. In nematodes, silencing Beclin-1, which is an essential autophagic modulator, abrogated RSV's beneficial effect on longevity.[132] In addition, deletion of Beclin-1/Atg6 prevented the induction of autophagy by RSV's activation of Sirt1.[132]

6.4.3 Adiposity

Obesity is a rapidly growing epidemic and originates from an imbalance between energy intake and energy expenditure.[40] In humans and other mammals, obesity is characterized by excessive accumulation of WAT mass, which functions as a major energy reservoir to store redundant energy in the form of triglyceride.[136] In obese persons, assays of protein and mRNA levels of essential autophagy markers indicated that autophagy was significantly enhanced compared with their lean counterparts and is positively related with degree of obesity.[137] In cellular and mice models, specific deletion of an essential protein (Atg5) for autophagy abolished adipogenesis and developed fewer subcutaneous mature adipocytes.[138] In mice adipose tissue, deletion of autophagy gene Atg7 improved metabolism, leading to a leanness phenotype and protection from diet-induced obesity.[41,136]

Further, in a study of obese men 30 days of RSV supplementation (150 mg day^{-1}) increased adipogenesis and prominently reduced the percentage of large adipocytes, resulting in contractible mean adipocyte size and improved insulin sensitivity.[15] Microarray analysis revealed that RSV administration induced signaling of lysosomal/phagosomal pathway and transcription factor EB (TFEB), suggesting an alternative approach of lipid droplet breakdown by autophagy (Figure 6.3).[15] TFEB is an essential regulator of autophagy and

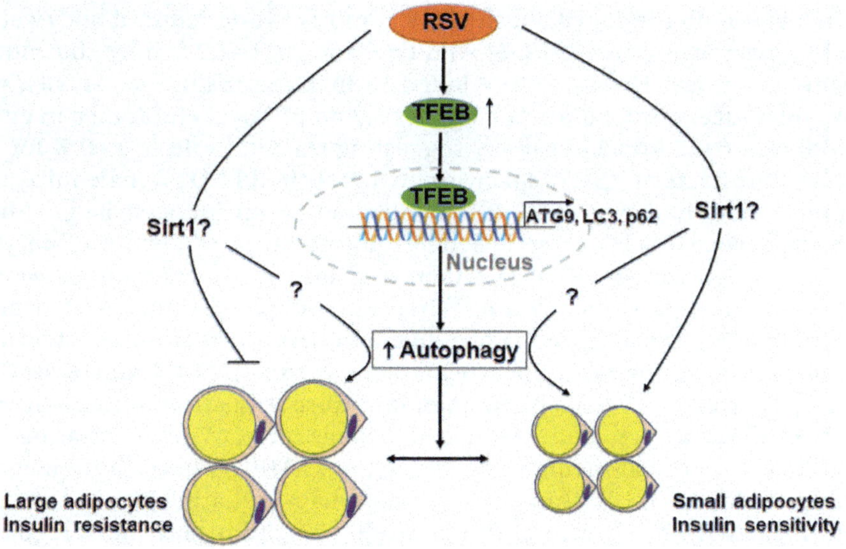

Figure 6.3 Resveratrol (RSV) regulates autophagy in adipocyte formation and expansion. RSV decreases the adipocyte size but paradoxically increases adipogenesis, improving insulin sensitivity in adipocytes. These changes were associated with activation of transcription factor EB (TFEB), a transcription factor regulating autophagy *via* mediating lysosomal and phagosomal pathways (*e.g.*, ATG9, LC3, and p62 gene expressions). Sirt1 was shown to suppress adipogenesis, whereas a direct role of Sirt1 in adipose autophagy and RSV-induced adipose changes remains to be defined (indicated by question marks).

subjects lysosomal biogenesis to autophagy.[139,140] During adipocyte differentiation, TFEB expression was simultaneously increased with autophagic activity.[139] Autophagy has been implicated in lipid droplet breakdown, which may explain the decreased adipocyte size by RSV.[41] However, a previous *in vitro* study of 3T3L1 preadipocytes indicated that RSV (12.5–50.0 μM) decreased adipogenesis and lipid accumulation.[141] In differentiated 3T3L1 mature adipocytes, RSV dose-dependently (0–100.0 μM) reduced lipid accumulation, which is dependent on Sirt1.[24] However, the opposite effect was observed in differentiating cells: when the concentrations of RSV were 1.0 μM and 10.0 μM, RSV increased adipogenic marker PPARγ expression and lipid accumulation[142] presumably because of differences in the dosage, duration, and stages of differentiation of RSV treatment. Further studies regarding how RSV regulates adipose autophagy at different dosage, duration, and stages of differentiation should be done to validate the effects of autophagy on adipose differentiation *in vitro* and *in vivo*.

Functionally and morphologically distinct from WAT, BAT is characterized by dense mitochondrial innervation, and vascular content, which is responsible for non-shivering thermogenesis, dissipating energy as heat.[25] Promoting BAT function is considered as a promising therapeutic approach to combat obesity.[143] The antiobesity effect of RSV partially contributes to its induction of browning of WAT.[144]

In high-fat diet-induced obesity, RSV administration reduced body weight and fat percentage in C57BL/6J wild-type and AMPKα2$^{-/-}$ mice, but not in AMPKα1$^{-/-}$ mice. However, RSV failed to increase insulin sensitivity, glucose tolerance, mitochondrial biogenesis, and physical endurance in either AMPK-α1 or -α2 knockout mice, suggesting the vital role of AMPK for the metabolic effects of RSV.[145] Likewise, in 5-month-old CD1 female mice with high-fat diet challenge, 0.1% RSV supplementation induced beige adipogenesis *in vivo*; this is characterized by induced beiging of white adipocytes, increased UCP1 expression, and increased fatty acid oxidation, as well as elevated AMPK phosphorylation.[144] In differentiated inguinal WAT stromal vascular cells, RSV induced brown-like adipocyte formation, as shown by overexpression of brown adipocyte markers, such as UCP1, Prdm16, PGC1α, and cytochrome c.[144] This phenomenon is absent in stromal vascular cells with AMPKα1 defects and AMPKα1 knockout mice,[146] demonstrating that AMPKα1 is a critical mediator of RSV action. AMPK induces autophagy by phosphorylation of Ulk1 (mammalian autophagy-initiating kinase 1) to activate its activity.[147] Therefore, future studies regarding how RSV regulates autophagy in obesity through AMPK-Ulk1 cascade will be of great value.

6.4.4 Cardiovascular Health

Cardiovascular diseases contribute to crucial morbidity and mortality in the developed countries, including the United States.[148] Therefore, it is urgent to develop preventive measures to lower the incidence of heart issues. Autophagy plays a vital role in the myocardium, among which it is active in basal cells to maintain contractile function of heart. In ischaemia-reperfusion injury, autophagy plays a protective role in the myocardium and cardiac cells.[149] Any fault or mutation of autophagic machinery gives rise to cardiac dysfunction that ultimately leads to heart failure.[150] It is presumably believed that the cardioprotective effect of RSV is not because of its direct action on the diseased heart; rather, it potentiates a preconditioning-like effect, a state of the art for cardioprotection.[131]

After hypoxia-reoxygenation or ischaemia-reperfusion of H_9C_2 cardiac myoblast cells, RSV (0.1 μM and 1.0 μM) induced cardiac autophagy.[149] By deactivating the mTOR pathway, RSV induced autophagy, which was correlated with improved cell survival and mitigated apoptosis (Figure 6.4A). Likewise, RSV and gamma-tocotrienol acted synergistically to induce autophagy, providing greater cardioprotective effects through the activation of the Akt-Bcl-2 pathway.[151] In left coronary artery ligation mice, high dosage (50 mg kg^{-1} day^{-1} for 2 weeks) but not low dose (5 mg kg^{-1} day^{-1}) of RSV ameliorated left ventricular dilation and significantly improved cardiac function with augmentation of autophagy, which is dependent on activating AMPK (Figure 6.4A).[152] However, how RSV regulates autophagy to combat DOX (a powerful antibiotic generally used in clinical cancer chemotherapy)-induced damage to hearts of mice is controversial. On the one hand, a study has shown that RSV (5.0 μM or 50.0 μM) or caloric restriction could induce autophagy (assessed by protein levels of Atg5, p62, and LC3-II/LC3I ratio) to combat DOX-induced damage to hearts of mice.[153] On the other hand, Xu *et al.* showed that RSV (> 15.0 μM) inhibited DOX-induced autophagy to reduce cardiomyocyte cell death.[154] Although it remains to be defined what caused the controversy, the methodology used to measure autophagy activity could be confounding. It is known that the steady-state levels of autophagy-related proteins are insufficient to reflect autophagy flux and may cause misinterpretation of autophagy activity. Therefore, it is important to block the degradation of autophagic vacuoles, leading to a time-dependent accumulation of autophagosomes in the cells or tissues to measure autophagic flux.[155]

The cardioprotective role of RSV *via* autophagy was further validated by specific autophagy inhibitor and gene-silencing technology. In H_9C_2 cardiac myoblast cells, the beneficial effects of RSV on improved cell survival and mitigated apoptosis were significantly attenuated by autophagy inhibitors, wortmannin and 3-MA (an inhibitor of autophagy's early steps).[149] Similarly, CQ (an autophagy inhibitor) elicited opposite effects to RSV, including exacerbation of cardiac remodeling associated with inhibition of autophagy.[152] However, the inhibitory effect of RSV on DOX-induced autophagy was abolished by Beclin-1 overexpression.[154]

6.4.5 Hepatic Metabolism

Over the years, extensive investigations into hepatic autophagy have indicated that autophagy machinery is not only important for maintaining liver homeostasis in physiologically normal conditions, but also for response to pathological states, such as metabolic dysregulation, infection, proteotoxicity, and

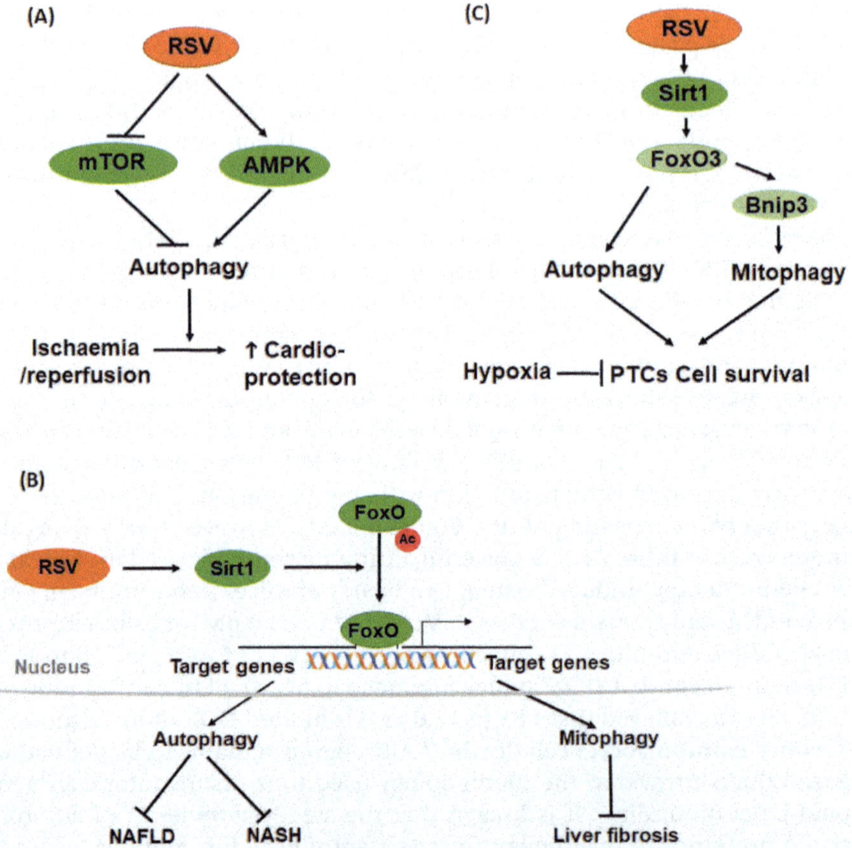

Figure 6.4 Resveratrol (RSV) regulates autophagy/mitophagy in heart, liver, and kidney. (A) RSV induces autophagy by deactivating mammalian target of rapamycin (mTOR) and activating AMP-activated protein kinase (AMPK), promoting cardiomyocyte cell survival from ischemia /reperfusion. (B) RSV activates Sirt1 to deacetylate FoxOs, inducing autophagy and mitophagy in the liver. The induced autophagy/mitophagy may protect against non-alcoholic fatty liver disease (NAFLD), non-alcoholic steatohepatitis (NASH), and liver fibrosis. (C) Under hypoxia conditions, RSV activates Sirt1 to deacetylate Forkhead box O3 (FoxO3), which induces autophagy and BCL2/adenovirus E1B 19-kDa interacting protein 3 (Bnip3)-dependent mitophagy to improve renal proximal tubular cell (PTC) survival.

even carcinogenesis. Dysfunction of autophagy has been implicated in the pathogenesis of several liver diseases, such as NAFLD, non-alcoholic steatohepatitis (NASH), viral hepatitis, and liver cancer (Figure 6.4B).[156] In NAFLD, hepatic steatosis was accompanied by decreased autophagy.[8] NASH could primarily progress into chronic liver disease and result in liver cirrhosis, fibrosis, and eventually hepatocellular carcinoma, which were accompanied by decreased autophagy.[157] A previous study showed that acute ethanol consumption is advantageous to liver health, which is ascribed to its induction of autophagy to reduce ethanol-induced liver injury.[10] All of this evidence suggests that pharmaceutical manipulation of autophagy might possess vast therapeutic meaning in liver diseases.

RSV has the competence to alleviate NAFLD, in which autophagy plays a protective role against NAFLD development.[8] In hepatocytes, RSV induced autophagy to ameliorate hepatic steatosis. Moreover, RSV increased the expression of Sirt1 in hepatocytes, as well as in HepG2 cells.[8] Incubation with Sirt1 inhibitor EX-527 or Sirt1 siRNA abrogated RSV-mediated autophagy. An *in vivo* study in mice with hepatic steatosis obtained parallel results.[8] In addition, RSV's role in NAFLD partially contributed to its action on the deacetylation activity of Sirt1 (Figure 6.4B).[158] In methionine/choline-deficient (MCD) diet-induced NASH animal model or medium-induced hepatic steatosis cell model, the phenotype of elevated levels of triglyceride, inflammatory factors TNFα, IL-1β, IL-6, alanine aminotransferase (ALT), and thiobarbituric acid reactive substances (TBARS) was ameliorated by RSV administration and accompanied by uplifted expressions of autophagy proteins, both *in vitro* and *in vivo*.[157] As to the beneficial effect of ethanol on liver, in cultured primary mouse hepatocytes and in liver ethanol significantly increased both mRNA and protein levels of many autophagy-related molecules.[10] At the same time, ethanol increased the nuclear translocation of FoxO3a and its target gene expression. These effects were diminished by overexpression of a dominant negative form of FoxO3a, indicating FoxO3a's positive role in regulating ethanol-induced autophagy and cell survival.[10] RSV, acting as Sirt1 agonist, further promoted ethanol-induced autophagy-related gene expressions through Sirt1's deacetylation of FoxO3a to increase its nuclear translocation (Figure 6.4B).[10]

GRX is an activated hepatic stellate cell model and is widely used to study the development and resolution of liver fibrosis. RSV (1.0–50.0 μM) treatment promoted cell death signals of GRX, along with improved mitochondrial dynamics and function. Accompanied with this, RSV treatment promoted autophagy/mitophagy, indicating that the cytotoxic effect of RSV on GRX may be due to the concomitant induction of apoptosis, autophagy/mitophagy, and mitochondrial biogenesis in GRX.[107]

Mitochondria are known to be involved in cholestatic liver injury. In a rat model of cholestasis developed by bile duct ligation surgery, RSV administration upregulated LC3-II levels, and reversed the decline in mitochondrial DNA copy number and downregulated Caspase 3 expression. In addition, RSV significantly decreased the numbers of apoptotic liver cells, indicating that early RSV treatment reverses impaired liver function, which may be partially through autophagy.[159]

To confirm the protective role of RSV on liver health *via* autophagy, pharmaceutical inhibition of autophagy proteins or genetic knockdown of autophagy genes has been done. In hepatocytes, the protective role of RSV on hepatic steatosis was abolished by 3-MA.[8] Similarly, CQ significantly mitigated RSV's protection against liver injury and inflammation, suggesting the indispensable role of autophagy through which RSV mitigated hepatic steatosis and inflammation in MCD-induced NASH.[157] *In vivo*, in wild-type and autophagic mediator ULK1 heterozygous knockout mice, 8 weeks of high-fat diet induced NAFLD. RSV (50 mg kg^{-1} daily for 4 weeks) ameliorated all measured features of NAFLD in wild-type mice, such as hepatic histology, fibrosis, oxidative status, and inflammation. However, in ULK1 heterozygous knockout mice the ameliorative effects were abrogated, indicating that the effects of RSV on NAFLD-caused hepatic injury improvement partially rely on regulating autophagic pathway.[160]

6.4.6 Renal Health

In kidney, plentiful studies suggest the vital role of autophagy in organ maintenance, disease, and aging. Hypoxic, ischemic, toxic, immunological, and oxidative insults would affect autophagy in renal epithelial cells to regulate the course of various kidney diseases.[161] In aged kidney, caloric restriction-induced mitochondrial autophagy was essential for cell survival under hypoxia.[162] Therefore, elucidating the essential role of autophagy in kidney physiology and diseases proposes unique strategies to kidney disease prevention and treatment.

RSV can protect against age-related or other types of renal diseases, including sepsis-related injury, drug-induced injury, diabetic nephropathy, ischemia-reperfusion injury, aldosterone-induced injury, and unilateral ureteral obstruction through its activation of Sirt1.[9] In those renal diseases, Sirt1 regulates several cellular events, including mitochondrial biogenesis, glucose/lipid metabolism, inflammation, apoptosis, autophagy, and adaptations to cellular stress, through the deacetylation of target proteins.[9] In aged kidney, long-term caloric restriction increased Sirt1 protein expression and attenuated mitochondrial and renal damage by hypoxia through Sirt1-mediated FoxO3 deacetylation, which was vital for Bnip3 expression and for subsequent autophagy in primary renal proximal tubular cells (Figure 6.4C). However, caloric restriction could not induce autophagy and followed improvement of damaged mitochondria in the kidneys of aged $Sirt1^{-/-}$ mice, highlighting the crucial role of the Sirt1–FoxO3 axis in cellular adaptation to hypoxia.[162]

6.5 Conclusion

Mitochondrial health underpins metabolic hemostasis in the cell.[163,164] The phytochemical RSV can activate the energy sensor Sirt1, which plays a key role in mitochondrial regulation, including biogenesis and autophagy. The Sirt1–mitochodnria axis is prevalent in metabolic regulation across tissues, and its dysregulation has been implicated in inflammation, aging, diabetes, and obesity. By contrast, RSV may reverse the changes by activating Sirt1 and restore metabolic homeostasis, demonstrating the potential of RSV to curb metabolic derangements. It should be noted, however, that controversies have arisen from studies that use different experimental models, RSV dosage, and treatment duration.[165–167] Future large-scale clinical trials are warranted to establish the effectiveness as well as toxicity of RSV.

Acknowledgments

This work is supported in part by the American Heart Association (Grant 18TPA34230082 to Z.C.), and the USDA National Institute of Food and Agriculture (Hatch project 1020373 to Z.C.).

References

1. Z. Cheng, E. M. Schmelz, D. Liu and M. W. Hulver, *Mol. Nutr. Food Res.*, 2014, **58**, 1739–1749.
2. Z. Cheng, S. Guo, K. Copps, X. Dong, R. Kollipara, J. T. Rodgers, R. A. Depinho, P. Puigserver and M. F. White, *Nat. Med.*, 2009, **15**, 1307.
3. K. Szkudelska and T. Szkudelski, *Eur. J. Pharmacol.*, 2010, **635**, 1–8.
4. R. Falchetti, M. P. Fuggetta, G. Lanzilli, M. Tricarico and G. Ravagnan, *Life Sci.*, 2001, **70**, 81–96.
5. M. A. Markus and B. J. Morris, *Clin. Interventions Aging*, 2008, **3**, 331.
6. W. Yu, Y. C. Fu and W. Wang, *J. Cell Biochem.*, 2012, **113**, 752–759.
7. D. C. Rubinsztein, G. Mariño and G. Kroemer, *Cell*, 2011, **146**, 682–695.
8. Y. Zhang, M. l. Chen, Y. Zhou, L. Yi, Y. X. Gao, L. Ran, S. H. Chen, T. Zhang, X. Zhou and D. Zou, *Mol. Nutr. Food Res.*, 2015, **59**, 1443–1457.
9. M. Kitada and D. Koya, *Oxid. Med. Cell Longev.*, 2013, **2013**, 568093.
10. H.-M. Ni, K. Du, M. You and W.-X. Ding, *Am. J. Pathol.*, 2013, **183**, 1815–1825.
11. F. Lang, Z. Qin, F. Li, H. Zhang, Z. Fang and E. Hao, *PloS One*, 2015, **10**, e0129196.
12. K.-F. Hsu, C.-L. Wu, S.-C. Huang, C.-M. Wu, J.-R. Hsiao, Y.-T. Yo, Y.-H. Chen, A.-L. Shiau and C.-Y. Chou, *Autophagy*, 2009, 5, 451–460.
13. Y. Feng, Y. Cui, J.-L. Gao, M.-H. Li, R. Li, X.-H. Jiang, Y.-X. Tian, K.-J. Wang, C.-M. Cui and J.-Z. Cui, *J. Mol. Med.Int. J. Mol. Med.*, 2016, **37**, 921–930.
14. H. Miki, N. Uehara, A. Kimura, T. Sasaki, T. Yuri, K. Yoshizawa and A. Tsubura, *Int. J. Oncol.*, 2012, **40**, 1020–1028.
15. E. Konings, S. Timmers, M. Boekschoten, G. Goossens, J. Jocken, L. Afman, M. Müller, P. Schrauwen, E. Mariman and E. Blaak, *Int. J. Obes.*, 2014, **38**, 470.
16. L. Ma, R. Fu, Z. Duan, J. Lu, J. Gao, L. Tian, Z. Lv, Z. Chen, J. Han and L. Jia, *Pathol.-Res. Pract.*, 2016, **212**, 310–318.
17. B. B. Aggarwal, A. Bhardwaj, R. S. Aggarwal, N. P. Seeram, S. Shishodia and Y. Takada, *Anticancer Res.*, 2004, **24**, 2783–2840.
18. B. Levine, N. Mizushima and H. W. Virgin, *Nature*, 2011, **469**, 323.
19. A. Kumar and S. Chauhan, *Eur. J. Med. Chem.*, 2016, **119**, 45–69.
20. J. Fang, A. Ianni, C. Smolka, O. Vakhrusheva, H. Nolte, M. Krüger, A. Wietelmann, N. G. Simonet, J. M. Adrian-Segarra and A. Vaquero, *Proc. Natl. Acad. Sci. U. S. A.*, 2017, **114**, E8352-E8361.

21. V. Lemos, R. M. De Oliveira, L. Naia, É. Szegö, E. Ramos, S. Pinho, F. Magro, C. Cavadas, A. C. Rego and V. Costa, *Hum. Mol. Genet.*, 2017, **26**, 4105–4117.
22. E. Jing, S. Gesta and C. R. Kahn, *Cell Metab.*, 2007, **6**, 105–114.
23. T. Shi, F. Wang, E. Stieren and Q. Tong, *J. Biol. Chem.*, 2005, **280**, 13560–13567.
24. F. Picard, M. Kurtev, N. Chung, A. Topark-Ngarm, T. Senawong, R. M. de Oliveira, M. Leid, M. W. McBurney and L. Guarente, *Nature*, 2004, **429**, 771.
25. L. Qiang, L. Wang, N. Kon, W. Zhao, S. Lee, Y. Zhang, M. Rosenbaum, Y. Zhao, W. Gu and S. R. Farmer, *Cell*, 2012, **150**, 620–632.
26. C. Cantó, L. Q. Jiang, A. S. Deshmukh, C. Mataki, A. Coste, M. Lagouge, J. R. Zierath and J. Auwerx, *Cell Metab.*, 2010, **11**, 213–219.
27. P. Gao, T.-T. Xu, J. Lu, L. Li, J. Xu, D.-L. Hao, H.-Z. Chen and D.-P. Liu, *J. Mol. Med.*, 2014, **92**, 347–357.
28. Z. Gerhart-Hines, J. T. Rodgers, O. Bare, C. Lerin, S. H. Kim, R. Mostoslavsky, F. W. Alt, Z. Wu and P. Puigserver, *EMBO J.*, 2007, **26**, 1913–1923.
29. Y. Cao, Y. Xue, L. Xue, X. Jiang, X. Wang, Z. Zhang, J. Yang, J. Lu, C. Zhang and W. Wang, *J. Hepatol.*, 2013, **59**, 1299–1306.
30. H.-S. Kim, C. Xiao, R.-H. Wang, T. Lahusen, X. Xu, A. Vassilopoulos, G. Vazquez-Ortiz, W.-I. Jeong, O. Park and S. H. Ki, *Cell Metab.*, 2010, **12**, 224–236.
31. Y. Li, S. Xu, A. Giles, K. Nakamura, J. W. Lee, X. Hou, G. Donmez, J. Li, Z. Luo and K. Walsh, *FASEB J.*, 2011, **25**, 1664–1679.
32. R. R. Alcendor, S. Gao, P. Zhai, D. Zablocki, E. Holle, X. Yu, B. Tian, T. Wagner, S. F. Vatner and J. Sadoshima, *Circ. Res.*, 2007, **100**, 1512–1521.
33. C.-P. Hsu, I. Odewale, R. R. Alcendor and J. Sadoshima, *Biol. Chem.*, 2008, **389**, 221–231.
34. Y. Ruan, C. Dong, J. Patel, C. Duan, X. Wang, X. Wu, Y. Cao, L. Pu, D. Lu and T. Shen, *Cell. Physiol. Biochem.*, 2015, **35**, 1116–1124.
35. I. Çakir, M. Perello, O. Lansari, N. J. Messier, C. A. Vaslet and E. A. Nillni, *PloS One*, 2009, **4**, e8322.
36. A. Z. Herskovits and L. Guarente, *Neuron*, 2014, **81**, 471–483.
37. H. Jeong, D. E. Cohen, L. Cui, A. Supinski, J. N. Savas, J. R. Mazzulli, J. R. Yates III, L. Bordone, L. Guarente and D. Krainc, *Nat. Med.*, 2012, **18**, 159.
38. D. J. Selkoe, *J. Biol. Chem.*, 1996, **271**, 18295–18298.
39. W. Qin, T. Yang, L. Ho, Z. Zhao, J. Wang, L. Chen, W. Zhao, M. Thiyagarajan, D. MacGrogan and J. T. Rodgers, *J. Biol. Chem.*, 2006, **281**, 21745–21754.
40. L. Liu, P. Zou, L. Zheng, L. E. Linarelli, S. Amarell, A. Passaro, D. Liu and Z. Cheng, *Cell Death Dis.*, 2015, **6**, e1586.
41. R. Singh, Y. Xiang, Y. Wang, K. Baikati, A. M. Cuervo, Y. K. Luu, Y. Tang, J. E. Pessin, G. J. Schwartz and M. J. Czaja, *J. Clin. Invest.*, 2009, **119**, 3329–3339.
42. S. Kajimura, B. M. Spiegelman and P. Seale, *Cell Metab.*, 2015, **22**, 546–559.
43. X. Liu, S. Wang, Y. You, M. Meng, Z. Zheng, M. Dong, J. Lin, Q. Zhao, C. Zhang and X. Yuan, *Endocrinology*, 2015, **156**, 2461–2469.

44. M. Chondronikola, E. Volpi, E. Børsheim, C. Porter, M. K. Saraf, P. Annamalai, C. Yfanti, T. Chao, D. Wong and K. Shinoda, *Cell Metab.*, 2016, **23**, 1200–1206.
45. A. Chalkiadaki and L. Guarente, *Cell Metab.*, 2012, **16**, 180–188.
46. J. N. Feige, M. Lagouge, C. Canto, A. Strehle, S. M. Houten, J. C. Milne, P. D. Lambert, C. Mataki, P. J. Elliott and J. Auwerx, *Cell Metab.*, 2008, **8**, 347–358.
47. A. S. Banks, N. Kon, C. Knight, M. Matsumoto, R. Gutiérrez-Juárez, L. Rossetti, W. Gu and D. Accili, *Cell Metab.*, 2008, **8**, 333–341.
48. R. Mayoral, O. Osborn, J. McNelis, A. M. Johnson, C. L. Izquierdo, H. Chung, P. Li, P. G. Traves, G. Bandyopadhyay and A. R. Pessentheiner, *Mol. Metab.*, 2015, **4**, 378–391.
49. H. Abdesselem, A. Madani, A. Hani, M. Al-Noubi, N. Goswami, H. B. Hamidane, A. M. Billing, J. Pasquier, M. S. Bonkowski and N. Halabi, *J. Biol. Chem.*, 2016, **291**, 2119–2135.
50. M. P. Gillum, M. E. Kotas, D. M. Erion, R. Kursawe, P. Chatterjee, K. T. Nead, E. S. Muise, J. J. Hsiao, D. W. Frederick and S. Yonemitsu, *Diabetes*, 2011, **60**, 3235–3245.
51. M. K. Nøhr, N. Bobba, B. Richelsen, S. Lund and S. B. Pedersen, *Int. J. Mol. Sci.*, 2017, **18**, 1006.
52. M. Boutant, M. Joffraud, S. S. Kulkarni, E. García-Casarrubios, P. M. García-Roves, J. Ratajczak, P. J. Fernández-Marcos, A. M. Valverde, M. Serrano and C. Cantó, *Mol. Metab.*, 2015, **4**, 118–131.
53. B. J. Gurd, *Appl. Physiol. Nutr. Metab.*, 2011, **36**, 589–597.
54. M. Suwa, H. Nakano, Z. Radak and S. Kumagai, *Metab. Clin. Exp.*, 2008, **57**, 986–998.
55. B. J. Gurd, C. G. Perry, G. J. Heigenhauser, L. L. Spriet and A. Bonen, *Appl. Physiol. Nutr. Metab.*, 2010, **35**, 350–357.
56. A. Philp and S. Schenk, *Exercise Sport Sci. Rev.*, 2013, **41**, 174.
57. R. Amat, A. Planavila, S. L. Chen, R. Iglesias, M. Giralt and F. Villarroya, *J. Biol. Chem.*, 2009, **284**, 21872–21880.
58. S. Nemoto, M. M. Fergusson and T. Finkel, *J. Biol. Chem.*, 2005, **280**, 16456–16460.
59. B. Chabi, P. J. Adhihetty, M. F. O'Leary, K. J. Menzies and D. A. Hood, *J. Appl. Physiol.*, 2009, **107**, 1730–1735.
60. A. Philp, A. Chen, D. Lan, G. A. Meyer, A. N. Murphy, A. E. Knapp, I. M. Olfert, C. E. McCurdy, G. R. Marcotte and M. C. Hogan, *J. Biol. Chem.*, 2011, **286**, 30561–30570.
61. B. J. Gurd, Y. Yoshida, J. Lally, G. P. Holloway and A. Bonen, *J. Physiol.*, 2009, **587**, 1817–1828.
62. Z. Shen, X. Liang, C. Q. Rogers, D. Rideout and M. You, *Am. J. Physiol.: Gastrointest. Liver Physiol.*, 2009, **298**, G364–G374.
63. H. Yin, M. Hu, X. Liang, J. M. Ajmo, X. Li, R. Bataller, G. Odena, S. M. Stevens and M. You, *Gastroenterology*, 2014, **146**, 801–811.
64. A. Purushotham, T. T. Schug, Q. Xu, S. Surapureddi, X. Guo and X. Li, *Cell Metab.*, 2009, **9**, 327–338.

65. R.-H. Wang, C. Li and C.-X. Deng, *Int. J. Biol. Sci.*, 2010, **6**, 682.
66. F. Xu, Z. Gao, J. Zhang, C. A. Rivera, J. Yin, J. Weng and J. Ye, *Endocrinology*, 2010, **151**, 2504–2514.
67. X. Q. Deng, L. L. Chen and N. X. Li, *Liver Int.*, 2007, **27**, 708–715.
68. Z. Gao, J. Zhang, I. Kheterpal, N. Kennedy, R. J. Davis and J. Ye, *J. Biol. Chem.*, 2011, **286**, 22227–22234.
69. S. E. Choi, S. Kwon, S. Seok, Z. Xiao, K.-W. Lee, Y. Kang, X. Li, K. Shinoda, S. Kajimura and B. Kemper, *Mol. Cell. Biol.*, 2017, **37**, e00006-17.
70. D. A. Guertin, D. M. Stevens, C. C. Thoreen, A. A. Burds, N. Y. Kalaany, J. Moffat, M. Brown, K. J. Fitzgerald and D. M. Sabatini, *Dev. Cell*, 2006, **11**, 859–871.
71. R.-H. Wang, H.-S. Kim, C. Xiao, X. Xu, O. Gavrilova and C.-X. Deng, *J. Clin. Invest.*, 2011, **121**, 4477–4490.
72. D. M. Erion, S. Yonemitsu, Y. Nie, Y. Nagai, M. P. Gillum, J. J. Hsiao, T. Iwasaki, R. Stark, D. Weismann and X. X. Yu, *Proc. Natl. Acad. Sci. U. S. A.*, 2009, **106**, 11288–11293.
73. E. D. B. Danz, J. Skramsted, N. Henry, J. A. Bennett and R. S. Keller, *Free Radic. Biol. Med.*, 2009, **46**, 1589–1597.
74. S. Oka, R. Alcendor, P. Zhai, J. Y. Park, D. Shao, J. Cho, T. Yamamoto, B. Tian and J. Sadoshima, *Cell Metab.*, 2011, **14**, 598–611.
75. M. Vinciguerra, M. P. Santini, C. Martinez, V. Pazienza, W. C. Claycomb, A. Giuliani and N. Rosenthal, *Aging Cell*, 2012, **11**, 139–149.
76. J. A. Baur, K. J. Pearson, N. L. Price, H. A. Jamieson, C. Lerin, A. Kalra, V. V. Prabhu, J. S. Allard, G. Lopez-Lluch and K. Lewis, *Nature*, 2006, **444**, 337.
77. C.-P. Hsu, P. Zhai, T. Yamamoto, Y. Maejima, S. Matsushima, N. Hariharan, D. Shao, H. Takagi, S. Oka and J. Sadoshima, *Circulation*, 2010, **122**, 2170–2182.
78. A. Planavila, R. Iglesias, M. Giralt and F. Villarroya, *Cardiovasc. Res.*, 2010, **90**, 276–284.
79. T. Kawashima, Y. Inuzuka, J. Okuda, T. Kato, S. Niizuma, Y. Tamaki, Y. Iwanaga, A. Kawamoto, M. Narazaki and T. Matsuda, *J. Mol. Cell. Cardiol.*, 2011, **51**, 1026–1036.
80. T.-M. Lu, J.-Y. Tsai, Y.-C. Chen, C.-Y. Huang, H.-L. Hsu, C.-F. Weng, C.-C. Shih and C.-P. Hsu, *J. Biomed. Sci.*, 2014, **21**, 57.
81. A. Prola, J. P. Da Silva, A. Guilbert, L. Lecru, J. Piquereau, M. Ribeiro, P. Mateo, M. Gressette, D. Fortin and C. Boursier, *Cell Death Differ.*, 2017, **24**, 343.
82. N. Hariharan, Y. Maejima, J. Nakae, J. Paik, R. A. DePinho and J. Sadoshima, *Circ. Res.*, 2010, **107**, 1470–1482.
83. A. Planavila, E. Dominguez, M. Navarro, M. Vinciguerra, R. Iglesias, M. Giralt, S. Lope-Piedrafita, J. Ruberte and F. Villarroya, *J. Mol. Cell. Cardiol.*, 2012, **53**, 521–531.
84. A. Kuno, Y. S. Hori, R. Hosoda, M. Tanno, T. Miura, K. Shimamoto and Y. Horio, *J. Biol. Chem.*, 2013, **288**, 5963–5972.
85. C. Zhang, Y. Feng, S. Qu, X. Wei, H. Zhu, Q. Luo, M. Liu, G. Chen and X. Xiao, *Cardiovasc. Res.*, 2011, **90**, 538–545.
86. G. Ramadori, C. E. Lee, A. L. Bookout, S. Lee, K. W. Williams, J. Anderson, J. K. Elmquist and R. Coppari, *J. Neurosci.*, 2008, **28**, 9989–9996.

87. H. Y. Cohen, C. Miller, K. J. Bitterman, N. R. Wall, B. Hekking, B. Kessler, K. T. Howitz, M. Gorospe, R. de Cabo and D. A. Sinclair, *Science*, 2004, **305**, 390–392.
88. D. Chen, J. Bruno, E. Easlon, S.-J. Lin, H.-L. Cheng, F. W. Alt and L. Guarente, *Genes Dev.*, 2008, **22**, 1753–1757.
89. M. Lu, D. A. Sarruf, P. Li, O. Osborn, M. Sanchez-Alavez, S. Talukdar, A. Chen, G. Bandyopadhyay, J. Xu and H. Morinaga, *J. Biol. Chem.*, 2013, **288**, 10722–10735.
90. M. Jiang, J. Wang, J. Fu, L. Du, H. Jeong, T. West, L. Xiang, Q. Peng, Z. Hou and H. Cai, *Nat. Med.*, 2012, **18**, 153.
91. D. Kim, M. D. Nguyen, M. M. Dobbin, A. Fischer, F. Sananbenesi, J. T. Rodgers, I. Delalle, J. A. Baur, G. Sui and S. M. Armour, *EMBO J.*, 2007, **26**, 3169–3179.
92. F. Ng, L. Wijaya and B. L. Tang, *Front. Cell. Neurosci.*, 2015, **9**, 64.
93. S. Hisahara, S. Chiba, H. Matsumoto, M. Tanno, H. Yagi, S. Shimohama, M. Sato and Y. Horio, *Proc. Natl. Acad. Sci. U. S. A.*, 2008, **105**, 15599–15604.
94. V. A. Rafalski, P. P. Ho, J. O. Brett, D. Ucar, J. C. Dugas, E. A. Pollina, L. M. Chow, A. Ibrahim, S. J. Baker and B. A. Barres, *Nat. Cell Biol.*, 2013, **15**, 614.
95. W. Guo, L. Qian, J. Zhang, W. Zhang, A. Morrison, P. Hayes, S. Wilson, T. Chen and J. Zhao, *J. Neurosci. Res.*, 2011, **89**, 1723–1736.
96. L. Zocchi and P. Sassone-Corsi, *Epigenetics*, 2012, 7, 695–700.
97. J. Gao, W.-Y. Wang, Y.-W. Mao, J. Gräff, J.-S. Guan, L. Pan, G. Mak, D. Kim, S. C. Su and L.-H. Tsai, *Nature*, 2010, **466**, 1105.
98. J.-S. Kim and J.-Y. Park, *J. Exerc. Nutr. Biochem.*, 2019, **23**, 7.
99. S. Davinelli, N. Sapere, M. Visentin, D. Zella and G. Scapagnini, *Immun. Ageing*, 2013, **10**, 28.
100. Q. Zhao, Z. Tian, G. Zhou, Q. Niu, J. Chen, P. Li, L. Dong, T. Xia, S. Zhang and A. Wang, *Theranostics*, 2020, **10**, 4822–4838.
101. M. M. Cao, X. Lu, G. D. Liu, Y. Su, Y. B. Li and J. Zhou, *Exp. Ther. Med.*, 2018, **15**, 576–584.
102. R. B. Devereux, M. J. Roman, M. Paranicas, M. J. O'Grady, E. T. Lee, T. K. Welty, R. R. Fabsitz, D. Robbins, E. R. Rhoades and B. V. Howard, *Circulation*, 2000, **101**, 2271–2276.
103. S. Ma, J. Feng, R. Zhang, J. Chen, D. Han, X. Li, B. Yang, X. Li, M. Fan and C. Li, *Oxid. Med. Cell Longev.*, 2017, **2017**, 4602715.
104. X. Zhang, Y. Feng, Y. Wang, J. Wang, D. Xiang, W. Niu and F. Yuan, *Life Sci.*, 2018, **207**, 234–245.
105. Y.-C. Chuang, S.-D. Chen, C.-Y. Hsu, S.-F. Chen, N.-C. Chen and S.-B. Jou, *Int. J. Mol. Sci.*, 2019, **20**, 998.
106. Y.-g. Li, W. Zhu, J.-p. Tao, P. Xin, M.-y. Liu, J.-b. Li and M. Wei, *Biochem. Biophys. Res. Commun.*, 2013, **438**, 270–276.
107. L. A. M. Martins, M. Q. Vieira, M. Ilha, M. de Vasconcelos, H. B. Biehl, D. B. Lima, V. Schein, F. Barbé-Tuana, R. Borojevic and F. C. R. Guma, *Cell Biochem. Biophys.*, 2015, **71**, 657–672.
108. S. Jäger, C. Handschin, J. St-Pierre and B. M. Spiegelman, *Proc. Natl. Acad. Sci. U. S. A.*, 2007, **104**, 12017–12022.

109. M. Lagouge, C. Argmann, Z. Gerhart-Hines, H. Meziane, C. Lerin, F. Daussin, N. Messadeq, J. Milne, P. Lambert and P. Elliott, *Cell*, 2006, **127**, 1109–1122.
110. K. Higashida, S. H. Kim, S. R. Jung, M. Asaka, J. O. Holloszy and D.-H. Han, *PLoS Biol.*, 2013, **11**, e1001603.
111. G. Ashrafi and T. Schwarz, *Cell Death Differ.*, 2013, **20**, 31.
112. A. M. Andres, A. Stotland, B. B. Queliconi and R. A. Gottlieb, *J. Mol. Cell. Cardiol.*, 2015, **78**, 62–72.
113. K. Palikaras, E. Lionaki and N. Tavernarakis, *Nature*, 2015, **521**, 525–528.
114. R. Sebori, A. Kuno, R. Hosoda, T. Hayashi and Y. Horio, *Oxid. Med. Cell Longev.*, 2018, **2018**, 9179270.
115. J. Wu, X. Li, G. Zhu, Y. Zhang, M. He and J. Zhang, *Exp. Cell Res.*, 2016, **341**, 42–53.
116. A. Kuno, R. Hosoda, R. Sebori, T. Hayashi, H. Sakuragi, M. Tanabe and Y. Horio, *Sci. Rep.*, 2018, **8**, 15555.
117. S. Cao, Z. Shen, C. Wang, Q. Zhang, Q. Hong, Y. He and C. Hu, *Food Funct.*, 2019, **10**, 344–354.
118. S. Goldman, Y. Zhang and S. V. Jin, *Autophagy*, 2010, **6**, 179–181.
119. B. Levine and G. Kroemer, *Cell*, 2008, **132**, 27–42.
120. M.-I. Chen, L. Yi, X. Jin, X.-Y. Liang, Y. Zhou, T. Zhang, Q. Xie, X. Zhou, H. Chang and Y.-j. Fu, *Autophagy*, 2013, **9**, 2033–2045.
121. X. Liao, J. C. Sluimer, Y. Wang, M. Subramanian, K. Brown, J. S. Pattison, J. Robbins, J. Martinez and I. Tabas, *Cell Metab.*, 2012, **15**, 545–553.
122. U. Švajger and M. Jeras, *Int. Rev. Immunol.*, 2012, **31**, 202–222.
123. Y. P. Chang, S. M. Ka, W. H. Hsu, A. Chen, L. K. Chao, C. C. Lin, C. C. Hsieh, M. C. Chen, H. W. Chiu and C. L. Ho, *J. Cell. Physiol.*, 2015, **230**, 1567–1579.
124. M. C. Haigis and D. A. Sinclair, *Annu. Rev. Pathol. Mechan. Dis.*, 2010, **5**, 253–295.
125. R. J. Colman, R. M. Anderson, S. C. Johnson, E. K. Kastman, K. J. Kosmatka, T. M. Beasley, D. B. Allison, C. Cruzen, H. A. Simmons and J. W. Kemnitz, *Science*, 2009, **325**, 201–204.
126. K. Pallauf and G. Rimbach, *Ageing Res. Rev.*, 2013, **12**, 237–252.
127. A. Meléndez, Z. Tallóczy, M. Seaman, E.-L. Eskelinen, D. H. Hall and B. Levine, *Science*, 2003, **301**, 1387–1391.
128. I. Bjedov, J. M. Toivonen, F. Kerr, C. Slack, J. Jacobson, A. Foley and L. Partridge, *Cell Metab.*, 2010, **11**, 35–46.
129. A. Del Roso, S. Vittorini, G. Cavallini, A. Donati, Z. Gori, M. Masini, M. Pollera and E. Bergamini, *Exp. Gerontol.*, 2003, **38**, 519–527.
130. F. Madeo, N. Tavernarakis and G. Kroemer, *Nature Cell Biol.*, 2010, **12**, 842.
131. G. Petrovski, N. Gurusamy and D. K. Das, *Ann. N. Y. Acad. Sci.*, 2011, **1215**, 22–33.
132. E. Morselli, M. Maiuri, M. Markaki, E. Megalou, A. Pasparaki, K. Palikaras, A. Criollo, L. Galluzzi, S. Malik and I. Vitale, *Cell Death Dis.*, 2010, **1**, e10.

133. M. Sugiyama, R. Kawahara-Miki, H. Kawana, K. Shirasuna, T. Kuwayama and H. Iwata, *J. Reprod. Dev.*, 2015, **61**, 251–259.
134. H. Zhao, S. Chen, K. Gao, Z. Zhou, C. Wang, Z. Shen, Y. Guo, Z. Li, Z. Wan and C. Liu, *Neuroscience*, 2017, **348**, 241–251.
135. G. Mariño, E. Morselli, M. V. Bennetzen, T. Eisenberg, E. Megalou, S. Schroeder, S. Cabrera, P. Bénit, P. Rustin and A. Criollo, *Autophagy*, 2011, **7**, 647–649.
136. Y. Zhang, S. Goldman, R. Baerga, Y. Zhao, M. Komatsu and S. Jin, *Proc. Natl. Acad. Sci. U. S. A.*, 2009, **106**, 19860–19865.
137. J. Kovsan, M. Blüher, T. Tarnovscki, N. Klöting, B. Kirshtein, L. Madar, I. Shai, R. Golan, I. Harman-Boehm and M. R. Schön, *J. Clin. Endocrinol. Metab.*, 2011, **96**, E268-E277.
138. R. Baerga, Y. Zhang, P.-H. Chen, S. Goldman and S. V. Jin, *Autophagy*, 2009, **5**, 1118–1130.
139. L. Liu, Z. Tao, L. D. Zheng, J. P. Brooke, C. M. Smith, D. Liu, Y. C. Long and Z. Cheng, *Cell Death Discovery*, 2016, **2**, 16066.
140. C. Settembre and A. Ballabio, *Autophagy*, 2011, **7**, 1379–1381.
141. S. Rayalam, J. Y. Yang, S. Ambati, M. A. Della-Fera and C. A. Baile, *Phytother. Res.*, 2008, **22**, 1367–1371.
142. P. Hu, L. Zhao and J. Chen, *Eur. J. Nutr.*, 2015, **54**, 569–579.
143. S. R. Farmer, *Nature*, 2009, **458**, 839.
144. S. Wang, X. Liang, Q. Yang, X. Fu, C. J. Rogers, M. Zhu, B. Rodgers, Q. Jiang, M. V. Dodson and M. Du, *Int. J. Obes.*, 2015, **39**, 967.
145. J.-H. Um, S.-J. Park, H. Kang, S. Yang, M. Foretz, M. W. McBurney, M. K. Kim, B. Viollet and J. H. Chung, *Diabetes*, 2010, **59**, 554–563.
146. S. Wang, X. Liang, Q. Yang, X. Fu, M. Zhu, B. Rodgers, Q. Jiang, M. V. Dodson and M. Du, *Mol. Nutr. Food Res.*, 2017, **61**, 1600746.
147. D. Accili and K. C. Arden, *Cell*, 2004, **117**, 421–426.
148. C. J. Murray and A. D. Lopez, *Lancet*, 1997, **349**, 1436–1442.
149. N. Gurusamy, I. Lekli, N. V. Gorbunov, M. Gherghiceanu, L. M. Popescu and D. K. Das, *J. Cell. Mol. Med.*, 2009, **13**, 373–387.
150. D. Fu, *Eur. Rev. Med. Pharmacol. Sci.*, 2015, **19**, e6.
151. I. Lekli, D. Ray, S. Mukherjee, N. Gurusamy, M. K. Ahsan, B. Juhasz, I. Bak, A. Tosaki, M. Gherghiceanu and L. M. Popescu, *J. Cell. Mol. Med.*, 2010, **14**, 2506–2518.
152. H. Kanamori, G. Takemura, K. Goto, A. Tsujimoto, A. Ogino, T. Takeyama, T. Kawaguchi, T. Watanabe, K. Morishita and M. Kawasaki, *Am. J. Pathol.*, 2013, **182**, 701–713.
153. D. Dutta, J. Xu, M. L. Dirain and C. Leeuwenburgh, *Free Radic. Biol. Med.*, 2014, **74**, 252–262.
154. X. Xu, K. Chen, S. Kobayashi, D. Timm and Q. Liang, *J. Pharmacol. Exp. Ther.*, 2012, **341**, 183–195.
155. C. Chen-Scarabelli, P. R. Agrawal, L. Saravolatz, C. Abuniat, G. Scarabelli, A. Stephanou, L. Loomba, J. Narula, T. M. Scarabelli and R. Knight, *J. Geriatr. Cardiol.*, 2014, **11**, 338.

156. J. L. Schneider and A. M. Cuervo, *Nat. Rev. Gastroenterol. Hepatol.*, 2014, **11**, 187–200.
157. G. Ji, Y. Wang, Y. Deng, X. Li and Z. Jiang, *Lipids Health Dis.*, 2015, **14**, 134.
158. Y. Colak, A. Yesil, H. H. Mutlu, O. T. Caklili, C. Ulasoglu, E. Senates, M. Takir, O. Kostek, Y. Yilmaz and F. Yilmaz Enc, *J. Gastrointest. Liver Dis.*, 2014, **23**, 311–319.
159. T.-K. Lin, L.-T. Huang, Y.-H. Huang, M.-M. Tiao, K.-S. Tang and C.-W. Liou, *Apoptosis*, 2012, **17**, 871–879.
160. L. Li, J. Hai, Z. Li, Y. Zhang, H. Peng, K. Li and X. Weng, *Food Chem. Toxicol.*, 2014, **63**, 166–173.
161. T. B. Huber, C. L. Edelstein, B. Hartleben, K. Inoki, M. Jiang, D. Koya, S. Kume, W. Lieberthal, N. Pallet and A. Quiroga, *Autophagy*, 2012, **8**, 1009–1031.
162. S. Kume, T. Uzu, K. Horiike, M. Chin-Kanasaki, K. Isshiki, S.-i. Araki, T. Sugimoto, M. Haneda, A. Kashiwagi and D. Koya, *J. Clin. Invest.*, 2010, **120**, 1043–1055.
163. Z. Cheng, Y. Tseng and M. F. White, *Trends Endocrinol. Metab.*, 2010, **21**, 589–598.
164. Z. Cheng, *Trends Endocrinol. Metab.*, 2019, **30**, 658–671.
165. S. Ramírez-Garza, E. Laveriano-Santos, M. Marhuenda-Muñoz, C. Storniolo, A. Tresserra-Rimbau, A. Vallverdú-Queralt and R. Lamuela-Raventós, *Nutrients*, 2018, **10**, 1892.
166. J. M. Pezzuto, *Biomol. Ther.*, 2019, **27**, 1.
167. A. P. Singh, R. Singh, S. S. Verma, V. Rai, C. H. Kaschula, P. Maiti and S. C. Gupta, *Med. Res. Rev.*, 2019, **39**, 1851–1891.

CHAPTER 7

The Role of Intestinal Microbiota and Microbial Metabolites in the Development of Host Metabolic Syndrome

VERONICA LAZAR,[a]* IULIA MAGDALENA VASILESCU,[a,b] ALINA MARIA HOLBAN[a] AND MARIANA CARMEN CHIFIRIUC[a]

[a]Department of Microbiology and Immunology, Faculty of Biology, University of Bucharest, and Research Institute of the University of Bucharest, Bucharest, Romania
[b]National Institute of Infectious Diseases Prof Dr Matei Bals, Bucharest, Romania
*Email: veronica.lazar2009@gmail.com

7.1 Introduction

The emerging role of the intestinal microbiota and microbial metabolites in the host health, as well as for the development of different pathological conditions, including the metabolic syndrome (MS), is one of the hottest current research subjects[1] and a huge potential for the development of novel diagnosis and therapeutic tools targeting gut microbiota is arising. However, the complex causal mechanisms, interactions between microbiota, host and environmental factors (particularly diet) as well as the involved bioactive metabolites

Nutritional Signaling Pathway Activities in Obesity and Diabetes
Food Chemistry, Function and Analysis No. 24
Edited by Zhiyong Cheng

Published by the Royal Society of Chemistry, www.rsc.org

have not been unveiled yet. Moreover, the majority of the performed studies have demonstrated this causality in animal models, and less in humans. Prospective studies, involving the careful monitoring of a selected panel of host and microbial markers through metagenomics, transcriptomics, proteomics and metabolomics assays are needed to establish if the changes in gut microbiota occur before or after the onset of disease.

7.2 Metabolic Syndrome

MS is defined by a set of risk factors (obesity, insulin resistance, high blood pressure, and dyslipidemia) that mutually enhance the risk of Type 2 diabetes and cardiovascular disease.[2] The reason we refer to these pathological states as a syndrome, and we do not consider them as independent risk factors, is that their pathophysiological mechanisms are interconnected.[3]

The spectrum of clinical manifestations in a patient at the time of diagnosis is variable, and may even vary between the members of the same family, making it difficult to establish the genetic determinism within this syndrome. It is considered that the MS represents a combination of cardiometabolic abnormalities that tend to associate, increasing the risk of cardiovascular events. Only a few genes have been indicated to have a sufficiently large impact to explain the association of several pathological features, characteristic for the MS. One such gene is Dyrk1B, responsible for the autosomal-dominant transmission of early-onset MS. Monogenic forms of obesity are linked to mutations in the melanocortin genes MC3R, MC4R (melanocortin receptors 3 and 4) and POMC (proopiomelanocortin produced by hypothalamic neurons), connected with leptin secreted by adipose tissue through a feedback mechanism. ADIPOQ gene mutations are associated with diabetes, high blood pressure, and dyslipidemia. Similarly, in the case of NR3C1, FOXC2, SREBP1, and GNB3 genes, several features of the MS are associated.[4]

Based on data collected by the National Cholesterol Education Program, it is estimated that 32–34% of US adults (31–34% of men and 33–35% of women) suffer from MS.[5,6] At a population of 310 million individuals, this is equivalent to over 100 million American adults affected by this pathology.

Currently, MS is diagnosed if at least three of the following five criteria, described by the National Cholesterol Education Program – Adult Treatment Panel III, are present in the same patient:[7] (1) abdominal circumference $\geq$102 cm in men or 88 cm in women; (2) plasma triglyceride level $\geq$150 mg dL^{-1}; (3) high-density lipoprotein (HDL)-cholesterol level $\leq$40 mg dL^{-1} in men or 50 mg dL^{-1} in women; (4) fasting blood glucose $\geq$110 mg dL^{-1}; and (5) systolic blood pressure $\geq$130 mmHg and/or diastolic blood pressure $\geq$85 mmHg.

7.2.1 Risk Factors

The intestinal microbiota is involved in the processes of regulation of the metabolic pathways of the host organism both directly, through the substances generated in the intestinal lumen, which pass to the circulatory system and produce effects in different target tissues, and indirectly, through some epigenetically active metabolites, which modulate the function of genes in hepatocytes or adipocytes. Under anaerobic conditions, species belonging to the genus *Bacteroides* and to the families Clostridiaceae and Lactobacillaceae, especially *Citrobacter* sp. and *Serratia* sp. strains, produce short-chain fatty acids (SCFA): acetate, propionate, and butyrate in molar ratios that vary from 3:1:1 to 10:2:1. Most SCFAs are metabolized to CO_2. Butyrate is used locally as an energy source for colonocytes, caliciform cells, or Paneth cells and regulates apoptosis, cell differentiation, and changes in the chemical structure of proteins and nucleic acids. In contrast, acetate and propionate pass into circulation, from where they are taken up by the liver and peripheral organs, where they are used as substrate for gluconeogenesis or lipogenesis.[8] Under physiological conditions, SCFA induces glucagon-like peptide (GLP-1 and GLP-2) secretion, and peptide tyrosine (PYY).[9] GLP1 stimulates pancreatic β cells to produce insulin, while PYY inhibits the absorption of nutrients from the intestinal lumen and controls appetite. The microbiota also contributes to fat storage by regulating the farnesoid nuclear X receptor (FXR), the bile acid receptor responsible for regulating bile acid synthesis and intrahepatic triglyceride accumulation.[10]

Recent studies have suggested that non-alcoholic fatty liver disease (NAFLD) is another important feature of the MS. NAFLD brings together a broader spectrum of hepatic disorders, from simple steatosis to steatohepatitis, with the complications caused by fibrosis – liver cirrhosis and increased risk of hepatocellular carcinoma. In the pathogenesis of both diseases, insulin resistance and hypersinsulinemia play a central role. It was reported that 90% of patients with NAFLD have at least one risk factor for MS, while 33% possess all the MS traits. In MS patients, the lipid content of the liver is increased independent of age, sex, and body mass index, and 30–100% of patients with NAFLD suffer from obesity; in obese people, steatosis occurs 4.6 times more frequently than in those of normal weight. There is an independent association between hepatic steatosis and carotid atherosclerotic plaques and endothelial dysfunction, which remains statistically significant even after the correction of risk factors for MS.[11]

Liver steatosis is commonly associated with obesity. Obesity, particularly the accumulation of visceral adiposity, is a component of the MS.[12] A study conducted by Ndumele *et al.*[13] suggests that there is an independent association between liver steatosis and systemic inflammation beyond that explained by the presence of obesity and MS. Increased levels of C-reactive protein (CRP) in patients with hepatic steatosis are considered as a marker of long-term cardiovascular risk and may be the basis of the association between hepatic steatosis and cardiovascular disease. The liver is the primary source of CRP and it has been shown that liver steatosis and histologically documented inflammation correlate with the systemic level of inflammatory biomarkers.[13] Abdominal obesity and MS predispose to hepatic steatosis, both by increasing the amount of free fatty acids that reach the liver and by intensifying the lipogenesis associated with hyperinsulinemia. Also, increased insulin resistance associated with hepatic steatosis exacerbates MS. The association between

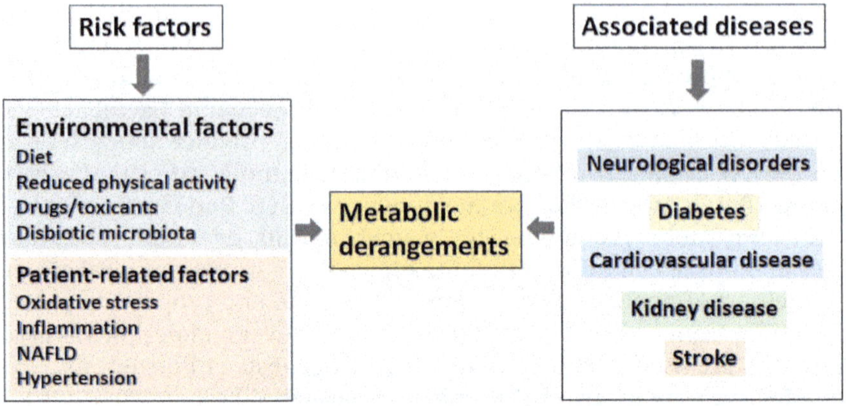

Figure 7.1 Risk factors and diseases associated with the metabolic derangements. Multiple environmental (*i.e.*, diet, reduced physical activity, drugs, toxicants, microbiota) and patient-dependent factors are key inducers and regulators of metabolic derangements. Another very important aspect in metabolic syndrome refers to the associated disorders, which significantly weigh on the outcome. Their impact is supported by increased mortality and morbidity rates in co-affected patients. NAFLD, non-alcoholic fatty liver disease.

liver steatosis, obesity, and cardiometabolic risk factors suggests that hepatic steatosis is another component of the MS. The combination of obesity, hepatic steatosis, and MS increases the probability of biological inflammatory syndrome in terms of CRP values. Figure 7.1 illustrates the main risk factors and diseases associated with MS.

7.2.2 Pathophysiological Mechanisms

The main factor responsible for the pathological consequences in the MS is hyperinsulinemia/insulin resistance, and the main cause of hyperinsulinemia is abdominal obesity and visceral lipid accumulation. Abdominal fat, also called "portal fat tissue," releases excess free fatty acids that reach the liver directly, where they stimulate triglyceride synthesis concomitantly with suppressing insulin clearance, resulting in hyperinsulinemia. The ability to release fatty acids from the portal fat tissue is increased due to the abundance of β-adrenergic receptors, not counterbalanced by the poorly represented α-adrenergic control. Visceral adiposity is, in turn, caused by a diet with high content of fats and refined sweets, under the conditions of a sedentary lifestyle. In the early stages of insulin resistance, the pancreas compensates by increasing insulin secretion, which causes β-cell hypertrophy. In fasting conditions, this basal compensation is sufficient to maintain blood glucose at normal values, but, after food ingestion, when glucose is rapidly absorbed and passes into circulation, this mechanism is exceeded and this is reflected in the modification of laboratory tests. The condition was called "glucose intolerance."[14]

The main insulin-responsive tissues are the liver, skeletal muscle, and adipose tissue, but the rest of the organs also respond to hyperinsulinemia in various ways. For example, in the kidneys sodium reabsorption is increased, which leads to increased intravascular volume and increased blood pressure. The cerebral nuclei respond by altering the natriuretic response to sodium retention and by adrenergic discharge, which amplifies the increase in blood pressure. The ovaries respond by hyperandrogenism, clinically manifested by hirsutism, oligomenorrhea, and polycystic ovary syndrome. In the liver, where large amounts of insulin are required to regulate glucose production, the mechanisms of lipogenesis do not exhibit insulin resistance and are overactivated, with increased very-low-density lipoprotein (VLDL) production, manifested by increased circulating triglyceride levels and decreased HDL-cholesterol concentrations. Thus, multiple manifestations of insulin resistance syndrome are, in fact, the result of a normal metabolic response to increased insulin concentration, induced in order to maintain a normal blood glucose level.[15]

A chronic condition of mild inflammation precedes and accompanies the manifestations of the MS, and because it does not have the features of acute and chronic inflammation that generate tissue injury, it has been called "low-grade inflammation" or "para-inflammation." This proinflammatory status is demonstrated by increased levels of CRP, which is used as a marker for coronary vascular disease in MS, but also as a predictor for NAFLD in correlation with biological analyses related to lipid and carbohydrate metabolism. One possible explanation would be that the adipose tissue of people with MS releases large amounts of cytokines that, in the liver, stimulate CRP synthesis.

Another possibility would be that insulin resistance is responsible for the release of a large cytokine amount. Far from being beneficial, this status is involved in the aggravation of the syndrome, oxidative stress, inflammatory reactions, hypertrophy of cells, and cellular organisms, generating vicious circles that amplify each other, leading to various dysfunctions.[16] It has been shown that chronic inflammation in morbid obesity is due to the continuous activation of the innate immune system. Chronic inflammation in morbid obesity is due to the continuous activation of the innate immune system. Some researchers use the term "metaflammation" to emphasize the presence of a metabolic inflammatory process. Inadequate irrigation of growing fatty tissue leads to inadequate oxygenation and contributes to inflammation. In those patients who cannot maintain benign obesity, hyperplasia and hypertrophy of adipocytes might appear, contributing to the proinflammatory status.

An increase in the level of non-esterified fatty acids reflects the inability of the adipose tissue to take up the entire amount of nutrients. During overload periods, the liver increases the production of particles containing apo-B, which transport triacylglycerols to the adipose tissue, forming low-density lipoproteins (LDLs). This occurs in visceral fat tissue with high efficiency and this form of deposit is ideal for releasing fats when needed. Subcutaneous adipose tissue has an increased capacity to store lipids, having a larger size. This explains the protective effect against MS in men, who, for genetic and hormonal reasons, have a smaller subcutaneous adipose compartment and tend to reach the limit at which deposition in visceral adipose tissue begins much faster. When the storage capacity of both deposits is exceeded, VLDL conversion is disrupted and hypertriglyceridemia occurs. Other tissues are thereafter used for lipid storage, such as the liver, muscles, pancreas, and heart. As these organs do not function specifically as fat stores, their accumulation disrupts their basic function, and lipotoxicity culminates in insulin resistance.[17]

Adipose tissue is the source of some signaling molecules, such as leptin, plasminogen activator inhibitor 1, tumor necrosis factor-α (TNFα), angiotensinogen, and interleukin-6 (IL-6), which, taken together, are called adipokines and contribute to oxidative stress and vascular inflammation. Leptin has procoagulant and antifibrinolytic properties and promotes thrombogenesis and atherogenesis by amplifying inflammation and vascular proliferation. TNFα and leptin may inhibit insulin-mediated glucose uptake. In contrast, adiponectin, also derived from adipose tissue, is low in obese patients and has inhibitory effects on inflammation and promotes insulin sensitivity. Adiponectin decreases the expression of adhesion molecules (intercellular adhesion molecule-1: ICAM-1), adhesion molecules to vascular cells (VCAM-1), P-selectin, on the endothelial surface and decreases the phagocytic activity of macrophages. Resistin, expressed mainly by macrophages and correlated with inflammatory markers, increases VCAM-1 and ICAM-1 expression together with monocyte chemoattractant-1 protein synthesis. Thus, adiponectin and resistin interfere with the adhesion of monocytes to the vascular endothelium, promoting their migration into the subendothelial space, one of the key elements of atherosclerosis.[18] The most important factors driving the endothelial dysfunction in MS are: hyperinsulinemia; hyperglycemia; and free fatty acids.[19]

7.3 Gut Microbiota and Metabolic Health

The human gastrointestinal tract, especially in the colon, is colonized by an abundant autochthonous microbiota, composed of a large number of bacteria, archaea, viruses, and unicellular eukaryotes,[20] exhibiting multiple roles in host physiology. The most predominant members of intestinal microbiota belong to five phyla (*i.e.*, Bacteroidetes, Firmicutes, Actinobacteria, Proteobacteria, and Cerrucomicrobia),[21] the most dominant being Firmicutes and Bacteroidetes.[22] In the healthy gut, anaerobic Bacteroidetes and Firmicutes represent >90% of the total bacterial species, with of course some interindividual variations, dictated by both host genome and environmental factors, such as antibiotic use, lifestyle, hygiene, and diet,[23–25] The gut microbiota contains >10 million genes,[26] by comparison with the eukaryotic cells, which express ~25 000 genes. The present hologenome theory considers that the evolutionary driving forces are not acting on individual organisms, but on the "holobiont," which includes the host organism and its microbiome (genes and metabolites).[27,28] If the host genome is well conserved and more stable, the microbiome is more able to change and adapt in accordance with the environment changes, whether due to novel bacteria intake, through a change of species proportion, or by horizontal gene transfer or mutations. It is well established that not all colonic species are equally represented; it is now known that about 40 bacterial species are dominant and represent approximately 75% of the human gut microbiota. But even if other species are less abundant, they cannot be neglected.[29] If a pressure factor (*e.g.*, antibiotics, a major change in diet) is acting, some dominant species can be drastically reduced, and some previously less abundant species in the new environmental conditions can rapidly adapt and multiply, becoming the new main actors/players leading to another issue, beneficial to the host or detrimental (diarrhea, fat storage, opportunistic infection, or even translocation).

Gut microbiota is involved in the development and regulation of intestinal mucosa permeability and immunity, acts as an anti-infectious barrier, regulates nutrient uptake control, and is involved in food digestion through saccharolytic (carbohydrate metabolism with the production of SCFA) or proteolytic (with the formation of both SCFA and other compounds, such as ammonia, amines, thiols, phenols, indoles) catabolic pathways.[30–34] Some of these microbial metabolites exhibit beneficial roles on the host, while others are potentially toxic (Figure 7.2).

The gut microbiota communicates with the host through direct and indirect means. The first microbial signals transmitted across the intestinal barrier are represented by microbial components, such as lipopolysaccharide (LPS) and peptidoglycans that act as pathogen-associated molecular patterns (PAMPs) and are recognized by the pattern recognition receptors (PRRs) from the intestinal mucosa cells.[35] Following recognition, they activate signaling processes in epithelial and endothelial host cells.[36,37] Following PAMP recognition, the epithelial cells secrete antimicrobial peptides (*e.g.*, as alpha, beta-defensin 2,5, lysozyme), but these defensins are reduced in obese patients.[38] The defensin level could be restored by probiotics which activate the intracellular TLR-MyD88 pathway to restore a certain level of defense.[39]

Gut microbiota can also modulate the host homeostasis indirectly, by the production of bioactive metabolites that can influence other organs directly or indirectly.[40] These metabolites can also interact with other endocrine hormones, such as leptin, ghrelin, glucagon-like peptide 1 (GLP-1), and PYY and stimulate the parasympathetic nervous system, thus affecting glucose homeostasis and other metabolic processes that could lead to MS development.[41–43]

An environmental change could influence both the host and the microbiome, leading to the development of metabolic disease. The host immunity, first through the innate, and second, though the adaptive immune system, represents the first line of adaptation to the changes of microbiome induced by different social, nutritional, chemical, and behavioral factors. Many studies highlight the association of the metabolic disease with a dysbiotic fecal microbiota.[44] Insulin resistance was associated with an abundance of *Prevotella copri* and *Bacteroides vulgatus*, associated with increased blood

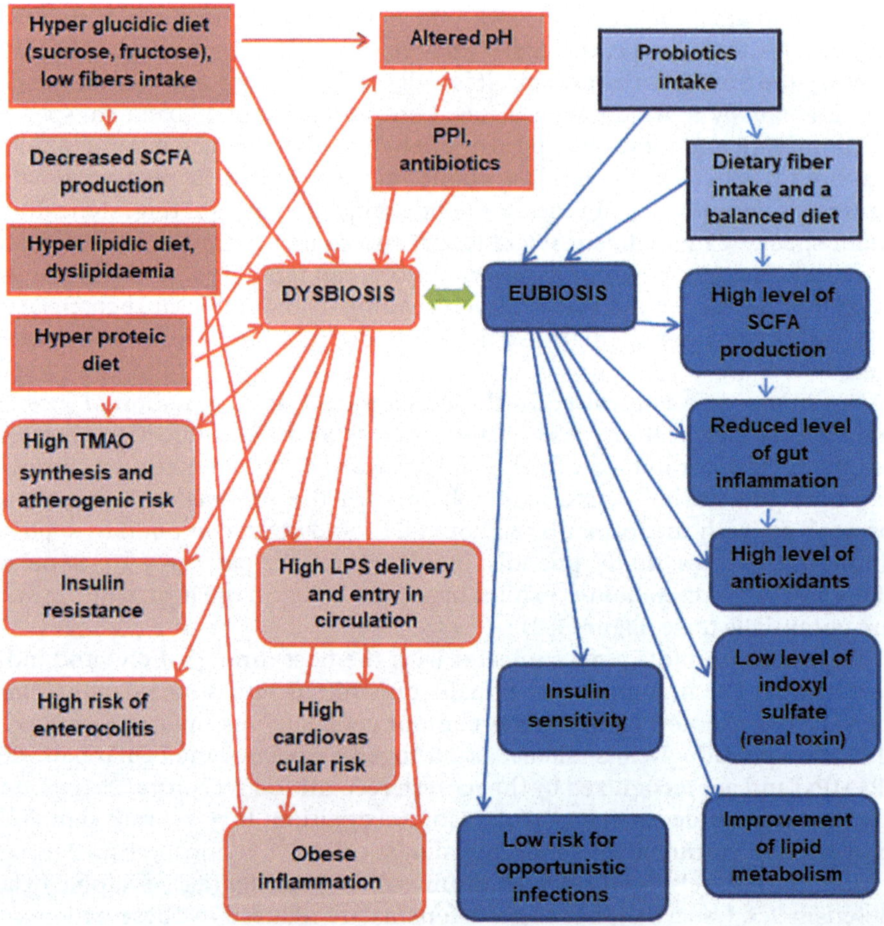

Figure 7.2 The effects of the main dietary substances on microbiota status (dysbiosis *versus* eubiosis) and on the host metabolic conditions which can evolve toward metabolic syndrome. The balance between disbiosis and eubiosis is fragile and strongly influenced by diet preferences which could control the diversity and proportion of gut microbiota species. This would subsequently impact on some physiopathological manifestations, such as insulin resistance, atherogenesis, enterocolitis, obesity, inflammation and cardiovascular diseases which are associated with the metabolic syndrome. Eubiosis could be improved or restored by individual interventions, such as dietary fiber intake and a balanced diet, but also by probiotics containing food intake (functional foods). These individual changes may induce the modulation of some physiopathological parameters, aiming to improve lipid metabolism and insulin sensitivity, to reduce gut inflammation and the risk for opportunistic infections. SCFA, short-chain fatty acids; TMAO, trimethylamine-*N*-oxide; PPI, proton pump inhibitors; LPS, lipopolysaccharides.

levels of branched-chain amino acids.[45] The obese microbiota correlates with a decreased abundance of *Bacteroides thetaiotaomicron* involved in glutamate conversion, leading to increased blood level of glutamate.[46] These changes in the composition of gut microbiota contribute to MS by increasing the energy harvest or by cross-talking with the immune, endocrine, and nervous system through different metabolites and signaling pathways.[47] The dysbiotic microbiota also favors the proliferation of some taxa (*e.g.*, Proteobacteria) and the subsequent translocation of bacterial components such as LPS and whole bacterial cells.[37,48,49] The translocated whole cells and bacterial fragments accumulate in tissues ("tissue microbiota"),[50] stimulate chronic inflammation and subsequently lead to insulin resistance, adipose tissue proliferation, and metabolic disease.

The correlation between gastrointestinal tract microbiota and the MS has been demonstrated by different studies showing that: (1) germ-free mice have a reduced adiposity and improved tolerance to glucose and insulin;[51] (2) genetically modified obese mice as well as obese people have an altered microbiota, consisting in reduced diversity,[52] while an improved metabolic health is associated with increased biodiversity;[53,54] and (3) microbiota transplantation from obese animals/people to lean animals promotes increased adiposity.[53,55–58]

The loss of microbiota biodiversity seems to be more frequent in Western populations, as a consequence of modern lifestyles, medical practices, and processed foods, and this shift in microbiota is potentially correlated with the high frequency of metabolic disease in industrialized countries. This could be explained among others by the alteration of microbial metabolic functions (*e.g.*, decreased production of SCFAs correlated with the extinction of the microbial taxa producing the glycoside hydrolases,[59,60] as a consequence of a decreased intake of dietary fiber. Metabolization of the complex dietary fibers requires numerous enzymes provided by the gut microbiota and takes place in the distal gut, where the transit is slowed down. The complex plant carbohydrates that can be metabolized by the microbiota are known as microbiota-accessible carbohydrates and are involved in gut microbiota modulation and the generation of microbial metabolites.[61] In the literature data regarding the "human gut microbiota in health and disease" topic, an omnipresent idea is to establish the influence of a long-term Western diet, rich in animal proteins, fats, and refined sugars, but there are studies reporting that even short-term diet changes have a certain influence on intestinal microbiota.[29] Due to the high level of cellular density and diversity, it is difficult to study microbiota modulation under different types of diet (with the exception of clinical trials on specific pathological conditions). The desideratum for personalized nutrition should be based on very precise studies on the influence of certain foods, in a certain amount, in certain combinations, with a certain rhythmicity of administration on healthy people, but it is difficult to enroll healthy people in such a scientific approach (not taking into account the temptation of the *forbidden fruit*). In this context, an educational effort would be needed regarding a balanced and diverse nutrition with nutritional principles able to satisfy the complex physiology of the host, but also a diverse, balanced, or eubiotic microbiota, which in turn is supported by its own metabolism in the physiology of the host; at the same time the problem of physical activity cannot be neglected, as it has a huge impact on host health (neuroendocrine equilibrium, immunity, weight control, microbiota, *etc.*).

7.4 Role of Microbial Metabolites in the Development of Host Metabolic Syndrome

Gut microbiota produces diverse primary and secondary metabolites,[62] which are dependent to a high degree on diet and can accumulate in the intestine or diffuse systemically, co-metabolized by the host, and then, be excreted through urine.[63]

7.4.1 Short-chain Fatty Acids

Among these metabolites, SCFAs seem to be involved in obesity and MS. SCFAs are the products of dietary fiber fermentation and represent (particularly butyrate) an important energetic source (adenosine triphosphate: ATP) for colonocytes. Type 2 diabetes was associated with a reduced abundance of butyrate-producing bacteria and an increased abundance of *Lactobacillus* spp.[64]

SCFAs (particularly propionate) can enter the intestinal gluconeogenesis pathway, which signals the central nervous system to protect the host from diet-induced obesity and associated glucose intolerance.[65]

SCFAs (butyrate and acetate) are involved in host signaling by inhibiting HDAC (histone deacetylase) or by activating G-protein-coupled receptors such as GPR41 and GPR43, affecting several important processes, including inflammation and enteroendocrine regulation,[66] by triggering the release of the hormone GLP-1.

Rectal and intravenous administration of acetate was associated with increased plasma concentration of PYY and GLP-1 (glucagon-like peptide-1) in humans.[67,68]

7.4.2 Amino Acids

The gut microbiota is involved in the metabolization of several amino acids originating from dietary and endogenous proteins.[69] The most abundant amino acid-fermenting bacterial groups in the small intestine of healthy individuals are *Clostridium*, *Bacillus–Lactobacillus–Streptococcus*, and Proteobacteria,[70] and in the large intestine *Clostridium* and *Peptostreptococcus*.[71] These amino acids can be utilized as precursors for the synthesis of microbial metabolites, including SCFAs. Elevated serum levels for some amino acids, including the branched-chain amino acids (BCAAs) (*i.e.*, leucine, valine, and isoleucine), and the aromatic amino acids (*i.e.*, phenylalanine and tyrosine), have been found in obese, insulin-resistant, and Type 2 diabetic individuals.[72] These amino acid alterations could be potentially involved in the development of metabolic disease, since the administration of BCAAs together with a high-fat diet has led to insulin resistance occurrence in rats,[73] while mice fed with a leucine-depleted diet have shown an improved insulin sensitivity.[74] Moreover, gastric bypass surgery induces a drastic decrease of plasma BCAA (phenylalanine and tyrosine).[75] The serum levels of different amino acids and SCFA were positively correlated with *Prevotella* (which is significantly enriched in obesity), *Alistipes* (also abundant in diabetic patients), and *Barnesiella* (having a pro-diabetogenic effect) and negatively correlated with *Bacteroides* and *Enterococcus*.[76a] We are tempted to believe that the "obese bacteria" are detrimental and must be fought, but, for instance, in the last trimester of pregnancy, these "bad" bacteria become abundant and are able to produce some metabolic and energy storage in fatty tissue changes, contributing to fetus growth and milk production in the maternal organism.[77]

Indole propionic acid (IPA) is a ligand for pregnane X receptor PXR). It has been shown that IPA inhibited the production of TNFα at intestinal level and increased the expression of the junctional protein-coding mRNAs, while PXR-deficient mice exhibited increased gut permeability and upregulation of the Toll-like receptor (TLR) signaling pathway. This data demonstrates that IPA regulates gut barrier integrity through PXR and TLR4 signaling pathway.[78]

Para-cresyl sulfate (PCS) results from tyrosine metabolization by the intestinal microbiota. Four-week administration of PCS to mice induced the occurrence of insulin resistance, loss of fat mass, and ectopic redistribution of lipid in muscle and liver. The mice showed altered insulin signaling in skeletal muscle through ERK1/2 activation. Treatment with the prebiotic arabinoxylo-oligosaccharide induced a decrease of PCS serum levels to the inhibition of *p*-cresol intestinal production, preventing the occurrence of insulin resistance and lipid metabolism alterations.[79]

7.4.3 Microbial Lipopolysaccharides and Inflammation in Metabolic Syndrome

The MS, obesity, and diabetes are all characterized by the occurrence of an inflammatory state, characterized by locally increased levels of proinflammatory cytokines produced by adipose tissue, liver, skeletal muscles, and pancreatic islets, as well as of circulating acute-phase proteins, pro-coagulant factors, cytokines, and chemokines. In MS, the low-grade inflammation revealed by macrophage infiltration, their functional polarization towards the inflammatory M1 phenotype and subsequent production of high levels of proinflammatory cytokines (TNFα, IL-1β, and IL-6)[80] can induce insulin resistance by reducing insulin signaling.[81] In exchange, in lean animal models, an anti-inflammatory cytokine profile dominated by IL-10 produced by M2 macrophages maintains insulin sensitivity.[82] The functional polarization of macrophages is induced by the cytokines produced by T and B cells; interferon-γ and TNFα induce polarization towards M1 while IL4–5–10–13, to M2.[83]

It has been shown that adipose tissue infiltration with CD8-T cells stimulates macrophage accumulation.[84] The proinflammatory state is also induced by the infiltration of IL-17 and IL-21/22-producing cells and other related Th17 populations, which proved to be increased in liver and adipose tissue.[85] Segmented filamentous bacteria may trigger a Th17 response, while *Clostridium* and *Bacteroides fragilis* induce the differentiation of regulatory T cells (Tregs), through butyrate-mediated pathways.[86,87] B cells can also contribute to metabolic inflammation, by presenting antigens, and subsequently activating the T cells.[83] The sIgA contribute essentially to maintenance of intestinal immune defense and influence microbiota composition.[88] It seems that metabolic tissue infiltration with adaptive immune cells is triggered by specific, but yet unknown, antigens, probably originating in the digestive tract.[89] The innate lymphoid cells (ILCs), which do not express specific T-cell receptors respond to bacterial aggression and tissue injury by activating the production of AMP and the overall local immune response, by the production of different cytokines (*e.g.*, IL-6, IL-17, IL-22, granulocyte–macrophage colony-stimulating factor, and TNF). ILCs are also regulating the Treg lymphocytes, which produce the anti-inflammatory cytokine IL-10.[90,91]

The innate lymphoid cells known as mucosa associated invariant T cells (MAIT) are also involved in the rapid response to microbiota,[92] generating a proinflammatory response (with an IL-17 profile) to the presence of certain microbial compounds, such as the riboflavin derivatives. Obese patients have shown a reduced number of circulating MAITs, but instead they have been found infiltrated in the adipose tissue.[93] Animal studies have shown that the αβ-T lymphocytes were very much decreased in the lamina propria of the ileum of high-fat diet-fed mice after only 1 week of diet administration,[94] probably as a consequence of a reduced phagocyte to lymphocyte co-activation.

Inflammation is involved in both the diabetogenic process and in the etiology of the chronic complications of diabetes[95] and targeting inflammatory pathways could represent a promising strategy for the prevention of diabetes and of its micro- and macrovascular complications.

Many studies suggest that alterations of gut microbiota (*e.g.*, the decrease of Gram-positive Firmicutes in favor of Gram-negative Bacteroidetes, overabundance or depletion of certain species, such as *Prevotella copri*) and the occurrence of subsequent metabolic changes are among the triggers of chronic inflammation, through the upregulation of TLR4, inflammatory pathways (c-Jun N-terminal kinase and nuclear factor-kappa B: NF-κB), intestinal permeability, acute-phase proteins and trimethylamine-*N*-oxide (TMAO) levels. Microbial LPS are incorporated up into chylomicrons which are formed from dietary saturated fats and subsequently they promote inflammation in the host that induces insulin resistance. Microbiota-based interventions with probiotic treatment exhibit an anti-inflammatory effect inflammation by restoring the microbiota eubiosis.[96]

L-Carnitine and phosphatidylcholine, found in red meat, are metabolized to TMAs, which in turn are oxidized by the flavin-containing monooxygenase 3 (FMO3) enzyme in the liver generating the atherogenic compound, TMAO.[97] TMAO is mainly eliminated from circulation by renal excretion; therefore, the renal function must be considered when evaluating the TMAO levels in the blood.[98] The TMAO proatherogenic effects are associated with a decrease of reverse cholesterol transport, alterations of tissue cholesterol and sterol metabolism, changes in bile acid composition, pool size and transport in the liver and gut.[99]

Biliary acids are formed by the microbiota from host cholesterol and facilitate the absorption of dietary fat and fat-soluble molecules. Primary biliary acids cholic acid (CA) and chenodeoxycholic acid (CDCA) are metabolized in the lower part of the small intestine and the colon to generate the secondary bile acids, deoxycholic acid (DCA) and lithocholic acid (LCA).[100] LCA activates host signaling through the G-protein-coupled bile acid receptor 1 (GPBAR1; also known as TGR5), with a major impact on host metabolism.[101] Tauro-β-muricholic acid (TβMCA), a naturally occurring FXR antagonist is deconjugated into β-muricholic acid, which prevents the inhibition of the FXR by TβMCA, having as consequences the induction of obesity, steatosis, and impaired tolerance to glucose and insulin.[102] FXR activation in the gut induces the expression of fibroblast growth factor 15, which inhibits the production of cholesterol 7 α-hydroxylase in the liver. Thus, the capacity to metabolize TβMCA links gut microbiota to obesity and insulin resistance.[103] Many studies state that the beneficial effects of bariatric surgery can be partially due to changes that occur in bile acid metabolism, affecting their signaling capacity.

7.5 Therapeutic Interventions Targeting the Gut Microbiota Detrimental Metabolic Activities

The interrelations among gut microbiota, its metabolites, and cardiometabolic diseases has highlighted the gut microbiome as a promising therapeutic target. Gut microbiota modulation can be achieved by many means, including: (1) a controlled diet (*e.g.*, fiber-rich diets that promote SCFA-producing microbiota);[104] (2) administration of prebiotics: oligosaccharides or complex saccharides that could improve glycemia and lipid profiles;[105] (3) administration of probiotics, that could act by pH modulation, antibacterial compound production, and competition with pathogens;[106] (4) microbial enzymes inhibitors, *e.g.*, 1,3 dimethylbutanol, a choline structural analog, that inhibited TMA lyase activity, decreasing TMA and TMAO plasma levels in mice and slowing down the atherosclerosis process;[107] (5) fecal microbiota transplantation, which proved to increase the abundance of butyrate-producing bacteria, such as *Roseburia*; however, the use of this intervention is limited due to the risk of endotoxins or infectious agent transfer, a potential alternative being: (6) the transplantation of a defined microbial consortium;[108] (7) design of a specific vaccine strategy for busting an efficient intestinal immune response to dysbiotic microbiota; and (8) genetic modulation of gut microbial metabolites. The most investigated therapeutic interventions targeting the gut microbiota in order to modulate their metabolic activities towards health and well-being are illustrated in Figure 7.3.

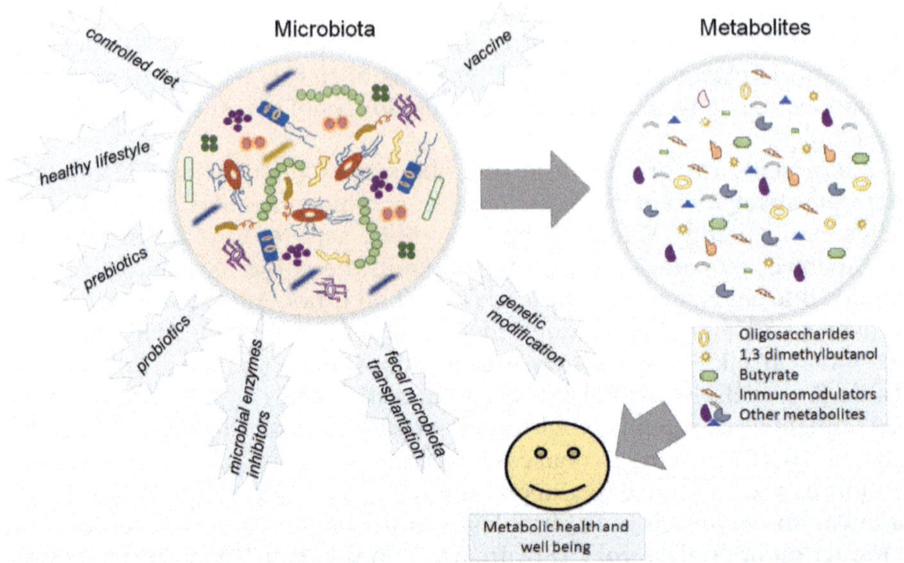

Figure 7.3 Therapeutic interventions targeting the gut microbiota to modulate microbial metabolite production and their influence on metabolic health and well-being. Microbiota composition may be modulated by patient-dependent (*i.e.*, controlled diet, healthy lifestyle, probiotic and prebiotic administration), but also by therapeutic means, such as the administration of new drugs (*i.e.*, microbial enzyme inhibitors), fecal transplantation, vaccines, and genetically modified microorganisms. Such new approaches aim to control the production of microbiota metabolites in a desired direction towards a balanced, healthy condition.

7.6 Conclusions

Intestinal microbiota metabolites exhibit both beneficial and detrimental effects on host metabolism and homeostasis, opening promising avenues for the treatment of metabolic diseases. However, the intimate mechanisms, the molecules, the metabolic/signaling pathways and the specific receptors involved in this complex cross-talk between microbes and host in the presence of varying dietary inputs need to be unveiled in order to be able to develop personalized therapeutic strategies with the capacity to modulate the gut microbiota and the host metabolism in a favorable direction. It must also be taken into account that the maintenance of a healthy microbiota is not an easy task! In order to favorably modulate the microbiota, the diet must meet the following conditions: (1) diversity, because it is the source of: nutrients, fibers, probiotics, prebiotics, vitamins and minerals, phytochemicals, antimicrobials, quorum-sensing molecules and inhibitors; (2) specific amounts depending on age and activity; (3) appropriate quality; and (4) healthy eating habits and lifestyle.

References

1. H. Herrema, R. G. Jzerman and M. Nieuwdorp, *Diabetologia*, 2017, **60**(4), 613–617.
2. A. S. Gami, B. J. Witt, D. E. Howard, P. J. Erwin, L. A. Gami, V. K. Somers and V. M. Montori, *J. Am. Coll. Cardiol.*, 2007, **49**, 403–414.
3. G. Reaven, *Arterioscler. Thromb. Vasc. Biol.*, 2012, **32**, 1754–1759.
4. M. D. Abou Ziki and A. Mani, *Curr. Opin. Lipidol.*, 2016, **27**(2), 162–171.
5. E. S. Ford, W. H. Giles and A. H. Mokdad, *Diabetes Care*, 2004, **27**, 2444–2449.
6. E. S. Ford, *Diabetes Care*, 2005, **28**, 2745–2749.
7. S. Swarup and R. Zeltser, *Metabolic Syndrome*, Start Pearls Publishing, Treasure Island, FL, 2019.
8. I. Rowland, G. Gibson, A. Heinken, *et al.*, *Eur. J. Nutr.* 2018, **57**(1), 1–24.
9. M. J. Zhang, Y. S. Sun, L. Q. Zhao, T. Chen, M. N. Fan, H. C. Jiao, J. P. Zhao, X. J. Wang, F. C. Li, H. F. Li and H. Lin, *Front. Microbiol.*, 2019, **10**, 2176.
10. E.-S. Lee, E.-J. Song and Y.-D. Nam, *J. Clin. Epigenet.*, 2017, **3**,2.
11. P. Paschos and K. Paletas, *Hippokratia*, 2009, **13**(1), 9–19.
12. A. M. Diehl, *Gut*, 2004, **53**(7), 923–924.
13. C. E. Ndumele, K. Nasir, R. D. Conceiçao, J. A. Carvalho, R. S. Blumenthal and R. D. Santos, *Arterioscler. Thromb. Vasc. Biol.*, 2011, **31**(8), 1927–1932.
14. C. K. Roberts, A. L. Hevener and R. J. Barnard, *Compr. Physiol.*, 2013, **3**(1), 1–58.
15. A. Ighbariya and R. Weiss, *J. Clin. Res. Pediatr. Endocrinol.*, 2017, **9**(Suppl 2), 49–57.
16. P. Sharma, *Indian J. Clin. Biochem.*, 2011, **26**(4), 317–318.
17. R. Monteiro and I. Azevedo, *Mediators Inflamm.*, 2010, **2010**, 289645.
18. R. Paoletti, C. Bolego, A. Poli and A. Cignarella, *Vasc. Health Risk Manag.*, 2006, **2**(2), 145–152.
19. A. Cozma, O. Orasan and D. Sampelean, *Rom. J. Intern. Med.*, 2009, **47**(2), 133–140.
20. A. S. Neish, *Gastroenterology*, 2009, **136**, 65–80.
21. T. W. H. Wilson, Y. Kitai and S. L. Hazen, *Circ. Res.*, 2017, **120**(7), 1183–1196.
22. J. L. Sonnenburg and F. Bäckhed, *Nature*, 2016, **535**(7610), 56–64.
23. J. Qin, R. Li, J. Raes, M. Arumugam, K. S. Burgdorf, C. Manichanh, *et al.*, *Nature*, 2010, **464**, 59–65.
24. S. Tamburini, N. Shen, H. C. Wu and J. C. Clemente, *Nat. Med.*, 2016, **22**, 713–722.
25. R. N. Carmody, G. K. Gerber, J. M. Luevano, Jr, D. M. Gatti, L. Somes, K. L. Svenson and P. J. Turnbaugh, *Cell Host Microbe*, 2015, **17**, 72–84.
26. J. Li, H. Jia and X. Cai, *Nat. Biotechnol.*, 2014, **32**, 834–841.
27. R. Burcelin, *Mol. Metab.*, 2016, **5**(9), 771–781.
28. V. Lazar, L.-M. Ditu, G. G. Pircalabioru, I. Gheorghe, C. Curutiu, A. M. Holban, A. Picu, L. Petcu and M. C. Chifiriuc, *Front. Immunol.*, 2018, **9**, 1830.

29. E. Rosenberg and I. Zilber-Rosenberg, *Microbiome*, 2018, **6**, 78.
30. I. Sekirov, S. L. Russell, L. C. Antunes and B. B. Finlay, *Physiol. Rev.*, 2010, **90**, 859–904.
31. F. Backhed, R. E. Ley, J. L. Sonnenburg, D. A. Peterson and J. I. Gordon, *Science,* 2005, **307**, 1915–1920.
32. S. Rakoff-Nahoum, J. Paglino, F. Eslami-Varzaneh, S. Edberg and R. Medzhitov, *Cell*, 2004, **118**, 229–241.
33. S. K. Mazmanian, C. H. Liu, A. O. Tzianabos and D. L. Kasper, *Cell*, 2005, **122**, 107–118.
34. V. Lazar, L. M. Ditu, G. G. Pircalabioru, A. Picu, L. Petcu, N. Cucu and M. C. Chifiriuc, *Front. Nutr.*, 2019, **6**, 21.
35. E. Larsson, V. Tremaroli, Y. S. Lee, O. Koren, I. Nookaew, A. Fricker, J. Nielsen, R. E. Ley and F. Backhed, *Gut*, 2012, **61**, 1124–1131.
36. P. D. Cani, S. Possemiers, T. Van de Wiele, Y. Guiot, A. Everard, O. Rottier, L. Geurts, D. Naslain, A. Neyrinck, D. M. Lambert, G. G. Muccioli and N. M. Delzenne, *Gut*, 2009, **58**, 1091–1103.
37. P. D. Cani, J. Amar, M. A. Iglesias, M. Poggi, C. Knauf, D. Bastelica, A. M. Neyrinck, F. Fava, K. M. Tuohy, C. Chabo, A. Waget, E. Delmee, B. Cousin, T. Sulpice, B. Chamontin, J. Ferrieres, J. F. Tanti, G. R. Gibson, L. Casteilla, N. M. Delzenne, M. C. Alessi and R. Burcelin, *Diabetes*, 2007, **56**, 1761–1772.
38. P. Vora, A. Youdim and L. S. Thomas, *J. Immunol.*, 2004, **173**, 5398–5405.
39. A. Menendez, B. P. Willing and M. Montero, *J. Innate Immun.*, 2013, **5**, 39–49.
40. R. Medzhitov, *Nature*, 2007, **449**, 819–826.
41. K. K. Ryan, V. Tremaroli, C. Clemmensen, P. Kovatcheva-Datchary, A. Myronovych, R. Karns, *et al.*, *Nature*, 2014, **509**, 183–188.
42. V. Tremaroli, F. Karlsson, M. Werling, M. Stahlman, P. Kovatcheva-Datchary, T. Olbers, *et al.*, *Cell Metab.*, 2015, **22**, 228–238.
43. R. J. Perry, L. Peng, N. A. Barry, G. W. Cline, D. Zhang, R. L. Cardone, *et al.*, *Nature*, 2016, **534**, 213–217.
44. T. Clavel, D. Borrmann and A. Braune, *Anaerobe*, 2006, **12**, 140–147.
45. H. K. Pedersen, V. Gudmundsdottir, H. B. Nielsen, T. Hyotylainen, T. Nielsen, B. A. Jensen, *et al.*, *Nature*, 2016, **535**, 376–381.
46. R. Liu, J. Hong, X. Xu, Q. Feng, D. Zhang, Y. Gu, *et al.*, *Nat. Med.*, 2017, **23**, 859–868.
47. X. Li, K. Watanabe and I. Kimura, *Front Immunol.*, 2017, **8**, 1882.
48. J. Amar, C. Chabo and A. Waget, *EMBO Mol. Med.*, 2011, **3**, 559–572.
49. M. Martinez-Medina, J. Denizot and N. Dreux, *Gut*, 2014, **63**, 116–124.
50. J. Lluch, F. Servant and S. Paisse, *PLoS One*, 2015, **10**, e0142334.
51. F. Bäckhed, H. Ding, T. Wang, L. V. Hooper, G. Y. Koh, A. Nagy, *et al.*, *Proc. Natl. Acad. Sci. U. S. A.*, 2004, **101**, 15718–15723.
52. P. J. Turnbaugh, M. Hamady, T. Yatsunenko, B. L. Cantarel, A. Duncan, R. E. Ley *et al.*, *Nature*, 2009, **457**, 480–484.
53. E. Le Chatelier, T. Nielsen, J. Qin, E. Prifti, F. Hildebrand, G. Falony *et al.*, *Nature*, 2013, **500**, 541–546.

54. A. Cotillard, S. P. Kennedy, L. C. Kong, E. Prifti, N. Pons, E. Le Chatelier, *et al.*, *Nature*, 2013, **500**, 585–588.
55. R. E. Ley, F. Bäckhed, P. Turnbaugh, C. A. Lozupone, R. D. Knight and J. I. Gordon, *Proc. Natl. Acad. Sci. U. S. A.*, 2005, **102**, 11070–11075.
56. P. J. Turnbaugh, R. E. Ley, M. A. Mahowald, V. Magrini, E. R. Mardis and J. I. Gordon, *Nature*, 2006, **444**, 1027–1031.
57. V. K. Ridaura, J. J. Faith, F. E. Rey, J. Cheng, A. E. Duncan, A. L. Kau, *et al.*, *Science*, 2013, **341**, 1241214.
58. J. K. Goodrich, J. L. Waters, A. C. Poole, J. L. Sutter, O. Koren, R. Blekhman, *et al.*, *Cell*, 2014, **159**, 789–799.
59. J. Qin, Y. Li, Z. Cai, S. Li, J. Zhu, F. Zhang, *et al.*, *Nature*, 2012, **490**, 55–60.
60. E. D. Sonnenburg, S. A. Smits, M. Tikhonov, S. K. Higginbottom, N. S. Wingreen and J. L. Sonnenburg, *Nature*, 2016, **529**(7585), 212–215.
61. E. D. Sonnenburg and J. L. Sonnenburg, *Cell Metab.*, 2014, **20**, 779–786.
62. M. S. Donia and M. A. Fischbach, *Science*, 2015, **349**, 1254766.
63. T. W. Meyer and T. H. Hostetter, *Kidney Int.*, 2012, **81**, 949–954.
64. F. H. Karlsson, V. Tremaroli, I. Nookaew, G. Bergstrom, C. J. Behre, B. Fagerberg, J. Nielsen and F. Backhed, *Nature*, 2013, **498**, 99–103.
65. F. De Vadder, P. Kovatcheva-Datchary, D. Goncalves, J. Vinera, C. Zitoun, A. Duchampt, *et al.*, *Cell*, 2014, **156**, 84–96.
66. K. M. Maslowski, A. T. Vieira, A. Ng, J. Kranich, F. Sierro, D. Yu, *et al.*, *Nature*, 2009, **461**, 1282–1286.
67. K. R. Freeland and T. M. Wolever, *Br. J. Nutr.*, 2010, **103**, 460–466.
68. H. Herrema, R. G. IJzerman and M. Nieuwdorp, *Diabetologia*, 2017, **60**(4), 613–617.
69. E. P. Neis, C. H. Dejong and S. S. Rensen, *Nutrients*, 2015, 7(4), 2930–2946.
70. Z. L. Dai, G. Wu and W. Y. Zhu, *Front. Biosci.*, 2011, **16**, 1768–1786.
71. C. Allison and G. T. Macfarlane, *Appl. Environ. Microbiol.*, 1989, **55**, 2894–2898.
72. R. Lin, W. Liu, M. Piao and H. Zhu, *Amino Acids*, 2017, **49**(12), 2083–2090.
73. F. Tremblay, M. Krebs, L. Dombrowski, A. Brehm, E. Bernroider, E. Roth, *et al.*, Diabetes, 2005, **54**, 2674–2684.
74. F. Xiao, Z. Huang, H. Li, J. Yu, C. Wang, S. Chen, *et al.*, *Diabetes*, 2011, **60**, 746–756.
75. B. Laferrere, D. Reilly, S. Arias, N. Swerdlow, P. Gorroochurn, B. Bawa, *et al.*, *Sci. Transl. Med.*, 2011, **3**, 80re2.
76. H. Zhang, J. K. DiBaise, A. Zuccolo, D. Kudrna, M. Braidotti, Y. Yu, P. Parameswaran, *et al.*, *Proc. Natl. Acad. Sci. U. S. A.*, 2009, **106**, 2365–2370.
77. O. Koren, J. K. Goodrich, T. C. Cullender, *et al.*, *Cell*, 2012, **150**, 470–480.
78. M. Venkatesh, S. Mukherjee, H. Wang, H. Li, K. Sun, A. P. Benechet, *et al.*, *Immunity*, 2014, **41**, 296–310.
79. L. Koppe, N. J. Pillon, R. E. Vella, *et al.*, *J. Am. Soc. Nephrol.*, 2013, **24**(1), 88–99.
80. S. E. Shoelson, J. Lee and A. B. Goldfine, *J. Clin. Invest.*, 2006, **116**, 1793–1801.

81. A. Bouloumie, C. A. Curat and C. Sengenes, *Curr. Opin. Clin. Nutr. Metab. Care*, 2005, **8**, 347–354.
82. C. N. Lumeng, J. L. Bodzin and A. R. Saltiel, *J. Clin. Invest.*, 2007, **117**, 175–184.
83. D. A. Winer, S. Winer and L. Shen, *Nat. Med.*, 2011, **17**, 610–617.
84. S. Nishimura, I. Manabe and M. Nagasaki, *Nat. Med.*, 2009, **15**, 914–920.
85. J. F. Cavallari, E. Denou and K. P. Foley, *Gut Microbes*, 2016, **7**, 82–89.
86. I. I. Ivanov, K. Atarashi, N. Manel, E. L. Brodie, T. Shima and U.Karaoz, *Cell*, 2009, **139**, 485–498.
87. K. Atarashi, T. Tanoue, T. Shima, A. Imaoka, T. Kuwahara and Y. Momose, *Science*, 2011, **331**, 337–341.
88. K. Suzuki, S. A. Ha, M. Tsuji and S. Fagarasan, *Semin. Immunol.*, 2007, **19**, 127–135.
89. B. Cousin, M. Andre and E. Arnaud, *Biochem. Biophys. Res.*, 2003, **301**, 1016–1022.
90. M. R. Hepworth, L. A. Monticelli and T. C. Fung, *Nature*, 2013, **498**, 113–117.
91. J. Qiu and L. Zhou, *Semin. Immunopathol.*, 2013, **35**, 657–670.
92. R. Salerno-Goncalves, T. Rezwan and M. B. Sztein, *Front. Immunol.*, 2014, **4**, 511.
93. I. Magalhaes, K. Pingris and C. Poitou, *J. Clin. Invest.*, 2015, **125**, 1752–1762.
94. L. Garidou, C. Pomie and P. Klopp, *Cell Metab.*, 2015, **22**, 100–112.
95. M. Moța, S. G. Popa, E. Mota, *et al.*, *J. Diabetes*, 2016, **8**(3), 336–344.
96. L. J. Kasselman, N. A. Vernice, J. DeLeon and A. B. Reiss, *Atherosclerosis*, 2018, **271**, 203–213.
97. Z. Wang, E. Klipfell, B. J. Bennett, R. Koeth, B. S. Levison, B. Dugar, *et al.*, *Nature*, 2011, **472**, 57–63.
98. W. H. Tang, Z. Wang, D. J. Kennedy, Y. Wu, J. A. Buffa, B. Agatisa-Boyle, X. S. Li, B. S. Levison and S. L. Hazen, *Circ. Res.*, 2015, **116**, 448–455.
99. R. A. Koeth, Z. Wang, B. S. Levison, J. A. Buffa, E. Org, B. T. Sheehy, E. B. Britt, X. Fu, Y. Wu, L. Li, J. D. Smith, J. A. DiDonato, J. Chen, H. Li, G. D. Wu, J. D. Lewis, M. Warrier, J. M. Brown, R. M. Krauss, W. H. Tang, F. D. Bushman, A. J. Lusis and S. L. Hazen, *Nat. Med.*, 2013, **19**, 576–585.
100. T. Midtvedt, *Am. J. Clin. Nutr.*, 1974, **27**, 1341–1347.
101. C. Thomas, R. Pellicciari, M. Pruzanski, J. Auwerx and K. Schoonjans, *Nature Rev. Drug Discov.*, 2008, **7**, 678–693.
102. A. Parséus, N. Sommer, F. Sommer, R. Caesar, A. Molinaro, M. Ståhlman, *et al.*, *Gut*, 2017, **66**(3), 429–437.
103. C. Jiang, C. Xie, F. Li, L. Zhang, R. G. Nichols, K. W. Krausz, J. Cai, Y. Qi, Z. Z. Fang, S. Takahashi, N. Tanaka, D. Desai, S. G. Amin, I. Albert, A. D. Patterson and F. J. Gonzalez, *J. Clin. Invest.*, 2015, **125**, 386–402.
104. O. T. Foye, I. F. Huang, C. C. Chiou, W. A. Walker and H. N. Shi, *FEMS Immunol. Med. Microbiol.*, 2012, **65**, 467–480.
105. P. D. Cani, A. M. Neyrinck, F. Fava, C. Knauf, R. G. Burcelin, K. M. Tuohy, G. R. Gibson and N. M. Delzenne, *Diabetologia*, 2007, **50**, 2374–2383.

106. V. Ojetti, E. C. Lauritano, F. Barbaro, A. Migneco, M. E. Ainora, L. Fontana, M. Gabrielli and A. Gasbarrini, *Expert Opin. Drug Metab. Toxicol.*, 2009, **5**, 675–682.
107. Z. Wang, A. B. Roberts, J. A. Buffa, B. S. Levison, W. Zhu, E. Org, X. Gu, Y. Huang, M. Zamanian-Daryoush, M. K. Culley, A. J. DiDonato, X. Fu, J. E. Hazen, D. Krajcik, J. A. DiDonato, A. J. Lusis and S. L. Hazen, *Cell*, 2015, **163**, 1585–1595.
108. J. C. Gregory, J. A. Buffa, E. Org, Z. Wang, B. S. Levison, W. Zhu, M. A. Wagner, B. J. Bennett, L. Li, J. A. DiDonato, A. J. Lusis and S. L. Hazen, *J. Biol. Chem.*, 2015, **290**, 5647–5660.

CHAPTER 8

Metabolomics in Nutritional Metabolism, Obesity, and Diabetes

L. BERVOETS[a] AND P. ADRIAENSENS[b]*

[a]Maastricht University, School of Nutrition and Translational Research in Metabolism (NUTRIM), Department of Medical Microbiology, P. Debyelaan 25, Maastricht, 6229 HX, the Netherlands
[b]Hasselt University, Institute for Materials Research, Applied and Analytical Chemistry, Agoralaan 1, Building D, B-3590 Diepenbeek, Belgium
*Email: peter.adriaensens@uhasselt.be

8.1 Introduction

Metabolomics comprises the study of the complete set of metabolites such as metabolic intermediates, hormones and other signaling molecules, and secondary metabolites, within a biological cell, tissue, organ or organism.[1] This collection of metabolites, also known as the metabolome, is the final product of biochemical reactions that take place within a biological system, starting from gene transcription and translation. In turn, metabolites can influence gene transcription and subsequent processes through many feedback loops. The metabolome is highly dynamic and provides a functional read-out of an organism's molecular phenotype at the time of sampling. It reflects the complex interplay between host genes, microbiota, drugs and environmental

Nutritional Signaling Pathway Activities in Obesity and Diabetes
Food Chemistry, Function and Analysis No. 24
Edited by Zhiyong Cheng

Published by the Royal Society of Chemistry, www.rsc.org

factors such as diet and ageing, and enables the determination of cellular processes altered by disease.[2] This, in combination with low costs on sample basis, has directed metabolomics research into the fields of nutrition,[3] microbiology,[4,5] toxicology[6] and drug discovery,[7] next to a large number of studies exploring the use of metabolomics in the field of disease diagnosis and prognosis.[8–10]

Human-based metabolomics has developed into a routine application for the measurement of endogenous (produced by the host organism, *e.g.* amino acids or lipids) and exogenous (originating from external sources, *e.g.* nutrients, microbiota or drugs) metabolites. The Human Metabolome Project (HMP) recently estimated that the human metabolome consists of more than 18 500 detected and quantified human metabolites,[11] a considerably higher number compared to the 883 in 2007.[12] In addition, 6777 metabolite–single-nucleotide polymorphism (SNP) interactions, 2497 metabolite–drug interactions, 25 570 metabolic pathway maps and 5498 metabolites associated with disease conditions were reported.[11] Unlike the human genome, which is now fully sequenced and freely available, metabolomics is still in its development stage. High-prevalence disease conditions such as diabetes, obesity and cardiovascular disorders are now being examined intensively by metabolomics research.[8,13,14]

Detection, identification and quantification of small-molecule metabolites in a biological sample (fluid, cell or tissue extracts) can be performed by mass spectrometry (MS) or nuclear magnetic resonance (NMR) spectroscopy. Both are well-established high-throughput analytical platforms for application in metabolomics research.[15]

The basic function of MS is to ionize chemical species and sort the ions based on their mass-to-charge ratio. MS can be connected to a chromatographic separation system such as gas- (GC-MS), liquid (LC-MS), capillary electrophoresis (CE-MS) or ion mobility spectrometer (IMS-MS) for optimal detection, or be used without chromatographic separation, *i.e.* direct-infusion electrospray ionization-mass spectrometry (ESI-MS). Although the sensitivity of MS measurements is higher than for NMR spectroscopy, sample preparation for different classes of substances (*i.e.* solvent extractions) lengthens its sample preparation time.[16]

An NMR spectrometer measures the frequency of the resonance of a given nucleus when subjected to a strong magnetic field. While carbon-13 (^{13}C) nuclei can be very useful for metabolite identification by NMR, the proton (^{1}H) nucleus is the most popular nucleus due to its high presence in metabolites, high relative sensitivity and natural abundance of 99.98%. ^{1}H NMR spectroscopy is a very robust, reproducible (> 98%) and non-destructive technique that requires little sample preparation. It can easily quantify metabolites, provide structural information and has low costs per sample. In a single run that only takes a few minutes, the recorded ^{1}H-spectrum of a sample provides information on the concentrations of a number of metabolites.[17]

Fundamentally, NMR spectroscopy and MS provide distinct spectral analysis of the same sample. Due to the lack of sensitivity or bias toward particular metabolites, no single platform is able to identify and quantify the entire range of metabolites within a biological matrix.[18] In order to gain the best metabolite coverage possible in a reasonable time, both techniques can be complementarily used, as has already been demonstrated by other studies.[13,19]

8.2 NMR-based Metabolomics

NMR-based metabolomics of biochemical changes which accompany diseased states originated in the 1980s.[20] In 1984, Nicholson *et al.*[20] showed that resonances for the ketone bodies 3-D-hydroxybutyrate, acetone and acetoacetate can be readily detected in serum, plasma and urine samples from fasting and diabetic subjects by 400-MHz ^{1}H NMR spectroscopy. Ever since, the refinement of methods for the measurement, analysis and interpretation of complex data sets as well as the number of applications of NMR-based metabolomics has expanded dramatically.[21]

^{1}H NMR-based metabolomics is typically performed according to a workflow starting from sample collection, sample preparation, NMR spectroscopic measurements and data analysis (Figure 8.1).[22] To minimize artefacts and

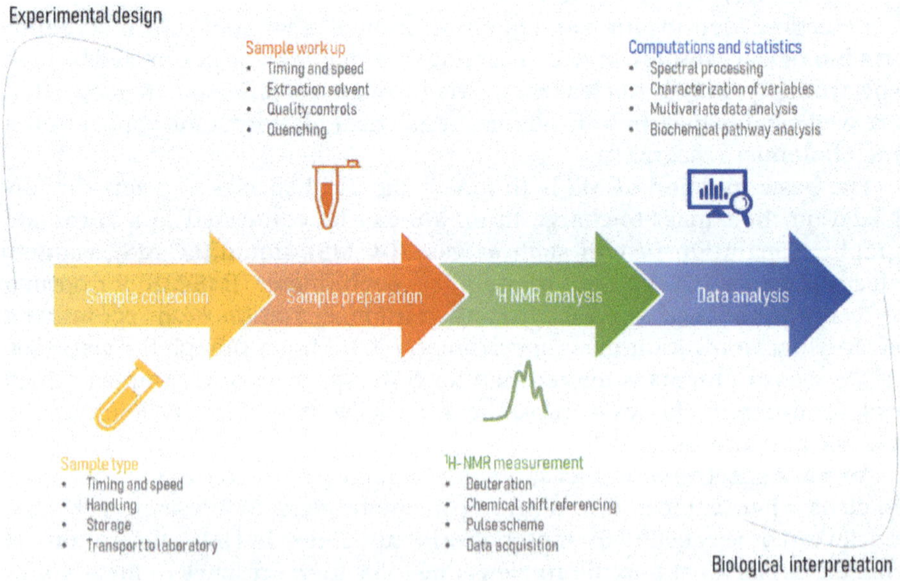

Figure 8.1 ^{1}H Nuclear Magnetic Resonance (NMR)-based metabolomics study workflow: starting from a biological question and experiment design, proceeding through sample collection and preparation, analytical experiments to acquire data, data processing and data analysis followed by biological interpretation. A suitable experimental design is needed for the correct biological interpretation with consideration of factors that can influence the ^{1}H NMR output, such as representativeness of the source material, recovery and preservation of metabolites, metabolome coverage and introduction of preparative and analytical bias.

bias due to biological and experimental variability, all experimental steps should be optimized carefully and performed according to standard operating procedures (SOPs).[23] Fundamentally, ^{1}H NMR spectroscopy demonstrates an inter- and intralaboratory coefficient of variation ranging between 0 and 10%, in contrast to MS, which is much more prone to experimental variation (8–250%).[24] For ^{1}H NMR spectroscopy of blood plasma, the detection limit and the number of metabolites detected are highly dependent on the sample preparation and analysis protocol. Estimates vary, but NMR spectroscopy can routinely detect up to 50–68 serum/plasma metabolites above concentrations of 2–10 µM.[25,26]

Although efforts are currently made within the field of metabolomics to move towards defining SOPs for preanalytical handling,[23] complete standardization of the preanalytical processing is not yet feasible between and within clinical settings. Alternatively, application of a preanalytical sample code that traces and manages these variations would allow sample harmonization. Hereto, the Standard PREanalytical Code (SPREC) was developed within the field of biobanking.[27,28] This is an easy-to-implement and comprehensive tool, consisting of seven elements that document the critical preanalytical details of biospecimens. However, its value within clinical metabolomics remains to be evaluated.

Next, we will describe and comment on specific precautions and improvements that should be taken into account at each experimental stage when performing NMR-based metabolomics research on plasma samples.

8.2.1 Sample Collection and Preparation

Differences in sample collection and preparation can induce preanalytical variation in the metabolic content of samples owing to differences in enzyme activity, reactivity of metabolites or degradation of metabolites. Gathering samples from multiple research centres, clinics or biobanks also adds to an increasing likelihood of discrepancies between sample handling.[29] In addition, since a high number of samples is needed for discovery and validation of metabolomics, automation of sample preparation is desirable. However, variations in NMR applications, sample types, total volume, sample viscosities, tube diameters and sample temperatures all need to be taken into account.[22]

8.2.1.1 Sample Type

NMR spectroscopy has proven to be a very powerful, robust and reproducible technology for metabolic profiling of different diagnostic biofluids. According to the Human Metabolome Database (HMDB), the number of expected and detected metabolites in human serum or plasma is the greatest, followed by urine, saliva, faeces, cerebrospinal fluid and sweat, respectively.[11] While serum traditionally makes up the majority of samples in clinical laboratories, plasma is preferentially used in laboratory medicine because it is more timesaving, has a higher yield and prevents coagulation-induced interferences.[30] Moreover, lithium-heparin is described as the most suitable anticoagulant for ^{1}H NMR analysis.[31]

Of particular interest for those involved in the field of paediatric research is the use of minimally invasive sample types such as urine and faeces, which can be routinely collected. Other minimally invasive samples gaining more attention these days are saliva, exhaled breath and mother's breastmilk. Minimally invasive sampling allows longitudinal sample sets with multiple collections to be obtained with relative ease. In addition, it enables direct capture of metabolite exposure and excretion information.

Although applications of NMR metabolomics are most prevalent in studies using biofluids, this approach has also been used to address a diverse range of applications involving other types of complex samples (*i.e.* cells, tissues, plants, food and beverages, and xenobiotics).[21]

8.2.1.2 Preanalytical Variation

Variations in the metabolite content of samples can have a major impact on the sample quality, impeding interpretation of analytical results, decreasing the credibility of metabolomics research outcomes and potentially affecting clinical implementation of metabolomics.[29,32] Consequently, variability should be under tight control in a clinical or research setting to ensure that differences in the metabolic profile are resulting from the physiological status and not from differences in preanalytical sampling conditions.

Blood measurements are the cornerstone of clinical testing. From an analytical point of view, however, blood plasma is a highly complex matrix. It contains many metabolites, of which albumin, lipoprotein particles and small-molecule metabolites such as glucose are prevailing. Since differences in sampling conditions can influence the concentration of these plasma metabolites, it is highly necessary to develop a well-designed and -controlled ^{1}H NMR study. To evaluate the impact of clinical sources of preanalytical bias on the plasma metabolome, we examined the effect of the following relevant preanalytical conditions on the plasma ^{1}H NMR metabolic profile in 20 healthy volunteers (12 females, median age: 32 years):[33]

- A double lithium-heparin concentration: occurs when blood collection tubes are underfilled, *e.g.* in a paediatric setting
- Short-term exposure to an oxidative atmosphere: occurs when the use of a vacuum collection tube holder cannot be applied, *e.g.* in a paediatric setting
- Haemolysis: occurs when blood drawing techniques are incorrectly applied
- Sample processing delay: 30 minutes, 3 hours or 8 hours
- Centrifugation temperature: at 4 °C or room temperature
- Freezing procedure: at −80 °C, dry ice or liquid nitrogen
- Storage duration: for 2 or 10 months.

Only increasing the sample processing delay from 30 minutes to 3 and 8 hours appeared to have a significant impact on the plasma concentration of pyruvate, lactate and glucose.[33] However, as the interindividual variation is still much larger than this preanalytical variation, the impact on multivariate group classification will be minimal. Nonetheless, it is recommended to keep the time gap between blood collection and centrifugation similar for all samples within a study. Preanalytical conditions that induce variation surpassing the interindividual variation should be avoided or samples collected under such conditions removed from the study.

Preparation of quality control samples can give insight into the amount of preanalytical variance introduced during a study. A pooled sample – made from equal amounts of each sample within a study – can serve as an internal quality control, whereas an external quality control can be produced with plasma samples obtained independently of the study. To ensure that the prepared buffer or samples are not being contaminated during the preparation process, buffer blanks have to be prepared and analysed too.[22]

It is highly recommended to follow a fixed and well-described protocol for sample collection, preparation and storage of ^{1}H NMR studies. Blood has to be drawn according to standard practices and correct techniques. Processing of samples has to be minimized to a maximum of 30 minutes and cooled centrifugation is preferred. Storage of samples at −80 °C for a maximum of 2 months is recommended; however, if the research setting does not allow this, the storage procedure needs to be similar for all samples within a study.[32,33] In order to minimize the impact of the dynamic changes in plasma metabolites induced upon meal consumption, plasma samples are preferably collected upon fasting (at least 8 hours).[34]

Finally, the preanalytical protocols applied in our study were encoded according to SPREC.[27,28] Implementation of SPREC within clinical metabolomics allows for an appropriate sample encoding and exclusion of samples subjected to unwanted, interfering preanalytical conditions. Without doubt, it will contribute to the validation of ^{1}H NMR metabolomics and assure the quality of samples used in clinical, biobank and multicentre research settings.

8.2.2 ^{1}H NMR Spectroscopic Analysis

Proton NMR spectroscopy of biofluids can provide one- or two-dimensional fingerprints suitable for identification and quantification of a large set of metabolites. The choice of suitable pulse schemes for acquisition of NMR spectra is crucial to obtain well-resolved patterns for detection of the largest possible numbers of metabolites.[35]

Most clinical NMR-based metabolomics studies rely on one-dimensional (1D) proton NMR spectra. The narrow chemical shift range of 1D NMR spectra increases the likelihood of overlapping proton resonance signals of specific metabolites. The appearance of signal overlap can be reduced by using stronger magnetic field strengths.[36] Using the appropriate sample preparation in combination with dedicated NMR pulse sequences can contribute to the attenuation of broader macromolecule signals overlying reasonably sharp resonances of smaller metabolites. The ^{1}H NOESY pulse sequence with water presaturation (NOESY-presat) is the most prevalent acquisition scheme to collect complete and quantitative metabolite fingerprints from biofluids. It is a simple sequence scheme that focuses on the reduction of the residual water "hump" and needs minimal parameter optimization.[37] The Carr–Purcell–Meiboom–Gill (CPMG) pulse sequence is mainly used to suppress broad signals of proteins and lipoproteins present in complex biofluids. This sequence scheme extracts broad background signals so that low-concentrated small metabolites can still be detected.[38,39]

In the studies discussed in this chapter, slightly T_2-weighted CPMG spectra (total spin-echo time of 32 ms; interpulse delay of 0.1 ms), preceded by 3 seconds' presaturation for water suppression was used.[33,40–42] Two-dimensional (2D) NMR (*e.g.* JRES, ^{1}H-^{1}H TOCSY or ^{1}H-^{13}C HSQC) spectra are sometimes recorded in parallel with 1D to aid in the assignment of metabolite signals by increasing signal dispersion and elucidating connectivities between signals.[43] However, in large studies the use of 2D NMR is rare because of the increased total instrument time required.[44]

8.2.2.1 Spectral Processing

To achieve optimal characterization of metabolites in the samples, NMR data sets require several steps of data pre-processing prior to multivariate statistics. First, all measured ^{1}H NMR spectra are subjected to phasing, baseline correction and chemical shift referencing – usually 4,4-dimethyl-4-silapentane-1-sulfonic acid (DSS) or trimethylsilylpropanoic acid (TSP) – and alignment using NMR software. Secondly, the spectrum has to be divided into regions of which the integration value (*i.e.* the area under the peak expressed in absolute units or relative to the total peak area, *i.e.* relative concentration) can be used as variable for further analysis. Separation of the NMR spectrum into integration regions can be challenging, in particular in crowded regions with severe signal overlap. Binning or bucketing is a commonly used data pre-processing technique in which the spectrum is segregated to produce such a reduced set of variables.[45] In point-wise or equal-width binning, the spectrum is divided into so-called equally sized bins, *e.g.* of 0.02 or 0.04 ppm. An important limitation of this method is the possible splitting of peaks, resulting in a loss of differentiating power and possibly data misinterpretation. To overcome this, smart bucketing (*i.e.* the segmentation into defined regions) can be applied based on spiking reference samples of a plasma pool with known metabolites (see below: 8.2.2.2 Spiking Methodology). After excluding the water region (4.7–5.2 ppm) and TSP (−0.3 to 0.3 ppm) all integration regions or variables for multivariate statistical analyses are obtained. Lastly, ^{1}H NMR data has to be scaled (*e.g.* Pareto, UV or mean-centred) and normalized (*e.g.* relative to the total integrated area of all regions except water and TSP, probabilistic quotient normalization (PQN), quantile or cubic-spline normalization).[46]

8.2.2.2 Spiking Methodology

The ^{1}H NMR spectral assignment of low-molecular-weight metabolites in biofluids is a time-consuming and complex procedure. It is generally based on the determination of chemical shifts, relative intensities and J-coupling constants of the proton resonance signals to provide structure-specific information for individual metabolites.[47] An NMR spectrometer detects hundreds of peaks, and all of these have to be identified and quantified in order to be able to map the number of metabolites or the final metabolome. To realize a complete and correct peak assignment, spiking experiments can be performed on specific biofluids. Here, we will explain a spiking methodology performed by Louis *et al.*[40]

^{1}H NMR spectra were recorded of several plasma samples prepared by spiking of reference plasma with a known metabolite. The reference plasma was obtained by combining plasma of several blood collections from a healthy person. Hereto, stock solutions of known and relevant metabolites were prepared by dissolving a relevant concentration of a known metabolite in the reference plasma. Next, a small amount of such stock solution was added to a reference plasma NMR sample (*e.g.* 10 μL stock solution to 200 μL reference plasma and 600 μL D_2O containing the TSP reference). This procedure was repeated for all metabolites of interest, using a fresh reference sample for each metabolite. The outcome of these spiking experiments subsequently allows an accurate identification and assignment of the chemical shifts and J-coupling patterns. Using a 400-MHz (9.4 T) NMR spectrometer, the described spiking method led to a segmentation of the NMR spectra in 110 well-defined integration regions. After integration and normalization (*e.g.* relative to the total integrated area, with exclusion of the contributions of TSP and water), these integration regions reflect relative metabolite concentrations and can now be used as variables for multivariate statistical analyses.[40]

8.2.3 Statistical Data Analyses

A significant challenge in metabolomics is the amount of data obtained from a single experiment. To overcome this, metabolomics combines analytical methods with multivariate projection methods. A substantial toolbox of multivariate statistical methods exist; however, we will here only focus on analyses discussed in this chapter. Statistics are mainly carried out by a combination of unsupervised and supervised analyses. Unsupervised principal component analysis (PCA) allows for the visualization of significant intrinsic clusters in the case–control data, identification of possible confounders and detection and removal of outliers in the training data set. Alongside, supervised orthogonal partial least squares discriminant analysis (OPLS-DA) is applied to construct (train) and validate a classification model which enables optimal discrimination between diseased and control subjects.[48]

To define variables strongly contributing to the group discrimination, we applied the following selection procedure, of which three selection criteria have to be fulfilled: significantly different in univariate statistics (a student *t*-test corrected for multiple testing by the Benjamini–Hochberg method);[49] an OPLS-DA absolute value of *p*(corr) (*i.e.* the loading scaled as a correlation coefficient) exceeding 0.5; and an OPLS-DA variable importance for the projection (VIP) value exceeding 0.7. In addition, the OPLS-DA S-line plot visualizes the covariance and correlation coefficient between the variables and the classification score in the model. Strongly discriminating variables combine a clear covariance (*p*[ctr]) with a high absolute value for the correlation scaled loading (abs[*p*(corr)]).[48]

Model characteristics such as the total explained intra- [R^2X(cum)] and intergroup [R^2Y(cum)] variation can be examined together with sensitivity, specificity and misclassification error (MCE) values in order to evaluate the strength performance of the OPLS-DA classifier. Predictive ability (Q^2(cum)) of the model can be demonstrated by cross-validation of the training set as well as by application of the model to an independent validation cohort.[48]

8.3 Application of Metabolomics in Obesity Research

Obesity is now recognized as a non-communicable chronic disease and often originates early in life.[50] Its aetiology is multifactorial and it is generally accepted that obesity is caused by an imbalance between energy intake and energy expenditure. It is a major public health problem with high prevalence numbers worldwide.[51] Especially in children, this disease has to be taken very seriously as children are in the developmental stage and therefore more vulnerable towards the impact of obesity on their health as compared to adults. Paediatric obesity is often accompanied by the development of comorbidities such as insulin resistance, metabolic syndrome, Type 2 diabetes mellitus (T2DM), coronary artery diseases and non-alcoholic fatty liver disease.[52] In addition, respiratory, orthopaedic, sleeping, psychosocial and emotional problems can have an enormous impact on a child's life. The need for effective and culturally appropriate population-based prevention strategies and novel interventions to manage paediatric obesity is now being recognized by the entire healthcare system.[53] As a side note, parents play an important role in facilitating a healthier lifestyle and life quality for their children.[54]

In order to develop effective and personalized prevention and treatment strategies, early screening for obesity and associated comorbidities is of utmost importance. It is well known that not all individuals with the same degree of obesity possess the same cardiometabolic health risks. Consequently, the concept of particular obesity subgroups, *i.e.* metabolically "healthy" obese (MHO), metabolically at-risk obese and metabolically unhealthy obese (MUO), was introduced by clinicians. MHO individuals are clinically obese, but do not exhibit traditional cardiometabolic risk factors.[55] Favourable cardiometabolic profiles are typically characterized by normal blood pressure, serum lipids, insulin and glucose, liver enzymes and hormonal profiles.[56] Previous research has shown that especially young individuals who maintain an MHO status during a follow-up of 5–10 years have a significantly lower risk of developing cardiometabolic complications.[57] Hence, closely monitoring and managing the metabolic health status of obese children over time could contribute to the prevention of future T2DM and cardiovascular disease.

As described above, the clinical definition of metabolic health status is based on traditional/routine clinical laboratory parameters. Focusing only on these clinical parameters may result in loss of information on other possibly important metabolites and their mutual dynamic biochemical interactions. In the area of metabolic health and disease, the field of clinical metabolomics has gained the most momentum and is now being conducted on a very large scale. It can serve as a complementary tool to traditional screening in order to more accurately and comprehensively identify metabolic risk profiles. Since metabolites can be captured in one measurement (especially true for NMR spectroscopy), it also enables fast detection. Furthermore, metabolomics can provide us with mechanistic insights into biomarker discovery and the comprehension of disease progression at metabolic levels. In the end, more targeted and effective screening and intervention strategies for obese metabolic phenotypes can be developed.

The paediatric age group is especially interesting to study because metabolic disturbances that occur in adulthood do often originate early in life. In the first stages of postnatal life, rapid and dynamic processes involving development and health occur. From a biochemical point of view, acquiring knowledge about this "neonatal window" could improve our understanding of various paediatric diseases. Moreover, the time evolution of the metabolic profile in neonates or children could be monitored from birth until later in life. Metabolomics can capture exposures that are notoriously difficult to quantify, and help to further refine phenotypes or specific groups of high-risk children, *e.g.* associated with excess adiposity and related metabolic sequelae over the life course. In this way, metabolomics has the potential to enhance our understanding of obesity/diabetes-related developmental origins of health and disease (DOHaD), to enable the timely implementation of curative interventions and to aid in the development of more individualized and effective treatments. In addition, metabolomics can be simple, safe and non-invasive with high sensitivity and specificity, making it a very suitable tool for application in the paediatric age group.

8.3.1 Metabolic Profiling of Paediatric Obesity

Since the metabolome provides a direct read-out of an organism's clinical phenotype (*i.e.* an indicator of both genetic and environmental perturbations), the interest in metabolomics as a platform to study the underlying biochemical mechanisms of obesity and related metabolic disorders is increasing.[58] Up to now, metabolomics studies focusing on obesity in childhood have been performed mostly with MS as analytical tool and show some conflicting results.[59–61]

Here, we will elaborate on previous research, which comprises the use of ^{1}H NMR spectroscopy as a profiling tool for studying the metabolite phenotype of obesity in children and adolescents. The application of NMR-based metabolomics to investigate biochemical pathways associated with childhood obesity is still underexplored. Possible reasons for this are that obesity was not recognized as a disease for a long time, the complex multifactorial origin of obesity which makes it difficult to study and the paediatric setting (lack of standardized definitions, invasive collection of biological samples and pubertal influences). Besides these clinical challenges, NMR spectroscopy offers excellent stability and integration accuracy (linear response between metabolite concentration and signal intensity/integration value), allowing not only to identify, but also to quantify the metabolites in biological samples. In addition, since there is no need for solvent extractions (as there is for MS), its reproducibility is very high.[17]

In order to profile the plasma metabolome of obese and normal-weight children (aged between 8 and 18 years), we used an untargeted ^{1}H NMR-based metabolomics approach. Detailed protocols regarding sample preparation and ^{1}H NMR analysis have been previously described.[42] A zoom-in of a typical plasma ^{1}H NMR spectrum of an obese and normal-weight child is illustrated in Figure 8.2.

Multivariate data analysis showed that a powerful OPLS-DA model could be built and was able to correctly classify 95.1% (58 out of 61, sensitivity) obese and 91.7% (33 out of 36, specificity) normal-weight individuals (Figure 8.3). The total amount of variation explained by the model *within* the groups was

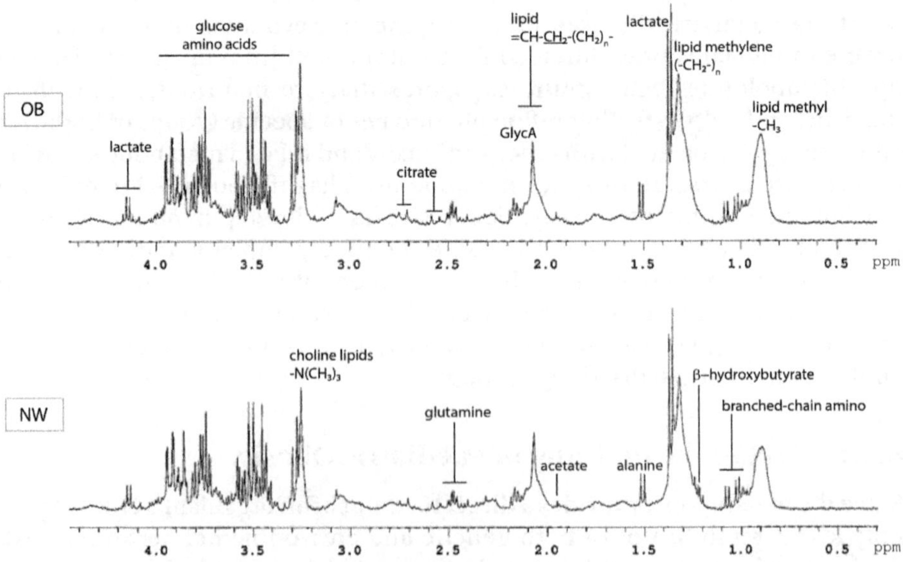

Figure 8.2 400-MHz Carr–Purcell–Meiboom–Gill (CPMG) ^{1}H nuclear magnetic resonance (NMR) spectra of the blood plasma of obese (OB, upper) and normal-weight (NW, lower) children, zoomed-in on the spectral region between 0.5 and 4.5 ppm. GlycA, *N*-acetyl peak from plasma glycoproteins.

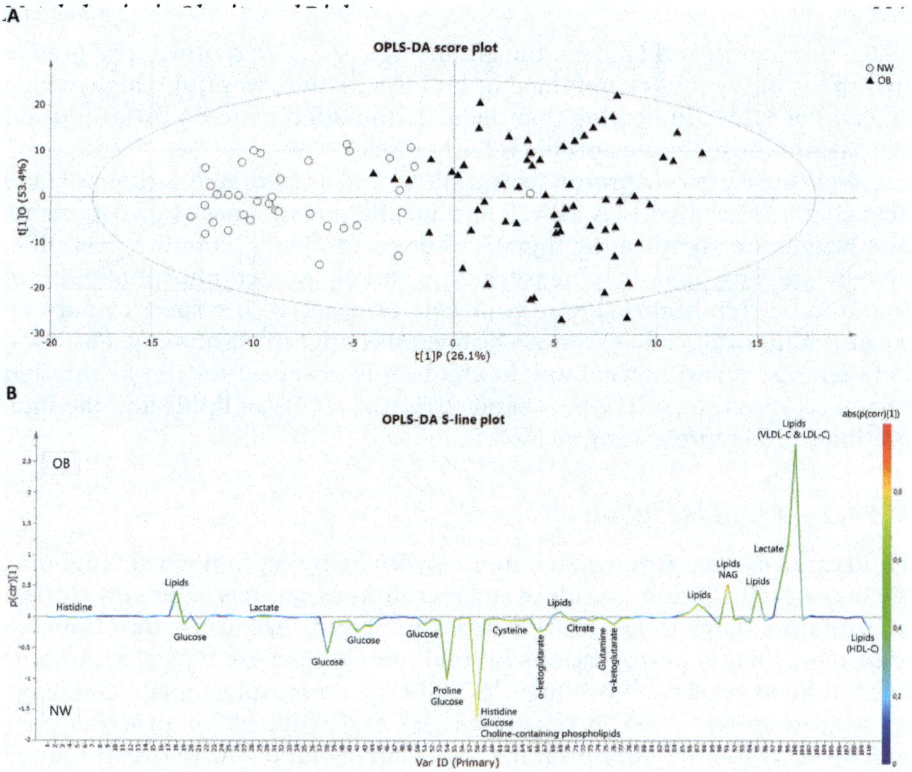

Figure 8.3 Orthogonal partial least squares discriminant analysis (OPLS-DA) score plot (**A**) and S-line plot (**B**) obtained to visualize differences between obese (OB, ▲) and normal-weight (NW, ○) children and adolescents. In the score plot (A), each participant is represented by his or her metabolic profile and visualized as a single symbol of which the location is determined by the contributions of the 110 variables in the ^{1}H nuclear magnetic resonance (NMR) spectrum. The OPLS-DA score plot shows the first predictive component (t[1]P: 26.1%), explaining the variation between the groups, *versus* the first orthogonal component (t[1]O: 53.4%) that explains the variation within the groups. The OPLS-DA S-line plot visualizes differences between obese (positive) and normal-weight subjects (negative). The left y-axis represents p(ctr)[1], the covariance between a variable and the classification score. It indicates if an increase or decrease of a variable is correlated with the classification score. The magnitude of the covariance is however difficult to interpret since covariance is scale-dependent. This means that a high value for the covariance does not necessarily imply a strong correlation, as the covariance is also influenced by the intensity of the signal with respect to the noise level. Therefore this measure will likely indicate variables with large signal intensities. The right y-axis shows p(corr)[1], the correlation coefficient between a variable and the classification score (*i.e.* the normalized covariance). It gives a linear indication of the strength of the correlation. As the correlation is independent of the intensity of the variable, it will be a better measure for the reliability of the variable in the classification process. In the S-line plot (B), the red color stands for the highest absolute value of the correlation coefficient. Strongly discriminating variables have a large intensity and large reliability of the variable in the classification process. Reproduced from reference 42 with permission from Future Science OA, Copyright 2018.

86.3% [R^2X(cum)] and *between* the groups was 72.5% [R^2Y(cum)]. The predictive ability of the model, obtained by cross-validation, was quite high, with a Q^2(cum) of 62.4%, indicating that the discrimination between the groups on the basis of the metabolic profile is fairly good.

Lipids (measured *via* protons on saturated and unsaturated carbons of fatty acid chains of triglycerides as well as phospholipids), *N*-acetyl glycoproteins and lactate are significantly higher, whereas proline, glutamine, histidine, cysteine, α-ketoglutarate, glucose, citrate and cholinated phospholipids are found to be significantly lower in plasma of obese with respect to normal-weight children and adolescents. The biggest change in plasma concentration between obese and normal-weight children is observed for the integration region representing $-C\underline{H}_2-C=O$ and $-C\underline{H}_2-CH=CH-$ of lipids and the integration region representing lactate.

8.3.1.1 Lipid Metabolism

Lipids are a diverse group of metabolites with many key biological functions, such as structural compounds of cell membranes, sources of energy storage and intermediates in signaling pathways.[62] Lipids are under tight homeostatic control and perturbations in lipid metabolism are typical symptoms of childhood obesity.[63] Although ^{1}H NMR spectroscopic signals challenge us to distinguish between several classes and types of lipoproteins, they empower us to study lipoproteins and their metabolism in greater detail than previously possible. Plasma ^{1}H NMR spectra of obese children show high concentrations of saturated and unsaturated fatty acid chains of triglycerides and phospholipids. Especially the CH_2 protons of saturated fatty acid chains, in particular of low-density lipoprotein (LDL) and very-LDL (VLDL), are increased.[64] The observed increase in saturated fatty acid chains can point to a preliminary increased hepatic production and secretion of triglyceride-containing VLDL particles, which are subsequently converted to LDL in the blood. This so-called proatherogenic/dyslipidaemic phenotype also proves to be strongly related to the intrahepatic lipid content in obese adolescents with normal glucose tolerance.[65] Hence, changes in concentrations of plasma lipid particles may reflect early disturbances in hepatic fat metabolism and might precede deregulation of glucose metabolism in obese children and adolescents. Indeed, hepatic steatosis is known to be frequently associated with development of Type 2 diabetes in both adolescents and adults.[66]

The NMR signal intensity corresponding to the nine protons of the three methyl groups of the choline head group of both sphingomyelins and phosphatidylcholines (referred to as cholinated phospholipids) was observed to be lower in the plasma of obese children as compared with normal-weight children. Our findings are substantiated by previous MS-based studies that have revealed increased levels of cholinated phospholipids in serum of obese children after substantial weight loss due to changes in diet and exercise.[67] Still, it is difficult to accurately interpret our findings since there is a lot of variation in sphingomyelin and phosphatidylcholine species concerning chain length and degree of saturation.[68] Also observing sphingomyelins and phosphatidylcholines in the same integration region makes it difficult to differentiate between both. Studying a fraction of the total cholinated phospholipids is only possible by applying more advanced techniques other than ^{1}H NMR spectroscopy (*e.g.* ^{31}P NMR or MS).

8.3.1.2 *Glucose Metabolism*

An accumulation of plasma lactate could reveal perturbations in hepatic glucose production and lipid accumulation.[69,70] Mitochondria-rich tissues will fail to metabolize lactate when their oxygen supply fails or if there are intrinsic abnormalities of oxidative pathways. The liver and adipose tissue will become lactate-producing rather than -consuming tissues under such circumstances.[70] Subcutaneous adipose tissue can be a significant source of whole-body lactate.[69] High levels of lactate in plasma of obese children can therefore be related to a failure of oxidative mechanisms and a large adipose mass.

Glucose was found to be lower in the plasma of obese as compared to normal-weight children. This is fully in agreement with an NMR-based study on Ningxiang growing pigs that showed lower glucose and higher *N*-acetyl glycoproteins in the serum of genetically obese compared with lean pigs consuming the same diet for 46 days.[71] Decreased glucose together with elevated *N*-acetyl glycoprotein levels could point to an obesity-associated activation of the hexosamine signaling pathway, as further described below.

Higher levels of *N*-acetyl glycoproteins may also indicate an increased production of acute-phase α-1 acid glycoproteins by the liver in response to proinflammatory cytokines.[72] It has been previously suggested that inflammatory mechanisms linking obesity to metabolic and cardiovascular complications are already activated in childhood.[73]

8.3.1.3 *Amino Acid Metabolism*

Concentrations of proline and glutamine were lower in the obese as compared with the normal-weight plasma metabolome. An MS-based metabolomic study has also shown that serum concentrations of glutamine and proline in obese children were lower when compared to their normal-weight counterparts.[59] In addition, after substantial weight loss, obese children appear to have increased levels of glutamine.[67] Although the exact mechanism behind reduced proline levels in obesity has not been unravelled yet, lower levels of glutamine could point to an obesity-associated activation of the hexosamine signaling pathway. The activation of the hexosamine signaling pathway is associated with insulin resistance in skeletal muscles,[74] deterioration of pancreatic β-cell function[75] and a reduction in energy expenditure.[76]

Dietary sources – such as beef, poultry, salmon, eggs, milk and walnuts – probably account for most of the histidine in the body. Histidine is a precursor for histamine and carnosine biosynthesis. The efficacy of histidine in protecting inflamed tissue is attributed to the capacity of the imidazole ring to scavenge reactive oxygen species (ROS) generated by cells during acute inflammatory response.[77] Serum histidine levels are lower and are negatively associated with inflammation and oxidative stress in obese women.[78] In addition, histidine supplementation has been shown to reduce insulin resistance, reduce body mass index and fat mass and suppress inflammation and oxidative stress in obese women with the metabolic syndrome. Histidine appears to suppress proinflammatory cytokine expression, possibly *via* the nuclear factor-κB (NF-κB) pathway, in adipocytes.[78]

Cysteine is a naturally occurring, non-essential sulphur-containing amino acid that is found in most proteins, although only in small quantities. It is a structural component of many tissues and hormones, it has antioxidant properties and is important in energy metabolism. As plasma concentrations of cysteine were found to be lower in obese as compared to normal-weight children, dysregulation of antioxidant and energy metabolism can take place in the obese phenotype. This hypothesis is further substantiated by the fact that cysteine may be essential for infants, the elderly and individuals with certain metabolic disease.[12]

8.3.1.4 TCA Cycle Intermediates

Citric acid is a weak acid that may be introduced with diet through consumption of citrus fruits, and is most concentrated in lemons and limes. Together with α-ketoglutarate, citrate (a derivative of citric acid) is a key molecule in the tricarboxylic acid (TCA) cycle, playing a fundamental role in determining the overall rate of this important metabolic process.[79] Reduced levels of α-ketoglutarate and citrate are observed in the obese plasma and can therefore be linked to an impaired TCA cycle, *i.e.* disturbed energy metabolism. Next to its key functions, α-ketoglutarate is a nitrogen scavenger and a source of glutamate and glutamine that stimulates protein synthesis and inhibits protein degradation in muscles. In particular, α-ketoglutarate can decrease protein catabolism and increase protein synthesis to enhance bone tissue formation in skeletal muscles.[79] Interestingly, enteric feeding of α-ketoglutarate supplements can significantly increase circulating plasma levels of hormones such as insulin, growth hormone and insulin-like growth factor-1.[79]

8.3.1.5 Dietary Influences on the Metabolome

Diet may play a key role in the establishment of the obese metabolome and development of obese phenotype. A study using ultra-high performance liquid chromatography-quadrupole time-of-flight MS (UPLC-Q-TOF-MS) and GC-MS showed abnormal lipid and energy metabolism in liver and blood samples of high-fat diet-induced obese mice.[80] Obese mice consuming a high-fat diet showed increased plasma levels of lipids (phosphatidylcholines, lysophosphatidylcholines and fatty acids) and decreased intermediates of lipid metabolism (betaine, carnitine and acylcarnitines). They also found decreased levels of glucose in high-fat diet-induced obesity that could point to a dysregulation of energy metabolism *via* β-oxidation, a catabolic process by which fatty acid molecules are broken down in the mitochondria to generate acetyl-CoA, which enters the citric acid cycle, and NADH and $FADH_2$, which are co-enzymes used in the electron transport chain. Moreover, the degree of insulin resistance – increased by obesity – was positively related to increased hepatic levels of mono- and disaccharides in obese mice, as expected, but was not related to the hepatic glucose level.[80]

8.3.2 Metabolic Subtyping of Paediatric Obesity

Metabolomics studies aiming to explore the MHO phenotype are still scarce, especially in the paediatric population. Nevertheless, early screening for the MHO phenotype could support the development of effective prevention and treatment strategies in the fight against metabolic complications due to childhood obesity. With this in mind, we have set up a pilot study in which plasma metabolic profiles of MHO and MUO children and adolescents were examined and compared, something that has – to the best of our knowledge – not been considered before.

MHO individuals were classified as being obese (defined according to the International Obesity Task Force criteria) but having none of the components of the metabolic syndrome (defined according to the criteria of the International Diabetes Federation (IDF) consensus for children older than 10 years).[81] Apart from obesity, the metabolic syndrome is defined as having two or more of the following components:

- High-density lipoprotein cholesterol <40 mg dL^{-1}; females 16 years or older: HDL-C <50 mg dL^{-1}
- Triglycerides ≥150 mg dL^{-1}
- Systolic blood pressure ≥130 mmHg or diastolic blood pressure ≥85 mmHg
- Fasting plasma glucose ≥100 mg dL^{-1} or T2DM.

MUO subjects are classified as being obese and having two or more components of the metabolic syndrome. A group that fell in between the classification of MHO and MUO was termed metabolically at-risk obese and was not included in the statistical analyses. A point of discussion is that there still exists no universally accepted definition for MHO in children, which complicates its implementation in current and future clinical research. In an attempt to achieve uniformity, we here suggest using the consensus-based paediatric IDF criteria.[81]

An untargeted ^{1}H NMR-based metabolomics approach was applied to profile the plasma metabolome of obese children and adolescents aged between 10 and 18. Detailed protocols regarding sample preparation and ^{1}H NMR analysis have been previously described.[42] The OPLS-DA model was able to explain a total amount of variation *within* the groups of 87.7% [R^2X(cum)] and *between* the groups of 50.5% [R^2Y(cum)]. The model allowed to classify 94.1% (16 out of 17, sensitivity) MUO and 88.9% (16 out of 18, specificity) MHO individuals correctly but still had a mediocre predictive ability of 24.9% [Q^2(cum)], most probably due to the limited number of participants.

Lipids and lactate levels were significantly decreased, whereas glutamine, histidine, lysine, valine, isoleucine, leucine and cholinated phospholipid concentrations were increased in the plasma of MHO with respect to MUO children and adolescents. The biggest change in plasma concentration between MHO and MUO children and adolescents is observed for the integration region representing cholinated phospholipids and the integration region representing lactate.

8.3.2.1 Lipid Metabolism

As already described above, cholinated phospholipids can be important in the regulation of lipid metabolism in obese subjects.[67] A higher level of cholinated phospholipids, as observed in MHO children, can point towards a healthier metabolic profile as compared to MUO subjects. However, the ^{1}H NMR experimental analyses of cholinated phospholipids have to be accompanied by more advanced techniques in order to better differentiate between sphingomyelins and phosphatidylcholine and consequently interpret our findings more deeply.

8.3.2.2 Amino Acid Metabolism

Increased plasma levels of glutamine in the MHO phenotype can be accountable for a reduced activation of the hexosamine signaling pathway. Hence, MHO individuals can be characterized by a reduced insulin resistance in skeletal muscles,[74] a better pancreatic β-cell function[75] and an increase in energy expenditure[76] as compared to MUO individuals. Histidine is related to the protection of inflamed tissue.[77] Increased levels of histidine in plasma of MHO can therefore be associated with less inflammation and less oxidative stress as compared to MUO individuals (as also discussed above). Lysine is an essential amino acid and is high in foods such as wheat germ, cottage cheese and chicken. Since lysine is highly concentrated in muscle compared to most other amino acids, it may support muscle metabolism, especially in MHO subjects.

8.3.2.3 Branched-chain Amino Acids

Nowadays, plasma concentrations of branched-chain amino acids (BCAAs: valine, isoleucine and leucine) are increasingly being investigated in the context of obesity, diabetes and cardiovascular disease. However, previous findings on BCAAs – studied in both obese children and adults – are inconsistent. In our study, BCAAs are found to be significantly reduced in plasma of MUO as compared with MHO children and adolescents. In line with our findings, Mihalik *et al.* demonstrated that adolescents with Type 2 diabetes showed lower concentrations of BCAAs as compared with obese as well as normal-weight individuals. They suggested that these lower concentrations result from an increased gluconeogenic drive.[82] In adults, however, higher circulating levels of BCAAs have been correlated with insulin resistance and future development of diabetes.[83] This finding most likely indicates different metabolic mechanisms for children and adolescents as compared with adults.

8.4 Application of Metabolomics in Diabetes Research

Diabetes mellitus is characterized by prolonged hyperglycaemia in blood due to dysfunction of insulin secretion or action. Type 1 diabetes mellitus (T1DM) is one of the most common paediatric diseases and its incidence is rising in many countries.[84] It results from a lack of pancreatic β-cell insulin production by autoimmune mechanisms.[85] T2DM is less prevalent in children and adolescents, and results from a combination of disordered insulin action and pancreatic β-cell secretion. Especially obese children with the metabolic syndrome are prone to develop T2DM. The risk for development of obesity and consequently T2DM depends on multiple factors, such as genetic susceptibility,[86] composition of the gut microbiota,[87] increased consumption of food rich in either saturated fat or refined carbohydrates, as well as reduced physical activity or exposure to chemicals.[88]

Insulin is a key hormone to maintain metabolic homeostasis, regulating carbohydrate, lipid and protein metabolism. Subsequently, insulin deficiency induces a variety of metabolic derangements.[13,89] In recent decades it has been shown that metabolites other than glucose play an important role in insulin deficiency and the development of diabetes. Given that pancreatic β-cell dysfunction has a central role in the initiation and progression of T1DM, the understanding of biochemical pathways associated with this dysfunction can provide more information on underlying pathways involved in diabetes. To this end, metabolomics can serve as a value-adding analytical tool to measure metabolic responses to diabetes.[90]

8.4.1 Metabolic Profiling of Type 1 Diabetes

Metabolomics research on paediatric study populations is still in its infancy. Up to now, only a few researchers investigated the plasma metabolic fingerprint of T1DM in children, using MS as analytical tool.[91–93] One study showed that children who later progressed to T1DM had reduced serum levels of succinic acid and phosphatidylcholine at birth, pointing towards a dysregulated metabolism preceding β-cell autoimmunity and T1DM.[91] In addition, methionine deficits in early childhood may lead to an increased risk of developing T1DM later in life.[92] When comparing the plasma metabolic profile of T1DM and healthy children, differences were observed in lipid metabolism (non-esterified fatty acids, lysophospholipids and other fatty acid derivatives) and some markers of differential activity of the gut microbiota (bile acids, *p*-cresol sulphate).[93]

We will here elaborate on our own research in which we investigated metabolic variations in the plasma of T1DM and non-diabetic matched children and adolescents using NMR spectroscopy-based metabolomics to gain insight into biochemical pathways that are altered in early stages of T1DM. The use of NMR-based metabolomics to obtain a deeper knowledge of the plasma metabolic profile of T1DM has not been fully explored in the paediatric population. With the aid of statistical pattern recognition, this can reveal novel insights into the biochemical consequences of early-onset diabetes.

An untargeted ^{1}H NMR-based metabolomics approach was applied to profile the plasma metabolome of T1DM and non-diabetic children and adolescents. Detailed protocols regarding subject characteristics, sample preparation and ^{1}H NMR analysis have been previously described elsewhere.[41] In general, all subjects were matched for gender, age and standardized body mass index. T1DM patients were diagnosed according to international consensus guidelines.[94] T1DM patients had diabetes for around 7 years and were treated with exogenous insulin. T1DM patients showed high fasting plasma glucose levels and haemoglobin A_{1c} (HbA_{1c}) concentrations above 6.5%. None of the subjects was using lipid-lowering drugs or other medication.

Multivariate OPLS-DA was used to build a model (statistical classifier) to differentiate between T1DM patients and controls (Figure 8.4). The total amount of variation *between* the groups that can be explained by the model is 81.0% [R^2Y(cum)] while the variation *within* the groups is 75.8% [R^2X(cum)]. The predictive ability of the model, obtained by cross-validation, is quite high with a Q^2(cum) of 50.7%, indicating that the discrimination between the groups on the basis of the metabolic phenotype is valid.

Besides the expected higher concentration of glucose, we found lower relative concentrations for lipids (triglycerides, phospholipids and cholinated phospholipids) and the amino acids serine, tryptophan and cysteine in plasma of T1DM children and adolescents as compared to non-diabetic controls.

8.4.1.1 Lipid Metabolism

Our findings of relatively lower levels of lipids in the plasma of T1DM as compared to controls are in agreement with other metabolomics studies.[13,89,91] In a prospective Finnish study, it was found that children who developed T1DM have reduced serum levels of phosphatidylcholine at birth, next to lower levels of multiple triglycerides and phospholipids throughout the follow-up.[91] In addition, it has been demonstrated that children developing T1DM before 4 years of age have low cord blood levels of phospholipids, mainly phosphatidylcholines.[95] It is suggested that T1DM progressors are choline-deficient at birth, and that choline deficiency is the key mechanism leading to lower serum triglyceride levels and their increased accumulation in the liver.[91] Choline metabolism also depends on the gut microbial composition,[96] making the

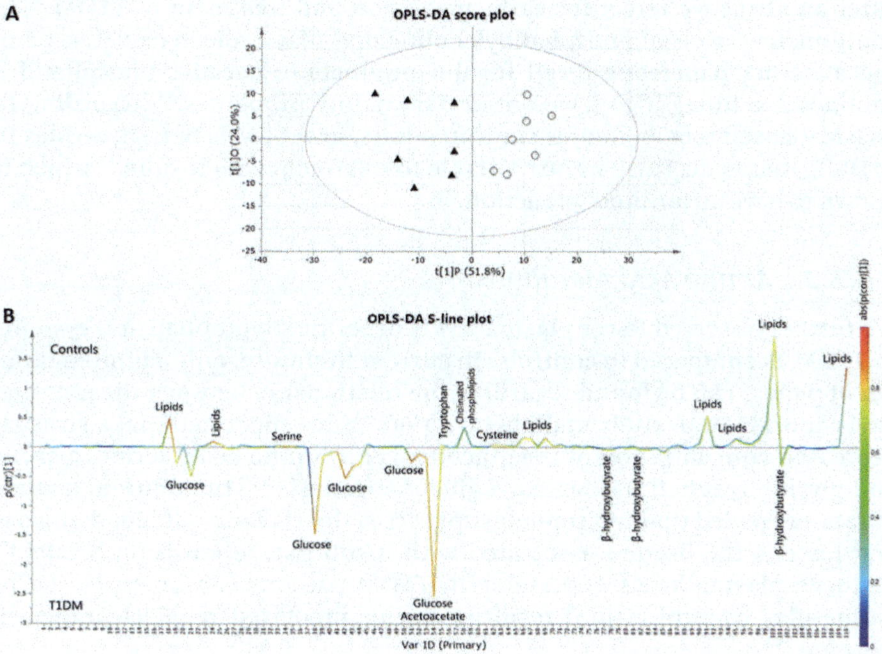

Figure 8.4 Orthogonal partial least squares discriminant analysis (OPLS-DA) score plot (**A**) and S-line plot (**B**) obtained to visualize differences between type 1 diabetes mellitus patients (T1DM, ▲) and healthy controls (○). In the score plot (A), each participant is represented by his or her metabolic profile and visualized as a single symbol, of which the location is determined by the contributions of the 110 variables in the ^{1}H nuclear magnetic resonance (NMR) spectrum. The OPLS-DA score plot shows the first predictive component (t[1]P: 51.8%), explaining the variation between the groups, *versus* the first orthogonal component (t[1]O: 24.0%) that explains the variation within the groups. The OPLS-DA S-line plot visualizes differences between T1DM patients (negative) and controls (positive). The left *y-axis* represents *p*(ctr)[1], the covariance between a variable and the classification score. It indicates if an increase or decrease of a variable is correlated to the classification score. The magnitude of the covariance is however difficult to interpret since covariance is scale-dependent. This means that a high value for the covariance does not necessarily imply a strong correlation, as the covariance is also influenced by the intensity of the signal with respect to the noise level. Therefore this measure will likely indicate variables with large signal intensities. The right *y*-axis shows *p*(corr)[1], the correlation coefficient between a variable and the classification score (*i.e.* the normalized covariance). It gives a linear indication of the strength of the correlation. As the correlation is independent of the intensity of the variable, it will be a better measure for the reliability of the variable in the classification process. In the S-line plot (**B**), the red color stands for the highest absolute value of the correlation coefficient. Strongly discriminating variables have a large intensity and large reliability of the variable in the classification process. Reproduced from reference 41, https://doi.org/10.1186/s13098-017-0246-9, under the terms of the CC BY 4.0 licence, https://creativecommons.org/licenses/by/4.0/.

latter an attractive target for early prevention and treatment of T1DM.[97] In addition, low levels of phosphatidylcholine may play a role in oxidative damage affecting pancreatic β-cell insulin production, because phosphatidylcholines are thought to have anti-inflammatory properties.[98] According to an NMR-based metabolomics study in adults, lower levels of triglycerides in T1DM patients may also be attributable to treatment with insulin,[89] which is known to have an antilipolytic action.[99]

8.4.1.2 Amino Acid Metabolism

We further observed lower plasma levels of serine, tryptophan and cysteine in T1DM as compared to controls. In current literature, only a limited number of papers can be found regarding the relationship between these amino acids and T1DM. A study in diabetic db^{-}/db^{-} mice suggested that a strongly decreased concentration of the gluconeogenic amino acids serine, alanine and glycine resulted in increased gluconeogenesis.[100] In addition, a study in rats suggested that tryptophan suppresses the elevation of blood glucose and lessens the burden associated with insulin secretion from β-cells.[101] Reduced plasma levels of cysteine in T1DM patients can be explained by glomerular hyperfiltration, resulting in an increased renal clearance of cysteine.[102]

8.5 Summary and Future Perspectives

In this chapter, we highlighted a series of metabolomic studies carried out in previous and recent years on the utility and application of NMR-based metabolomics in an attempt to understand pathophysiological mechanisms in obesity and diabetes. Remarkably, a number of metabolites have been identified, including specific lipids, carbohydrates, amino acids and TCA cycle intermediates that could play an important role in the underlying disease progression of obesity and diabetes in children. Focusing on childhood gives the advantage of examining early subtle changes in the metabolic profile that develop in premature disease stages, that can occur even before changes in traditional measurements, as observed by clinical practitioners. In addition, metabolic adaptation can take place in children, leading to minimized dysregulation in disease metabolism compared to adulthood. In this way, primary changes in the metabolic profile may be used as a biomarker to eventually define the best intervention/therapeutic strategy to reverse or combat the development of metabolic disease. To give an example, restoring or improving the plasma metabolic profile, *e.g.* by re-establishing lipid and amino acid availability or by modulating gut microbial composition, might prevent β-cell destruction and delay obesity or T1DM progression in children and adolescents.

To conclude, NMR-based metabolomics can be used to search for subtle changes in the plasma of children with obesity or diabetes, and therefore, it might become an important tool for the early diagnosis and prognosis of these diseases. Continuing technological improvements will expand our knowledge of the causes and progression of obesity and diabetes, and will direct future research to higher levels.

Acknowledgements

The authors thank all the volunteers who participated in these studies. They also gratefully acknowledge colleagues at Hasselt University, Jessa Hospital and University Biobank Limburg for their contributions. This work was supported by a grant from ZonMW/JPI HDHL Intestinal Microbiomics (grant 50-52905-98-599) and by Hasselt University and the Research Foundation Flanders (FWO Vlaanderen; Hercules project AUHL/15/2- GOH3816N).

References

1. O. Fiehn, *Plant Mol. Biol.*, 2002, **48**, 155–171.
2. R. Goodacre, *J. Nutr.*, 2007, **137**, 259S–266S.
3. D. P. Jones, Y. Park and T. R. Ziegler, *Annu. Rev. Nutr.*, 2012, **32**, 183–202.
4. G. Falony, S. Vieira-Silva and J. Raes, *Annu. Rev. Microbiol.*, 2015, **69**, 305–321.
5. S. Paramsothy, S. Nielsen, M. A. Kamm, N. P. Deshpande, J. J. Faith, J. C. Clemente, R. Paramsothy, A. J. Walsh, J. van den Bogaerde, D. Samuel, R. W. L. Leong, S. Connor, W. Ng, E. Lin, T. J. Borody, M. R. Wilkins, J. F. Colombel, H. M. Mitchell and N. O. Kaakoush, *Gastroenterology*, 2019, **156**, 1440–1454 e1442.
6. T. Ramirez, M. Daneshian, H. Kamp, F. Y. Bois, M. R. Clench, M. Coen, B. Donley, S. M. Fischer, D. R. Ekman, E. Fabian, C. Guillou, J. Heuer, H. T. Hogberg, H. Jungnickel, H. C. Keun, G. Krennrich, E. Krupp, A. Luch, F. Noor, E. Peter, B. Riefke, M. Seymour, N. Skinner, L. Smirnova, E. Verheij, S. Wagner, T. Hartung, B. van Ravenzwaay and M. Leist, *ALTEX*, 2013, **30**, 209–225.
7. D. S. Wishart, *Nat. Rev. Drug Discov.*, 2016, **15**, 473–484.
8. J. R. Ussher, S. Elmariah, R. E. Gerszten and J. R. Dyck, *J. Am. Coll. Cardiol.*, 2016, **68**, 2850–2870.
9. Y. Zhou, M. Oresic, M. Leivonen, P. Gopalacharyulu, J. Hyysalo, J. Arola, A. Verrijken, S. Francque, L. Van Gaal, T. Hyotylainen and H. Yki-Jarvinen, *Clin. Gastroenterol. Hepatol.*, 2016, **14**, 1463–1472 e1466.
10. E. R. Kim, H. N. Kwon, H. Nam, J. J. Kim, S. Park and Y. H. Kim, *Sci. Rep.*, 2019, **9**, 4786.
11. D. S. Wishart, Y. D. Feunang, A. Marcu, A. C. Guo, K. Liang, R. Vazquez-Fresno, T. Sajed, D. Johnson, C. Li, N. Karu, Z. Sayeeda, E. Lo, N. Assempour, M. Berjanskii, S. Singhal, D. Arndt, Y. Liang, H. Badran, J. Grant, A. Serra-Cayuela, Y. Liu, R. Mandal, V. Neveu, A. Pon, C. Knox, M. Wilson, C. Manach and A. Scalbert, *Nucleic Acids Res.*, 2018, **46**, D608–D617.
12. D. S. Wishart, D. Tzur, C. Knox, R. Eisner, A. C. Guo, N. Young, D. Cheng, K. Jewell, D. Arndt, S. Sawhney, C. Fung, L. Nikolai, M. Lewis, M. A. Coutouly, I. Forsythe, P. Tang, S. Shrivastava, K. Jeroncic, P. Stothard, G. Amegbey, D.

Block, D. D. Hau, J. Wagner, J. Miniaci, M. Clements, M. Gebremedhin, N. Guo, Y. Zhang, G. E. Duggan, G. D. Macinnis, A. M. Weljie, R. Dowlatabadi, F. Bamforth, D. Clive, R. Greiner, L. Li, T. Marrie, B. D. Sykes, H. J. Vogel and L. Querengesser, *Nucleic Acids Res.*, 2007, **35**, D521–526.
13. I. R. Lanza, S. Zhang, L. E. Ward, H. Karakelides, D. Raftery and K. S. Nair, *PLoS One*, 2010, 5, e10538.
14. E. T. Cirulli, L. Guo, C. Leon Swisher, N. Shah, L. Huang, L. A. Napier, E. F. Kirkness, T. D. Spector, C. T. Caskey, B. Thorens, J. C. Venter and A. Telenti, *Cell Metab.*, 2019, **29**, 488–500 e482.
15. W. B. Dunn, N. J. Bailey and H. E. Johnson, *Analyst*, 2005, **130**, 606–625.
16. C. S. Clendinen, M. E. Monge and F. M. Fernandez, *Analyst*, 2017, **142**, 3101–3117.
17. A. C. Dona, M. Kyriakides, F. Scott, E. A. Shephard, D. Varshavi, K. Veselkov and J. R. Everett, *Comput. Struct. Biotechnol. J.*, 2016, **14**, 135–153.
18. K. Bingol and R. Bruschweiler, *Curr. Opin. Biotechnol.*, 2017, **43**, 17–24.
19. Z. Pan, H. Gu, N. Talaty, H. Chen, N. Shanaiah, B. E. Hainline, R. G. Cooks and D. Raftery, *Anal. Bioanal. Chem.*, 2007, **387**, 539–549.
20. J. K. Nicholson, P. J. Sadler, J. R. Bales, S. M. Juul, A. F. MacLeod and P. H. Sonksen, *Lancet*, 1984, 2, 751–752.
21. C. K. Larive, G. A. Barding, Jr. and M. M. Dinges, *Anal. Chem.*, 2015, **87**, 133–146.
22. A. C. Dona, B. Jimenez, H. Schafer, E. Humpfer, M. Spraul, M. R. Lewis, J. T. Pearce, E. Holmes, J. C. Lindon and J. K. Nicholson, *Anal. Chem.*, 2014, **86**, 9887–9894.
23. P. Bernini, I. Bertini, C. Luchinat, P. Nincheri, S. Staderini and P. Turano, *J. Biomol. NMR*, 2011, **49**, 231–243.
24. M. E. Dumas, E. C. Maibaum, C. Teague, H. Ueshima, B. Zhou, J. C. Lindon, J. K. Nicholson, J. Stamler, P. Elliott, Q. Chan and E. Holmes, *Anal. Chem.*, 2006, **78**, 2199–2208.
25. R. A. Wevers, U. Engelke and A. Heerschap, *Clin. Chem.*, 1994, **40**, 1245–1250.
26. N. Psychogios, D. D. Hau, J. Peng, A. C. Guo, R. Mandal, S. Bouatra, I. Sinelnikov, R. Krishnamurthy, R. Eisner, B. Gautam, N. Young, J. Xia, C. Knox, E. Dong, P. Huang, Z. Hollander, T. L. Pedersen, S. R. Smith, F. Bamforth, R. Greiner, B. McManus, J. W. Newman, T. Goodfriend and D. S. Wishart, *PLoS One*, 2011, **6**, e16957.
27. F. Betsou, S. Lehmann, G. Ashton, M. Barnes, E. E. Benson, D. Coppola, Y. DeSouza, J. Eliason, B. Glazer, F. Guadagni, K. Harding, D. J. Horsfall, C. Kleeberger, U. Nanni, A. Prasad, K. Shea, A. Skubitz, S. Somiari, E. Gunter, B. International Society for Biological and Environmental Repositories Working Group on Biospecimen Science, *Cancer Epidemiol. Biomarkers Prev.*, 2010, **19**, 1004–1011.
28. S. Lehmann, F. Guadagni, H. Moore, G. Ashton, M. Barnes, E. Benson, J. Clements, I. Koppandi, D. Coppola, S. Y. Demiroglu, Y. DeSouza, A. De Wilde, J. Duker, J. Eliason, B. Glazer, K. Harding, J. P. Jeon, J. Kessler, T. Kokkat, U. Nanni, K. Shea, A. Skubitz, S. Somiari, G. Tybring, E. Gunter,

F. Betsou, B. International Society for Biological and Environmental Repositories Working Group on Biospecimen Science, *Biopreserv. Biobank*, 2012, **10**, 366–374.

29. O. Teahan, S. Gamble, E. Holmes, J. Waxman, J. K. Nicholson, C. Bevan and H. C. Keun, *Anal. Chem.*, 2006, **78**, 4307–4318.
30. *World Health Organization*, Use of anticoagulants in diagnostic laboratory investigations, WHO/DIL/LAB/99.1 Rev.2, WHO, Geneva, 2002.
31. S. Deprez, B. C. Sweatman, S. C. Connor, J. N. Haselden and C. J. Waterfield, *J. Pharm. Biomed. Anal.*, 2002, **30**, 1297–1310.
32. J. Pinto, M. R. Domingues, E. Galhano, C. Pita, C. Almeida Mdo, I. M. Carreira and A. M. Gil, *Analyst*, 2014, **139**, 1168–1177.
33. L. Bervoets, E. Louis, G. Reekmans, L. Mesotten, M. Thomeer, P. Adriaensens and L. Linsen, *Metabolomics*, 2015, **11**, 1197–1207.
34. P. Yin, R. Lehmann and G. Xu, *Anal. Bioanal. Chem.*, 2015, **407**, 4879–4892.
35. O. Beckonert, H. C. Keun, T. M. Ebbels, J. Bundy, E. Holmes, J. C. Lindon and J. K. Nicholson, *Nat. Protoc.*, 2007, **2**, 2692–2703.
36. E. Louis, F. X. Cantrelle, L. Mesotten, G. Reekmans, L. Bervoets, K. Vanhove, M. Thomeer, G. Lippens and P. Adriaensens, *Magn. Reson. Chem.*, 2017, **55**, 706–713.
37. R. T. McKay, *Concept Magn. Reson. A*, 2011, **38a**, 197–220.
38. H. Y. Carr and E. M. Purcell, *Phys. Rev.*, 1954, **94**, 630–638.
39. S. Meiboom and D. Gill, *Rev. Sci. Instrum.*, 1958, **29**, 688.
40. E. Louis, L. Bervoets, G. Reekmans, E. De Jonge, L. Mesotten, M. Thomeer and P. Adriaensens, *Metabolomics*, 2015, **11**, 225–236.
41. L. Bervoets, G. Massa, W. Guedens, E. Louis, J. P. Noben and P. Adriaensens, *Diabetol. Metab. Syndr.*, 2017, **9**, 48.
42. L. Bervoets, G. Massa, W. Guedens, G. Reekmans, J. P. Noben and P. Adriaensens, *Future Sci. OA*, 2018, **4**, FSO310.
43. I. A. Lewis, S. C. Schommer, B. Hodis, K. A. Robb, M. Tonelli, W. M. Westler, M. R. Sussman and J. L. Markley, *Anal. Chem.*, 2007, **79**, 9385–9390.
44. J. Kruk, M. Doskocz, E. Jodlowska, A. Zacharzewska, J. Lakomiec, K. Czaja and J. Kujawski, *Appl. Magn. Reson.*, 2017, **48**, 1–21.
45. A. H. Emwas, E. Saccenti, X. Gao, R. T. McKay, V. Dos Santos, R. Roy and D. S. Wishart, *Metabolomics*, 2018, **14**, 31.
46. S. M. Kohl, M. S. Klein, J. Hochrein, P. J. Oefner, R. Spang and W. Gronwald, *Metabolomics*, 2012, **8**, 146–160.
47. E. M. Lenz and I. D. Wilson, *J. Proteome Res.*, 2007, **6**, 443–458.
48. L. Eriksson, T. Byrne, E. Johansson, J. Trygg and C. Vikström, *Multi- and Megavariate Data Analysis*, Umetrics Academy, Malmö, Sweden, 3rd edn, 2013.
49. Y. Benjamini and Y. Hochberg, *J. R. Statist. Soc., Ser. B*, 1995, **57**, 289–300.
50. D. Haire-Joshu and R. Tabak, *Annu. Rev. Public Health*, 2016, **37**, 253–271.
51. M. Ng, T. Fleming, M. Robinson, B. Thomson, N. Graetz, C. Margono, E. C. Mullany, S. Biryukov, C. Abbafati, S. F. Abera, J. P. Abraham, N. M. Abu-Rmeileh, T. Achoki, F. S. AlBuhairan, Z. A. Alemu, R. Alfonso, M. K. Ali, R.

Ali, N. A. Guzman, W. Ammar, P. Anwari, A. Banerjee, S. Barquera, S. Basu, D. A. Bennett, Z. Bhutta, J. Blore, N. Cabral, I. C. Nonato, J. C. Chang, R. Chowdhury, K. J. Courville, M. H. Criqui, D. K. Cundiff, K. C. Dabhadkar, L. Dandona, A. Davis, A. Dayama, S. D. Dharmaratne, E. L. Ding, A. M. Durrani, A. Esteghamati, F. Farzadfar, D. F. Fay, V. L. Feigin, A. Flaxman, M. H. Forouzanfar, A. Goto, M. A. Green, R. Gupta, N. Hafezi-Nejad, G. J. Hankey, H. C. Harewood, R. Havmoeller, S. Hay, L. Hernandez, A. Husseini, B. T. Idrisov, N. Ikeda, F. Islami, E. Jahangir, S. K. Jassal, S. H. Jee, M. Jeffreys, J. B. Jonas, E. K. Kabagambe, S. E. Khalifa, A. P. Kengne, Y. S. Khader, Y. H. Khang, D. Kim, R. W. Kimokoti, J. M. Kinge, Y. Kokubo, S. Kosen, G. Kwan, T. Lai, M. Leinsalu, Y. Li, X. Liang, S. Liu, G. Logroscino, P. A. Lotufo, Y. Lu, J. Ma, N. K. Mainoo, G. A. Mensah, T. R. Merriman, A. H. Mokdad, J. Moschandreas, M. Naghavi, A. Naheed, D. Nand, K. M. Narayan, E. L. Nelson, M. L. Neuhouser, M. I. Nisar, T. Ohkubo, S. O. Oti, A. Pedroza, D. Prabhakaran, N. Roy, U. Sampson, H. Seo, S. G. Sepanlou, K. Shibuya, R. Shiri, I. Shiue, G. M. Singh, J. A. Singh, V. Skirbekk, N. J. Stapelberg, L. Sturua, B. L. Sykes, M. Tobias, B. X. Tran, L. Trasande, H. Toyoshima, S. van de Vijver, T. J. Vasankari, J. L. Veerman, G. Velasquez-Melendez, V. V. Vlassov, S. E. Vollset, T. Vos, C. Wang, X. Wang, E. Weiderpass, A. Werdecker, J. L. Wright, Y. C. Yang, H. Yatsuya, J. Yoon, S. J. Yoon, Y. Zhao, M. Zhou, S. Zhu, A. D. Lopez, C. J. Murray and E. Gakidou, *Lancet*, 2014, **384**, 766–781.
52. R. Lakshman, C. E. Elks and K. K. Ong, *Circulation*, 2012, **126**, 1770–1779.
53. A. Pandita, D. Sharma, D. Pandita, S. Pawar, M. Tariq and A. Kaul, *Diabetes Metab. Syndr. Obes.*, 2016, **9**, 83–89.
54. D. A. Kornet-van der Aa, T. M. Altenburg, C. H. van Randeraad-van der Zee and M. J. Chinapaw, *Obes. Rev.*, 2017, **18**, 581–593.
55. R. L. Prince, J. L. Kuk, K. A. Ambler, J. Dhaliwal and G. D. Ball, *Diabetes Care*, 2014, **37**, 1462–1468.
56. G. V. Denis and M. S. Obin, *Mol. Aspects Med.*, 2013, **34**, 59–70.
57. S. L. Appleton, C. J. Seaborn, R. Visvanathan, C. L. Hill, T. K. Gill, A. W. Taylor, R. J. Adams and North West Adelaide Health Study Team, *Diabetes Care*, 2013, **36**, 2388–2394.
58. A. Zhang, H. Sun and X. Wang, *Obes. Rev.*, 2013, **14**, 344–349.
59. S. Wahl, Z. Yu, M. Kleber, P. Singmann, C. Holzapfel, Y. He, K. Mittelstrass, A. Polonikov, C. Prehn, W. Romisch-Margl, J. Adamski, K. Suhre, H. Grallert, T. Illig, R. Wang-Sattler and T. Reinehr, *Obes. Facts*, 2012, **5**, 660–670.
60. W. Perng, M. W. Gillman, A. F. Fleisch, R. D. Michalek, S. M. Watkins, E. Isganaitis, M. E. Patti and E. Oken, *Obesity (Silver Spring)*, 2014, **22**, 2570–2578.
61. N. F. Butte, Y. Liu, I. F. Zakeri, R. P. Mohney, N. Mehta, V. S. Voruganti, H. Goring, S. A. Cole and A. G. Comuzzie, *Am. J. Clin. Nutr.*, 2015, **102**, 256–267.
62. E. Fahy, S. Subramaniam, H. A. Brown, C. K. Glass, A. H. Merrill, Jr., R. C. Murphy, C. R. Raetz, D. W. Russell, Y. Seyama, W. Shaw, T. Shimizu, F. Spener, G. van Meer, M. S. VanNieuwenhze, S. H. White, J. L. Witztum and E. A. Dennis, *J. Lipid Res.*, 2005, **46**, 839–861.
63. S. R. Daniels, *Am. J. Clin. Nutr.*, 2011, **94**, 699–700.

64. P. Soininen, A. J. Kangas, P. Wurtz, T. Tukiainen, T. Tynkkynen, R. Laatikainen, M. R. Jarvelin, M. Kahonen, T. Lehtimaki, J. Viikari, O. T. Raitakari, M. J. Savolainen and M. Ala-Korpela, *Analyst*, 2009, **134**, 1781–1785.
65. A. M. Cali, T. L. Zern, S. E. Taksali, A. M. de Oliveira, S. Dufour, J. D. Otvos and S. Caprio, *Diabetes Care*, 2007, **30**, 3093–3098.
66. E. D'Adamo and S. Caprio, *Diabetes Care*, 2011, **34 Suppl 2**, S161–165.
67. T. Reinehr, B. Wolters, C. Knop, N. Lass, C. Hellmuth, U. Harder, W. Peissner, S. Wahl, H. Grallert, J. Adamski, T. Illig, C. Prehn, Z. Yu, R. Wang-Sattler and B. Koletzko, *Eur. J. Nutr.*, 2015, **54**, 173–181.
68. C. J. DeLong, Y. J. Shen, M. J. Thomas and Z. Cui, *J. Biol. Chem.*, 1999, **274**, 29683–29688.
69. P. A. Jansson, A. Larsson, U. Smith and P. Lonnroth, *J. Clin. Invest.*, 1994, **93**, 240–246.
70. B. Phypers and J. M. T. Pierce, *Contin. Educ. Anaesthesia Crit. Care Pain*, 2006, **6**, 128–132.
71. Q. He, P. Ren, X. Kong, Y. Wu, G. Wu, P. Li, F. Hao, H. Tang, F. Blachier and Y. Yin, *J. Nutr. Biochem.*, 2012, **23**, 133–139.
72. C. Gabay and I. Kushner, *N. Engl. J. Med.*, 1999, **340**, 448–454.
73. H. S. Schipper, R. Nuboer, S. Prop, H. J. van den Ham, F. K. de Boer, C. Kesmir, I. M. H. Mombers, K. A. van Bekkum, J. Woudstra, J. H. Kieft, I. E. Hoefer, W. de Jager, B. Prakken, M. van Summeren and E. Kalkhoven, *Diabetologia*, 2012, **55**, 2800–2810.
74. M. E. Patti, A. Virkamaki, E. J. Landaker, C. R. Kahn and H. Yki-Jarvinen, *Diabetes*, 1999, **48**, 1562–1571.
75. H. Kaneto, G. Xu, K. H. Song, K. Suzuma, S. Bonner-Weir, A. Sharma and G. C. Weir, *J. Biol. Chem.*, 2001, **276**, 31099–31104.
76. S. Obici, J. Wang, R. Chowdury, Z. Feng, U. Siddhanta, K. Morgan and L. Rossetti, *J. Clin. Invest.*, 2002, **109**, 1599–1605.
77. J. W. Peterson, I. Boldogh, V. L. Popov, S. S. Saini and A. K. Chopra, *Lab. Invest.*, 1998, **78**, 523–534.
78. R. N. Feng, Y. C. Niu, X. W. Sun, Q. Li, C. Zhao, C. Wang, F. C. Guo, C. H. Sun and Y. Li, *Diabetologia*, 2013, **56**, 985–994.
79. N. Wu, M. Yang, U. Gaur, H. Xu, Y. Yao and D. Li, *Biomol. Ther. (Seoul)*, 2016, **24**, 1–8.
80. H. J. Kim, J. H. Kim, S. Noh, H. J. Hur, M. J. Sung, J. T. Hwang, J. H. Park, H. J. Yang, M. S. Kim, D. Y. Kwon and S. H. Yoon, *J. Proteome Res.*, 2011, **10**, 722–731.
81. P. Zimmet, K. G. Alberti, F. Kaufman, N. Tajima, M. Silink, S. Arslanian, G. Wong, P. Bennett, J. Shaw, S. Caprio and IDF Consensus Group, *Pediatr. Diabetes*, 2007, **8**, 299–306.
82. S. J. Mihalik, S. F. Michaliszyn, J. de las Heras, F. Bacha, S. Lee, D. H. Chace, V. R. DeJesus, J. Vockley and S. A. Arslanian, *Diabetes Care*, 2012, **35**, 605–611.
83. C. B. Newgard, J. An, J. R. Bain, M. J. Muehlbauer, R. D. Stevens, L. F. Lien, A. M. Haqq, S. H. Shah, M. Arlotto, C. A. Slentz, J. Rochon, D. Gallup, O. Ilkayeva, B. R. Wenner, W. S. Yancy, Jr., H. Eisenson, G. Musante, R. S. Surwit, D. S. Millington, M. D. Butler and L. P. Svetkey, *Cell Metab.*, 2009, **9**, 311–326.

84. C. Patterson, L. Guariguata, G. Dahlquist, G. Soltesz, G. Ogle and M. Silink, *Diabetes Res. Clin. Pract.*, 2014, **103**, 161–175.
85. M. A. Atkinson, G. S. Eisenbarth and A. W. Michels, *Lancet*, 2014, **383**, 69–82.
86. F. R. Day and R. J. Loos, *J. Nutrigenet. Nutrigenomics*, 2011, **4**, 222–238.
87. G. M. Barlow, A. Yu and R. Mathur, *Nutr. Clin. Pract.*, 2015, **30**, 787–797.
88. S. Bouret, B. E. Levin and S. E. Ozanne, *Physiol. Rev.*, 2015, **95**, 47–82.
89. L. Brugnara, R. Mallol, J. Ribalta, M. Vinaixa, S. Murillo, T. Casserras, M. Guardiola, J. C. Vallve, S. G. Kalko, X. Correig and A. Novials, *PLoS One*, 2015, **10**, e0136348.
90. J. R. Bain, *Diabetes*, 2013, **62**, 349–351.
91. M. Oresic, S. Simell, M. Sysi-Aho, K. Nanto-Salonen, T. Seppanen-Laakso, V. Parikka, M. Katajamaa, A. Hekkala, I. Mattila, P. Keskinen, L. Yetukuri, A. Reinikainen, J. Lahde, T. Suortti, J. Hakalax, T. Simell, H. Hyoty, R. Veijola, J. Ilonen, R. Lahesmaa, M. Knip and O. Simell, *J. Exp. Med.*, 2008, **205**, 2975–2984.
92. M. Pflueger, T. Seppanen-Laakso, T. Suortti, T. Hyotylainen, P. Achenbach, E. Bonifacio, M. Oresic and A. G. Ziegler, *Diabetes*, 2011, **60**, 2740–2747.
93. C. Balderas, F. J. Ruperez, E. Ibanez, J. Senorans, J. Guerrero-Fernandez, I. G. Casado, R. Gracia-Bouthelier, A. Garcia and C. Barbas, *Electrophoresis*, 2013, **34**, 2882–2890.
94. M. E. Craig, C. Jefferies, D. Dabelea, N. Balde, A. Seth, K. C. Donaghue, P. International Society for Pediatric and Adolescent Diabetes, *Pediatr. Diabetes*, 2014, **15 Suppl 20**, 4–17.
95. D. La Torre, T. Seppanen-Laakso, H. E. Larsson, T. Hyotylainen, S. A. Ivarsson, A. Lernmark, M. Oresic and the DiPiS Study Group, *Diabetes*, 2013, **62**, 3951–3956.
96. M. E. Dumas, R. H. Barton, A. Toye, O. Cloarec, C. Blancher, A. Rothwell, J. Fearnside, R. Tatoud, V. Blanc, J. C. Lindon, S. C. Mitchell, E. Holmes, M. I. McCarthy, J. Scott, D. Gauguier and J. K. Nicholson, *Proc. Natl. Acad. Sci. U. S. A.*, 2006, **103**, 12511–12516.
97. C. He, Y. Shan and W. Song, *Nutr. Res.*, 2015, **35**, 361–367.
98. I. Treede, A. Braun, R. Sparla, M. Kuhnel, T. Giese, J. R. Turner, E. Anes, H. Kulaksiz, J. Fullekrug, W. Stremmel, G. Griffiths and R. Ehehalt, *J. Biol. Chem.*, 2007, **282**, 27155–27164.
99. B. Verges, *Diabetes Metab.*, 2009, **35**, 353–360.
100. E. Altmaier, S. L. Ramsay, A. Graber, H. W. Mewes, K. M. Weinberger and K. Suhre, *Endocrinology*, 2008, **149**, 3478–3489.
101. T. Inubushi, N. Kamemura, M. Oda, J. Sakurai, Y. Nakaya, N. Harada, M. Suenaga, Y. Matsunaga, K. Ishidoh and N. Katunuma, *J. Nutr. Sci. Vitaminol. (Tokyo)*, 2012, **58**, 415–422.
102. F. Wollesen, L. Brattstrom, H. Refsum, P. M. Ueland, L. Berglund and C. Berne, *Kidney Int.*, 1999, **55**, 1028–1035.

CHAPTER 9

Proteomics in Nutrition, Obesity and Diabetes Research

P. RUIZ-LIMON,[a] M. BALONGO,[a] M. INSENSER,[b*]
F. J. TINAHONES[a,c] AND M. MURRI[a,c*]

[a]Unidad de Gestión Clínica de Endocrinología y Nutrición, Instituto de Investigación Biomédica de Málaga (IBIMA), Hospital Clínico Virgen de la Victoria, Málaga, Spain
[b]Diabetes, Obesity and Human Reproduction Research Group, Department of Endocrinology and Nutrition, Hospital Universitario Ramón y Cajal, Universidad de Alcalá and Instituto Ramón y Cajal de Investigación Sanitaria (IRYCIS), Centro de Investigación Biomédica en Red de Diabetes y Enfermedades Metabólicas Asociadas (CIBERDEM), Madrid, Spain
[c]CIBER Fisiopatología de la Obesidad y Nutrición (CIBEROBN), Instituto de Salud Carlos III, Málaga, Spain
*Email: mariainsen@gmail.com or moramurri@gmail.com

9.1 Introduction

Nutritional interventions impact on the pathogenesis of obesity and diabetes at multiple levels. The shift in dietary behaviours, lifestyle patterns and prevalence of diabetes and obesity are spreading worldwide. The growing interest in nutritional research-related topics is continuously demanding the development and implementation of analytical techniques able to address the influence of diet in health and disease and the application to clinical

Nutritional Signaling Pathway Activities in Obesity and Diabetes
Food Chemistry, Function and Analysis No. 24
Edited by Zhiyong Cheng

Published by the Royal Society of Chemistry, www.rsc.org

medicine and diagnostics. Today, it is imperative to provide solid evidence of the mechanisms through which food can promote human health and well-being.[1] Despite significant advances in our understanding of nutritional signaling, a global strategy is required for the integration of advanced analytical techniques ("omics" platforms) and bioinformatics analysis.

Proteins are the key players in many cellular processes. Protein dysfunction impacts on the pathogenesis of obesity and diabetes. The study at protein level of nutritional interventions has the potential to explain molecular mechanisms and metabolic pathways of the effect of diets and supplements on health status in healthy, diabetic and obese subjects.[2]

Among "omics" techniques, proteomics has a high potential for linking genotype and phenotype, because proteins, rather than nucleic acids, represent the functional output of a cell. Proteomics allows the identification and quantification of complex protein mixtures with high accuracy. The rapid development of proteomics technologies has contributed to the characterization of molecular and cellular changes in protein abundance.[3] Proteomics technologies have opened new insights into the identification of potential biomarkers toward patient stratification and the development of new therapeutic targets focused on nutrition response associated with diabetes and obesity (Figure 9.1). There are several studies of proteomics in diabetes and obesity;

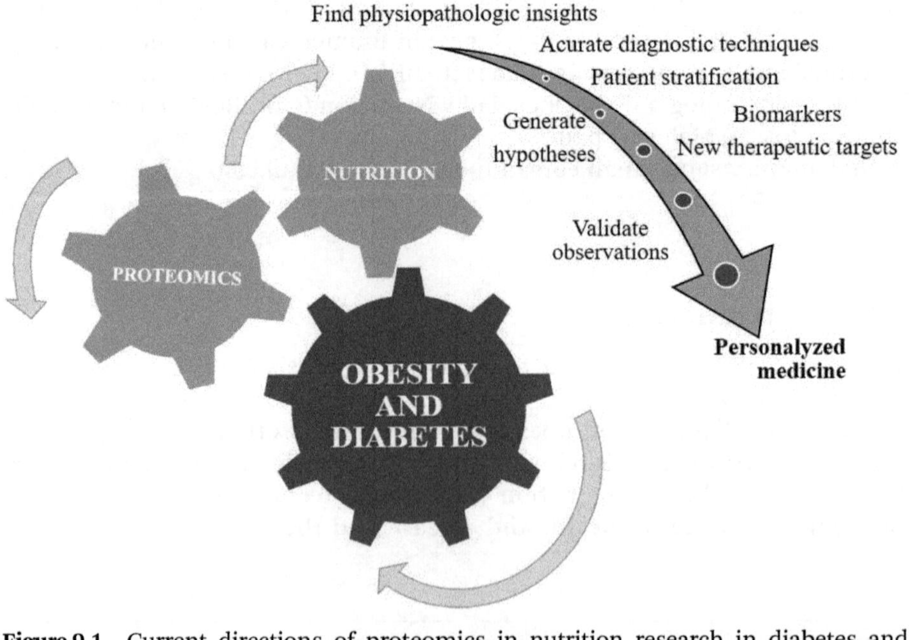

Figure 9.1 Current directions of proteomics in nutrition research in diabetes and obesity. The application of proteomics approaches is needed to understand the aetiological aspects of diabetes and obesity, the identification of potential biomarkers and the discovery of novel therapeutic targets. The data obtained from proteomics research has a potential role to validate clinical observations and generate new hypotheses.

however, literature related to nutritional signaling is scarce due to the higher complexity of proteomes as compared to transcriptomes and metabolomes. Proteomics applied to nutritional research aims to understand at molecular level how nutrients interact with the body, analyse the different responses to dietary interventions and find novel biomarkers for health or disease that are sensitive to such interventions.[4]

This chapter summarizes the evidence relating to the effects of nutrition on human health status, diabetes and obesity through proteomics analysis to offer an integrative vision of the complex relationships of the nutrient signaling network. First, an overview is given of the gel-based and gel-free approaches for the separation and characterization of protein profiles. The next sections review studies on the effects of food supplements on health status and nutritional signaling activities in obesity and diabetes. As a future perspective, further proteomics studies are needed to validate the discoveries made so far and to emphasize the importance of identifying integrative solutions to nutritional intervention outcomes.

9.2 Proteomics Technologies for Nutritional Signaling

Proteomics is an ideal technique for the high-throughput discovery of protein changes in response to dietary interventions in a biological fluid or tissue of interest. The first step in a proteomics biomarker study is the separation and quantification of proteins by comparing the proteome before and after nutritional stimuli. As shown in Figure 9.2, the main strategies commonly used to decrease sample complexity are based on either gel electrophoresis (GE) or liquid chromatography (LC). After applied high-resolution separation techniques, samples are analysed in a mass spectrometer.

Mass spectrometry (MS) is a highly selective analytical technique that separates the components of a sample by their mass and electrical charge. A mass spectrometer ionizes chemical compounds to generate charged molecules or molecule fragments and separates them on the basis of their mass-to-charge ratio (m/z value). A mass spectrometer consists of an ionization source, a mass analyser and a detector. The ionization techniques include matrix-assisted laser desorption ionization (MALDI) and electrospray ionization (ESI) because they cause little or no fragmentation of the molecule during the ionization and desorption processes. Mass analysers include time-of-flight (TOF) and quadrupole TOF (QTOF) instruments, ion traps and linear quadrupole ion trap-Fourier transform (LTQFTs). Typical detectors include electron multipliers (in ion traps), microchannel plates (in TOF) and an alternating-current image that is Fourier-transformed to an m/z spectrum (in LTQFT and Orbitrap). Another identification method is tandem MS (MS/MS). This analysis involves a first phase where selected peptides are isolated and subjected to further fragmentation in a gas-containing collision cell, and the second phase of MS analysis to provide amino acid sequence information.

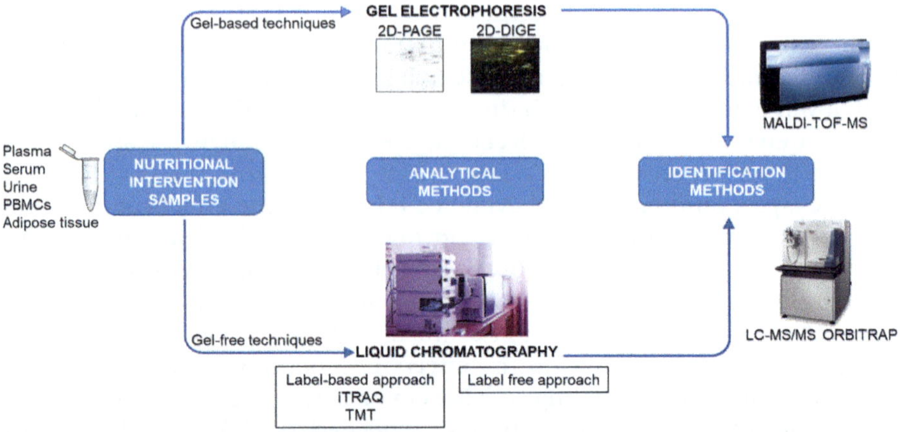

Figure 9.2 Schematic workflow of the gel-free and gel-based proteomics approaches. Samples collected after nutritional intervention can be analysed in two ways: with gel-based techniques to separate complex protein mixtures using electrophoresis (2D-DIGE and 2D-PAGE) or with gel-free techniques using a liquid chromatography separation method, which may use a marker (iTRAQ and TMT) or quantify based on the spectra data for quantifying the differences. After applied analytical separation techniques, samples are analysed in a mass spectrometer (MALDI-TOF-MS or LC-MS/MS ORBITRAP). 2D-DIGE, two-dimensional difference gel electrophoresis; 2D-PAGE, two-dimensional polyacrylamide gel electrophoresis; iTRAQ, isobaric tags for relative and absolute quantitation; LC-MS, liquid chromatography-mass spectrometry; MALDI-TOF-MS, matrix-assisted laser desorption/ionization-time-of-flight-mass spectrometry; MS, mass spectrometry; PBMCs, peripheral blood mononuclear cells; TMT, tandem mass tag.

In this section, we briefly explain the different platforms currently applied in nutrition research. No single method will allow the detection and identification of a whole proteome and a combination of several analytical techniques might increase the depth of proteome coverage. In addition, the proper integration of proteomics studies with other "omics" platforms in nutrition, obesity and diabetes research will help to translate findings from clinical practice into effective personalized nutritional strategies.[2]

9.2.1 Gel-based Techniques Used in Nutrition, Obesity and Diabetes Research

The classical approaches used in proteomics to separate complex proteins mixtures are based on two-dimensional polyacrylamide gel electrophoresis (2D-PAGE). Proteins are separated on a polyacrylamide gel based on their isoelectric point (the pH at which a protein is neutral or has no net charge) and molecular weight. The gel with the proteins is stained, digitalized and quantified by the assessment of spot intensities. Post-translational modifications, such as phosphorylation, or different protein isoforms can be easily detected because of the resulting shifts in protein mass or isoelectric point.

Differential GE (2D-DIGE) technology uses different fluorescent dyes (Cy2, Cy3, Cy5) to label protein samples before electrophoresis.[5] This method reduces inter-gel variation and permits the quantification of protein abundance with high accuracy and statistical certainty because pooled samples are used as an internal standard for quantification.

Gel-based proteomics techniques are widely used in nutrition research, although they were not free of limitations. Gel-based proteomics is a time-consuming and labour-intensive method and fails to efficiently separate proteins with extreme molecular weight and pH, or very hydrophobic proteins. Nevertheless, several proteomics studies have identified proteins with different relative abundance between obese or diabetic patients compared with a control group and between different nutritional interventions in both diseases. Protein spots with differential abundance are excised from the gel for subsequent identification by MS. The excised spots are digested with a protease (usually trypsin) and the resulting mixture of peptides is then analysed by MS. Most of the gel-based approaches reviewed here used the MALDI matrix in order to promote peptide ionization for MS analysis. This involves separation of peptides according to their TOF, which is dependent upon the m/z ratio. The resultant peptide m/z "fingerprint" is then compared against databases of theoretical peptide fingerprints of all proteins to identify the protein of interest.

Data collected from patients with obesity and/or diabetes allows the non-targeted discovery of novel molecules and metabolic pathways associated with these complex metabolic disorders. Moreover, the effect of nutritional interventions in these patients allows us the identification of potential biomarkers toward the patient stratification and the development of new therapeutic targets focused on nutrition response.

9.2.2 Gel-free Techniques Used in Nutrition, Obesity and Diabetes Research

Research in proteomics applied to nutrition has moved in a relatively short time from classical gel-based methodologies to more specific, faster and higher-throughput gel-free techniques based on conventional separation techniques such as liquid chromatography coupled with MS (LC-MS).[6]

Complex mixture proteins derived from biological fluids or tissue samples are digested into peptides with proteases like trypsin. Trypsin cleaves, with fairly high specificity, protein-peptide bonds at the C-termini of arginine and lysine residues.[7] To reduce the complexity of the proteolytic mixture, tryptic peptides can be fractionated into multiple samples based on properties such as charge, size, polarity or hydrophobicity and then separated using LC.[8] Chromatographic separation consists of dissolving the sample into a solvent and passing this mobile phase through a stationary phase; a chromatographic column containing surfaces with specific interaction chemistries allows the separation of the dissolved peptides. Each isolated peptide goes to a mass spectrometer producing a mass spectrum based on the m/z ratio of ions, which can be interpreted to determine the identity of the protein in the sample. With the resolving power of current mass analysers, it is not possible to identify peptides solely based on their intact masses. However, it is possible to fragment peptide ions in a mass spectrometer at the weakest bonds (usually the peptide bond between amino acids) by colliding them with inert gases. The resulting fragment ions are analysed in an MS/MS spectrum, which can be used to identify the amino acid sequence.[8]

MS-based quantification allows different nutritional interventions to be compared with each other. Relative quantification can be performed using two strategies: label-based and label-free methods.

In the label-based approach, peptides are chemically or metabolically linked with various tags in which ion signals corresponding to relative peptide abundance in the analysed sample. These approaches enable sample multiplexing, resulting in high quantification accuracy and precision based on the relative intensities of the reporter ions in fragment spectra. The most common labelling strategies include stable isotope labelling by amino acids in cell culture (SILAC), isotope-coded affinity tag (ICAT), isobaric tags for relative and absolute quantification (iTRAQ) and tandem mass tags (TMTs).[9–14]

By contrast, in label-free analysis each sample is analysed separately, and proteins are quantified on the basis of precursor ion intensity or the number of associated fragment spectra. Higher dynamic range and simpler sample preparation make such approaches popular compared with labelling techniques. The improvement of the label-free approach consists of the advance of a high-resolution mass spectrometer such as Orbitrap as well as TOF instruments. Label-free proteomics involves running different samples consecutively. Relative label-free methods can roughly be subdivided into counting the number of peptide-to-spectrum matches (PSMs; spectral counting) obtained for each protein[15] or measurement of peptide signal intensities by using the extracted ion chromatogram (XIC).[16–18] Spectral counting is based on determination and comparison of the number of MS/MS fragment-ion spectra of peptides between different samples without the consideration of physicochemical peptide properties. The second label-free technique is based on the measurement of MS signal intensity of the area under the chromatographic peak of the peptide precursor ion.

A limitation of label-free approaches is that they do not support multiplexing of samples and that every sample is handled and measured separately, which negatively affects the reproducibility of label-free approaches. These methods require highly reproducible workflows and an appropriate normalization. Furthermore, the impact of sample heterogeneity and contaminants in samples has to be considered.[19,20]

9.3 Proteomics Studies on the Effect of Food Supplement in Health Status

Food supplements are defined by the European Food Safety Authority (EFSA) as a concentrated source of nutrients or other substances with a nutritional or physiological effect (such as vitamins, minerals, amino acids, essential fatty acids and various plants and herbal extracts) that are marketed in "dose" form. Food supplements are proposed to correct nutritional deficiencies, maintain an adequate intake of certain nutrients or to support specific physiological functions.

In this section, proteomics responses to dietary supplement have been analysed (Table 9.1), such as alpha-tocopherol, cranberries, flaxseed, glucosamine and chondroitin supplementation, olive oil, orange juice and omega-3 polyunsaturated fatty acids (ω-3 PUFAs).

Alpha-tocopherol is one of the eight natural forms of vitamin E and is well established as an efficient antioxidant that protects membranes from free radical-initiated oxidation.[21–23] Aldred *et al.*[24] carried out proteomics analysis in plasma of healthy subjects who took 268 mg day^{-1} of alpha-tocopherol after 14 and 28 days. Results showed an increase in proteins involved in the regulation of the transcription process, oxygen transport and lipid metabolism; of the last one, particularly, the protein proapolipoprotein A1 is significantly increased. Proapolipoprotein A1 is enzymatically processed to obtain the mature apolipoprotein A1 (ApoA-1)[24] and ApoA-1 is associated with high-density lipoprotein (HDL) and has protective capacities against cardiovascular diseases.

Bekiares *et al.*[25] studied the effect of sweetened dried cranberry during 2 weeks on urinary proteome in healthy humans and found that proteins significantly decreased, mainly in processes such as blood coagulation, inflammation, cell adhesion and lipid transport. Cranberries have unique phenolic compounds which may be beneficial for reducing urinary tract infections,[26] providing antioxidant protection,[27,28] improving the lipid profile[29,30] and improving vasodilation to reduce cardiovascular disease (CVD) risk.[31] Also, cranberry phenolics have been implicated in improved insulin resistance in a rat model[32] and reduced rat hepatocyte glycogen phosphorylase activity.[33] Effects in the rat model are suggestive of hypoglycaemic effects for diabetic humans.[34]

Navarro *et al.*[35] performed proteomics analysis on the plasma of healthy subjects who took 1500 mg day^{-1} of glucosamine sulphate (GS) and 1200 mg day^{-1} of chondroitin sulphate (CS) supplementation for 28 days. Results showed altered proteins involved in the regulation of the transcription process, blood coagulation and apoptotic process. Also, this dietary supplement significantly decreased proteins involved in cell adhesion and inflammation. GS is a key building block for glycosaminoglycans in the extracellular matrix of cartilage and other connective tissues. CS is a complex polysaccharide, which is a structural component of cartilage, providing resistance to compression. Oral GS and CS supplements are thought to have anti-inflammatory and antiapoptotic effects.[36,37]

Table 9.1 Characteristics of the proteomics studies in nutritional signaling activities[†]

Study (ref.)	Sample type	Sample size	Analytic method	Identification method	Protein difference abundance: Increased	Protein difference abundance: Decreased	BMI before ($kg\ m^{-2}$)	BMI after ($kg\ m^{-2}$)	BMI ($kg\ m^{-2}$)	Obesity NW	Diabetes/MetS
24	Plasma	N:32 (21F/11M)	2D-PAGE	MALDI-TOF MS	4					NW	–
52	PBMCs	N:12 (7F/5M)	TMT	LC-MS/MS Orbitrap	HFHC + W:41 HFHC + G:26 HFHC + OJ:22	HFHC + W:7 HFHC + G:5 HFHC + OJ:16			20–25	NW	–
38	PBMCs	N:7 (7M)	2D-PAGE	MALDI-TOF MS	7	6				NW	–
54	Serum	N:81 fish oil:42 sunflower oil:39	2D-PAGE	MALDI-TOF MS LC-MS/MS		Fish oil:8				NW	–
47	Urine	N:63 LPOO:34 (19F/15M) HPOO:29 (17F/12M)	CE	ESI-TOF MS	LPOO: 3 HPOO: 6		LPOO: 23.9 ± 1.2 HPOO: 24.2 ± 4.8	LPOO: 24.6 ± 5.1 HPOO: 24.4 ± 4.8	< 25	NW	–
25	Urine	N:10 (8F/2M)	Label-free	Orbitrap Q-Exactive	1	21	24.07 ± 2.31		20.5–28.7	NW + OW	–
43	HDL proteome	N:6 (6M)	2D-DIGE	MALDI-TOF MS LC-MS/MS Orbitrap	6	6	26.1	26.3	24–28	OW	–
35	Plasma	N:18 (9F/9M)	Fluorescent labelling	Microarrays	49	51	F: 29 ± 2.4 M: 27 ± 2.8		25–32.5	OW	–
127	Plasma	N:14 (8F/6M)	iTRAQ	LC-MS/MS	46	41	37.9 ± 1.4	30.0 ± 1.1	>30	OW	Diabetes
62	Urine	N:5 (0F/5M)	2D-DIGE	MALDI-TOF MS	5	7	36 ± 8	31 ± 5	≥30	OW/Obese	–
157	Subcutaneous adipose tissue	N:8 (4F/4M)	2D-PAGE	MALDI-TOF MS LC-MS/MS	26	14	32.6 ± 1.1	29.5 ± 1.2	≥27	OW/Obese	–

65	Plasma	N : 781	Label-free	UHPLC-MS/MS	70	201			27–45	OW/Obese	–
64	Nuclear and cytoplasmic fraction of PBMCs	N: 39	2D-PAGE	MALDI-TOF MS	HSFA: 5 HMUFA: 3 HPUFA: 4 HPUFA ω-3: 3	HSFA: 5 HMUFA: 3 HPUFA: 1 HPUFA ω-3: 2	HSFA: 6 ± 1.2 HMUFA: 34.5 ± 1.2 HPUFA: 35.7 ± 1 HPUFA ω-3: 35 ± 1.2		≥25	OW/Obese	MetS
61	Plasma	N: 932 (620F/312M)	Fluorescently labelled polynucleotide aptamers	Multiplexed aptamer-based proteomics technology	53	51	F: 33.8/M: 33.8		≥25	OW/Obese	–
60	Plasma	N: 52 (44F/8M)	Label-free	LC-MS/MS Orbitrap	30	63	34.52 ± 0.55	30.27 ± 0.7	30–40	OW/Obese	–
58	Plasma	M: 473 (307F/166M)	TMT	LC-MS/MS Orbitrap	24	15	F: 34.3 ± 5.0 M: 34.2 ± 4.7	F: 30.58 ± 4.5 M: 30.61 ± 4.3	30–40	OW/Obese	–
63	Nuclear fraction PBMCs – cytoplasmic fraction PBMCs	N: 24 (6 per diet, 3F/3M)	2D-PAGE	MALDI-TOF MS	HSFA: 3 HMUFA: 1 LFHCC: 1 LFHCC ω-3: 4	HSFA: 1 HMUFA: 5 LFHCC: 5 LFHCC ω-3: 6	HSFA: 36.8 ± 2.4 HMUFA: 33.7 ± 4.3 LFHCC: 35.2 ± 3.4 LFHCC ω-3: 34.3 ± 3.8			OW/Obese	MetS

† 2D-DIGE, two-dimensional fluorescence difference gel electrophoresis; 2D-PAGE, two-dimensional gel electrophoresis; BMI, body mass index; CE, capillary electrophoresis; MALDI-TOF MS, matrix-assisted laser desorption/ionization-time of flight mass spectrometry; ESI-TOF MS, electrospray ionization-time of flight mass spectrometry; F, female; G, glucose; HDL, high-density lipoprotein; HFHC, high-fat high-carbohydrate; HMUFA, high monounsaturated fatty acid; HPOO, high-phenolic olive oil; HPUFA n-3, high-polyunsaturated fatty acid n-3; HPUFA, high-polyunsaturated fatty acid; HSFA, high-saturated-fatty-acid; iTRAQ, isobaric tags for relative and absolute quantification; LC-MS/MS, liquid chromatography mass spectrometry/mass spectrometry; LFHCC: low-fat, high carbohydrate diets supplemented with placebo; LFHCC n-3, low-fat, high-carbohydrate diets supplemented with long chain n-3 polyunsaturated fatty acids; LPOO, low-phenolic olive oil; M, male; MetS, metabolic syndrome; NW, normal weight; OJ, orange juice; OW, overweight; PBMCs, peripheral blood mononuclear cells; TMT, tandem mass tag; UHPLC-MS, ultra-high-pressure liquid chromatography-mass spectrometry; W,n water.

The study of Fuchs *et al.*[38] was performed in healthy subjects who consumed every morning 0.4 g of flaxseed per kg of body weight per day for 7 days. Flaxseed (*Linum usitatissimum*) is composed of enterolignan precursors, primarily secoisolariciresinol diglucoside,[39] total fibres, omega-3 fatty acids and alpha-linolenic acid. It has been shown to decrease lipid profile parameters, plasma trans fats,[40] atherogenicity,[41,42] glycaemia and pro-inflammatory oxylipins.[39] Proteome analysis of peripheral blood mononuclear cells (PBMCs) from healthy subjects after the intervention showed that flaxseed consumption increased levels of proteins involved in antioxidant pathways and decreased levels of proteins involved in cell adhesion and oxidant pathways that reduced reactive oxygen species in cells and consequently diminished the proinflammatory state. Likewise, the have observed an augmentation in proteins involved in stress response.

Burillo *et al.*[43] studied HDL proteome of healthy overweight volunteers who took 2 g day^{-1} of ω-3 PUFAs for 5 weeks. This study showed an alteration in proteins involved in blood coagulation, immunity and cell adhesion. Likewise, this work demonstrated a significant increase in the protein, Apo-A1, involved in lipid metabolism. This finding is in line with studies mentioned above, because Apo-A1 is the major component of HDL, and Apo-A1 protects from atherosclerosis development by its participation in cholesterol efflux and through its anti-inflammatory and antioxidant properties.[43] Different observational and intervention studies have confirmed a beneficial effect of diets enriched in ω-3 PUFAs on health status. Other potential beneficial effects of ω-3 PUFAs include antithrombogenic and antioxidant effect;[44] anti-inflammatory effect;[45] promotion of endothelial relaxation and a mild hypotensive effect.[46]

Silva *et al.*[47] evaluated the impact of olive oil (OO) supplementation, both low and high in phenolics on urinary proteomics biomarkers related to coronary artery disease (CAD). For that purpose, healthy participants were supplied with 20 mL of OO as a supplement (not heated or cooked) for 6 weeks. The proteomics study was carried out in urine. The results obtained in this urine proteomics study showed elevated abundance of proteins involved in extracellular matrix organization, among which the authors reported that eight of the 12 sequenced peptides were significantly regulated toward healthy scoring, including four collagen α-1 (I) chain, one α-2 (1) chain, one α-2 (V) chain, and one α-2 (VI) chain fragments. In this line, several studies have identified specific polypeptides to CAD panel biomarkers and certain collagens were the most abundant peptides sequenced.[48,49] Specifically, Delles *et al.*[50] found decreased urinary excretion of certain collagen types 1 and 3 alpha-chain fragments but an increase of specific collagen types 1 and 3 alpha-chain fragments with a C-terminal motif, related to CAD.[50] Also, Silva *et al.*[47] have shown that the consumption of both OOs improved the proteomics CAD score at endpoint compared with baseline. OOH is a mixture of fatty acids, such as unsaturated fatty acids, saturated fatty acids and other minor compounds, such as polyphenols, hydrocarbons, phytosterols and triterpenes.[51]

Other work in healthy participants with orange juice supplements was accomplished by Chaves *et al.*[52] These authors studied the proteome response at 3 and 5 hours in PBMCs after the ingestion of a high-fat and high-carbohydrate meal with 500 mL orange juice or water or an isocaloric drink. Orange juice (*Citrus sinensis*) has a high content of vitamin C, phenolic compounds and carotenoids,[53] and also is high in antioxidants and ascorbic acid. Briefly, when the meal was consumed with orange juice compared with water or isocaloric drink, it resulted in an upregulation of proteins involved in cell cycle and signal transduction and also a downregulation in proteins involved in immune response, showing an anti-inflammatory effect.

De Roos *et al.*[54] studied healthy volunteers who received 3.5 g day^{-1} of fish oil for 6 weeks and the proteomics study was performed in serum samples. Fish oil is a rich source of long-chain (ω-3) PUFAs. ω-3 PUFAs are precursors of anti-inflammatory molecules and provide benefits against chronic inflammatory conditions.[55] De Roos *et al.*[54] reported a decrease in proteins involved in cell adhesion, inflammation, blood coagulation, immune response and lipid metabolism. The proteins involved in lipid metabolism were ApoA-1 and apolipoprotein L-1 (ApoL-1). The first was the major component of HDL but the authors observed only decreased levels of one positional variant of the four other variants of ApoA-1; for that reason fish oil supplementation did not change overall levels of ApoA-1in serum. Levels of the second, ApoL-1, were positively associated with plasma triglycerides in normolipidaemic and dyslipidaemic humans,[56,57] which means that decreased levels in serum ApoL-1 might play a key role in the hypotriglyceridaemic effects of fish oil.

9.4 Proteomics Studies on Nutritional Signaling Activities in Obesity

Several proteomics studies have been performed on the effects of different diets in obese subjects (Tables 9.1 and 9.2).

A study by Oller-Moreno *et al.*[58] involved adults following an 8-week low-calorie diet (Modifast, Nutrition et Santé, France) consisting of 800 kcal day^{-1}. Adults whose weight loss was ≥8% after 8 weeks were randomly assigned to ad-libitum diets for 26 weeks as maintenance intervention. The diets differing in protein content and glycaemic index: (1) low protein, low glycaemic index; (2) low protein, high glycaemic index; (3) high protein, low glycaemic index; (4) high protein, high glycaemic index; and (5) control diet according to accepted national dietary guidelines.[59]

In a study by Geyer *et al.*,[60] participants followed a weekly supervised very-low-calorie powder diet (800 kcal day^{-1}; Cambridge Weight Plan, Corby, UK) for 8 weeks to achieve a weight loss of at least 7.5% of their initial body weight. During the weight maintenance phase, the calorie intake of the study participants was restricted to the estimated daily energy needs subtracted by 600 kcal. In the case of weight gain, up to two meals a day during the weight maintenance period were allowed to be replaced by Cambridge Weight Plan products to ensure weight maintenance.

Participants in the study of Carayol *et al.*[61] followed 8 weeks of weight loss intervention on a low-calorie formula diet (800 kcal, Modifast).

Table 9.2 Identified proteins under the effect of different diets in obesity[‡]

Proteins	ID protein	Abundance difference	Sample type	Diets	Duration	Study
Afamin	P43652	Decreased	Plasma	8 weeks LCD 800 kcal day^{-1} + 26 weeks ad-libitum diet	34 weeks	58
		Decreased	Plasma	VLCD 800 kcal day^{-1}	8 weeks	60
		Increased	Plasma	LCD 800 kcal day^{-1}	8 weeks	61
Albumin	P02768	Increased	Urine	LCD 1500 kcal day^{-1}	3 months	62
		Increased	Plasma	VLCD 800 kcal day^{-1}	8 weeks	60
Alpha-2-antiplasmin	P08697	Increased	Plasma	LCD 800 kcal day^{-1}	8 weeks	61
		Increased	Plasma	VLCD 800 kcal day^{-1}	8 weeks	60
Annexin A2	P07355	Decreased	Nuclear fraction PBMCs	HMUFA	12 weeks	63
		Increased	Subcutaneous adipose tissue	5 weeks VLCD 500 kcal day^{-1} + 1 week n + 2 weeks m	8 weeks	65
Antithrombin 3	P01008	Decreased	Plasma	LCD 800 kcal day^{-1}	8 weeks	61
		Increased	Plasma	8 weeks LCD 800 kcal day^{-1} + 26 weeks ad-libitum diet	34 weeks	58
		Increased	Plasma	VLCD 800 kcal day^{-1}	8 weeks	60
Beta-Ala-His dipeptidase	Q96KN2	Decreased	Plasma	VLCD 800 kcal day^{-1}	8 weeks	60
		Increased	Plasma	LCD 800 kcal day^{-1}	8 weeks	61
Capping protein muscle Z-line, beta	P47756	Decreased	Cytoplasmatic fraction PBMCs	HPUFA n-3	Postprandial	64
		Decreased	Cytoplasmatic fraction PBMCs	HMUFA	12 weeks	63
CD5 antigen-like	O43866	Decreased	Plasma	8 weeks LCD 800 kcal day^{-1} + 26 weeks ad-libitum diet	34 weeks	58
		Increased	Plasma	VLCD 800 kcal day^{-1}	8 weeks	60
Ceruloplasmin	P00450	Increased	Plasma	8 weeks LCD 800 kcal day^{-1} + 26 weeks ad-libitum diet	34 weeks	58
		Increased	Plasma	VLCD 800 kcal day^{-1}	8 weeks	60
Coagulation factor 10	P00742	Decreased	Plasma	VLCD 800 kcal day^{-1}	8 weeks	60
		Increased	Plasma	LCD 800 kcal day^{-1}	8 weeks	61

Complement C1q subcomponent subunit C	P02747	Decreased	Plasma	LCD 800 kcal day^{-1}	8 weeks	61
		Increased	Plasma	8 weeks LCD 800 kcal day^{-1} + 26 weeks ad-libitum diet	34 weeks	58
Complement C3	P01024	Decreased	Plasma	VLCD 800 kcal day^{-1}	8 weeks	60
		Increased	Plasma	LCD 800 kcal day^{-1}	8 weeks	61
Complement component C6	P13671	Increased	Plasma	8 weeks LCD 800 kcal day^{-1} + 26 weeks ad-libitum diet	34 weeks	58
		Increased	Plasma	VLCD 800 kcal day^{-1}	8 weeks	60
		Decreased	Plasma	LCD 800 kcal day^{-1}	8 weeks	61
Complement component C7	P10643	Increased	Plasma	8 weeks LCD 800 kcal day^{-1} + 26 weeks ad-libitum diet	34 weeks	58
		Increased	Plasma	VLCD 800 kcal day^{-1}	8 weeks	60
Complement component C9	P02748	Decreased	Plasma	LCD 800 kcal day^{-1}	8 weeks	61
		Increased	Plasma	8 weeks LCD 800 kcal day^{-1} + 26 weeks ad-libitum diet	34 weeks	58
		Increased	Plasma	VLCD 800 kcal day^{-1}		
Complement factor 1	P05156	Decreased	Plasma	8 weeks LCD 800 kcal day^{-1} + 26 weeks ad-libitum diet	34 weeks	58
		Decreased	Plasma	VLCD 800 kcal day^{-1}	8 weeks	60
		Increased	Plasma	LCD 800 kcal day^{-1}	8 weeks	61
Corticosteroid-binding globulin	P08185	Increased	Plasma	8 weeks LCD 800 kcal day^{-1} + 26 weeks ad-libitum diet	34 weeks	58
		Increased	Plasma	VLCD 800 kcal day^{-1}	8 weeks	60
		Increased	Plasma	LCD 800 kcal day^{-1}	8 weeks	61
Cysteine-rich secretory protein 3	P54108	Increased	Plasma	8 weeks LCD 800 kcal day^{-1} + 26 weeks ad-libitum diet	34 weeks	58
Extracellular matrix protein 1	Q16610	Increased	Plasma	8 weeks LCD 800 kcal day^{-1} + 26 weeks ad-libitum diet	34 weeks	58
		Increased	Plasma	VLCD 800 kcal day^{-1}	8 weeks	60

(*continued*)

Table 9.2 (*Continued*)

Proteins	ID protein	Abundance difference	Sample type	Diets	Duration	Study
Fatty acid transporter protein type 3	P05413	Decreased	Plasma	LCD 800 kcal day^{-1}	8 weeks	61
		Decreased	Subcutaneous adipose tissue	5 weeks VLCD 500 kcal day^{-1} + 1 week n + 2 weeks m	8 weeks	65
Fibrinogen alpha chain	P02671	Decreased	Nuclear fraction PBMCs	HMUFA	12 weeks	63
		Increased	Plasma	VLCD 800 kcal day^{-1}	8 weeks	60
Fibrinogen beta chain	P02675	Decreased	Nuclear fraction PBMCs	LFHCC	12 weeks	63
		Increased	Nuclear fraction PBMCs	LFHCC n-3	12 weeks	63
		Decreased	Cytoplasmatic fraction PBMCs	HSFA	Postprandial	64
		Increased	Plasma	VLCD 800 kcal day^{-1}	8 weeks	60
Ficolin 3	O75636	Decreased	Plasma	VLCD 800 kcal day^{-1}	8 weeks	60
		Increased	Plasma	LCD 800 kcal day^{-1}	8 weeks	61
Galectin-3-binding protein	Q08380	Decreased	Plasma	8 weeks LCD 800 kcal day^{-1} + 26 weeks ad-libitum diet	34 weeks	58
		Decreased	Plasma	VLCD 800 kcal day^{-1}	8 weeks	60
		Increased	Plasma	LCD 800 kcal day^{-1}	8 weeks	61
Gelsolin	P06396	Increased	Cytoplasmatic fraction PBMCs	LFHCC + n3	12 weeks	63
		Decreased	Nuclear fraction PBMCs	HMUFA	12 weeks	63
		Increased	Plasma	8 weeks LCD 800 kcal day^{-1} + 26 weeks ad-libitum diet	34 weeks	58
Haptoglobin	P00738	Increased	Plasma	LCD 800 kcal day^{-1}	8 weeks	61
		Decreased	Plasma	VLCD 800 kcal day^{-1}	8 weeks	60

Heat shock 70-kDa protein 1A	P0DMV8	Decreased	Cytoplasmatic fraction PBMCs	HSFA	Postprandial	64
		Increased	Cytoplasmatic fraction PBMCs	HMUFA	Postprandial	64
		Decreased	Plasma	LCD 800 kcal day^{-1}	8 weeks	61
Heparin cofactor 2	P05546	Decreased	Plasma	VLCD 800 kcal day^{-1}	8 weeks	60
		Increased	Plasma	LCD 800 kcal day^{-1}	8 weeks	61
Insulin-like growth factor-binding protein 2	P18065	Decreased	Plasma	LCD 800 kcal day^{-1}	8 weeks	61
		Increased	Plasma	8 weeks LCD 800 kcal day^{-1} + 26 weeks ad-libitum diet	34 weeks	58
Inter-alpha-trypsin inhibitor heavy chain H2	P19823	Increased	Plasma	VLCD 800 kcal day^{-1}	8 weeks	60
		Increased	Plasma	8 weeks LCD 800 kcal day^{-1} + 26 weeks ad-libitum diet	34 weeks	58
Inter-alpha-trypsin inhibitor heavy chain H3	Q06033	Increased	Plasma	8 weeks LCD 800 kcal day^{-1} + 26 weeks ad-libitum diet	34 weeks	58
		Increased	Plasma	VLCD 800 kcal day^{-1}	8 weeks	60
Kininogen-1	P01042	Increased	Plasma	LCD 800 kcal day^{-1}	8 weeks	61
		Increased	Plasma	VLCD 800 kcal day^{-1}	8 weeks	60
Lysozyme	P61626	Decreased	Plasma	LCD 800 kcal day^{-1}	8 weeks	61
		Increased	Plasma	VLCD 800 kcal day^{-1}	8 weeks	60
Mannan-binding lectin serine protease 2	O00187	Decreased	Plasma	8 weeks LCD 800 kcal day^{-1} + 26 weeks ad-libitum diet	34 weeks	58
		Decreased	Plasma	VLCD 800 kcal day^{-1}	8 weeks	60
Platelet glycoprotein Ib alpha chain	P07359	Decreased	Plasma	LCD 800 kcal day^{-1}	8 weeks	61
		Increased	Plasma	VLCD 800 kcal day^{-1}	8 weeks	60
Protein disulphide isomerase	P07237	Increased	Cytoplasmatic fraction PBMCs	HMUFA	Postprandial	64

(*continued*)

Table 9.2 (*Continued*)

Proteins	ID protein	Abundance difference	Sample type	Diets	Duration	Study
		Decreased	Cytoplasmatic fraction PBMCs	HSFA	Postprandial	64
		Increased	Plasma	LCD 800 kcal day^{-1}	8 weeks	61
Selenoprotein P	P49908	Increased	Plasma	8 weeks LCD 800 kcal day^{-1} + 26 weeks ad-libitum diet	34 weeks	58
		Increased	Plasma	VLCD 800 kcal day^{-1}	8 weeks	60
Serum amyloid A-1 protein	P0DJI8	Increased	Plasma	LCD 800 kcal day^{-1}	8 weeks	61
		Decreased	Plasma	VLCD 800 kcal day^{-1}	8 weeks	60
Serum amyloid P-component	P02743	Increased	Plasma	LCD 800 kcal day^{-1}	8 weeks	61
		Decreased	Plasma	VLCD 800 kcal day^{-1}	8 weeks	60
Sex hormone-binding globulin	P04278	Increased	Plasma	8 weeks LCD 800 kcal day^{-1} + 26 weeks ad-libitum diet	34 weeks	58
		Increased	Plasma	VLCD 800 kcal day^{-1}	8 weeks	60
		Decreased	Plasma	LCD 800 kcal day^{-1}	8 weeks	61
Thrombospondin 1	P07996	Increased	Cytoplasmatic fraction PBMCs	HSFA	Postprandial	64
		Decreased	Nuclear fraction PBMCs	LFHCC	12 weeks	63
Vitronectin	P04004	Decreased	Plasma	8 weeks LCD 800 kcal day^{-1} + 26 weeks ad-libitum diet	34 weeks	58
		Decreased	Plasma	VLCD 800 kcal day^{-1}	8 weeks	60

‡ 5 weeks VLCD 500 kcal day^{-1} + 1 week n + 2 weeks m, 5 weeks very-low-calorie diet 500 kcal day^{-1} + 1 week normalizing + 2 weeks maintenance; 8 weeks LCD 800 kcal day^{-1} + 26 weeks ad-libitum diet, 8 weeks low-calorie diet 800 kcal day^{-1} + 26 weeks assigned ad-libitum diet; HMUFA, high monounsaturated fatty acid; HPUFA n-3, high-polyunsaturated fatty acid n-3; HSFA, high-saturated fatty acid; LCD 1500 kcal day^{-1}, low-calorie diet 1500 kcal day^{-1}; LCD 800 kcal day^{-1}, low-calorie diet 800 kcal day^{-1}; LFHCC n-3, low-fat, high-carbohydrate diet supplemented with long-chain n-3 polyunsaturated fatty acids; LFHCC, low-fat, high-carbohydrate diet supplemented with placebo; PBMCs, peripheral blood mononuclear cells; VLCD 800 kcal day^{-1}, very-low-calorie powder diet 800 kcal day^{-1}.

Subjects in the study of Alfadda *et al.*[62] received a folder listing different combinations of meals and snacks that collectively would provide 1500 kcal day^{-1}. They then used their food lists to self-select three meals and two snacks a day during the 3-month duration of the study.

Several diets were used in a study by Rangel-Zúñiga *et al.*[63] for a period of 12 weeks, described as follows: high-saturated fatty acid (HSFA); high-monounsaturated fatty acid (HMUFA); low-fat high-complex carbohydrate (LFHCC); and low-fat, high-complex carbohydrate diet supplemented with n-3 polyunsaturated fatty acids (LFHCC n-3).

Camargo *et al.* evaluated the postprandial effect of several diets,[64] including HSFA, HMUFA and two high-polyunsaturated fatty acid (HPUFA) meals supplemented with n-3 PUFA or placebo.

The intervention of Bouwman *et al.*[65] was composed of a 5-week very-low-calorie diet (VLCD, 500 kcal day^{-1}, Modifast, Nutrition et Santé, France). During week 6, the VLCD was gradually replaced by normal ad-libitum meals with protein or carbohydrate supplements. Afterwards, there was a weight maintenance period of a low-fat ad-libitum diet with dietary consulting. For this study, the maintenance period was set for 2 weeks.

9.4.1 Protein Description

Those proteins that were found to be altered in at least two studies were included in Table 9.2, and many of them have been described in detail according to their function in this section.

9.4.1.1 Extracellular Matrix Proteins

Extracellular matrix (ECM) proteins represent the non-cellular components that surround the cells within all tissues and organs, providing them with mechanical support but also regulating key cellular processes in physiological and pathological conditions.[66] The composition is organ-specific; however, the major constituents are fibrous proteins (collagen, elastin, fibronectin, laminin) and highly acidic and hydrated glycans (glycosaminoglycan and proteoglycan).[66] ECM proteins are involved in inflammation and immunity. It has been reported that ECM macromolecules have the capacity to trigger or to inhibit complement activation, or act either as proinflammatory damage-associated molecular patterns or generate anti-inflammatory signals.[66]

9.4.1.1.1 Extracellular Matrix Protein 1 Low-calorie diets increased plasma ECM protein 1 (ECM1).[58,60] ECM1 is a multifunctional protein that interacts with the majority of other ECM proteins. Through these interactions, ECM1 enhances ECM protein binding and mediates a variety of processes such as extracellular matrix formation, cell adhesion, cell signaling, cell-mediated immune responses and regulation of tissue differentiation and/or maturation.[67,68]

9.4.1.1.2 Inter-α-trypsin Inhibitors Low-calorie diets increased plasma inter-α-trypsin inhibitor heavy-chain H2 and H3.[58,60] The inter-α-trypsin inhibitors are a family of structurally related plasma serine protease inhibitors involved in ECM stabilization and remodelling,[69] which are critical in normal processes such as tissue morphogenesis, differentiation and wound healing. The inter-α-trypsin inhibitor complex is a macromolecular arrangement of structurally related heavy-chain proteins (H1, H2 and H3) covalently cross-linked to the chondroitin sulphate chain of the proteoglycan bikunin. The inter-α-trypsin inhibitor complex is abundant in plasma and associated with inflammation, kidney diseases, cancer and diabetes.[70]

9.4.1.1.3 Thrombospondin 1 HSFA diet increased cytoplasm fraction PBMC thrombospondin 1 (THBS1),[64] while LFHCC diet decreased its abundance in nuclear fraction PBMCs.[63] THBS1 is a multifunctional glycoprotein released from various types of cells, including platelets, macrophages and adipocytes.[71] THBS1 interacts with a cohort of target molecules through its multiple functional domains and participates in a wide range of physiological and pathological processes such as cell–cell and cell–matrix interactions, cell proliferation and migration, tissue remodelling, wound healing, angiogenesis and inflammation.[67,72] THBS1 is also known to be a major regulator of transforming growth factor-β_1, which acts as a central regulator of the inflammatory response and also participates in regulating cell proliferation and ECM production. It has been reported that THBS1 is increased in adipose tissue and serum of obese rodents and obese and insulin-resistant subjects[71,73] and serum THBS1 levels are increased in response to high-fat diet challenge.[71]

9.4.1.1.4 Vitronectin VLCD and low-calorie diets decreased plasma vitronectin (VN).[58,60] VN, an abundant plasma and matrix glycoprotein, plays an important role in cell adhesion and motility. Vitronectin interacts with glycosaminoglycans and proteoglycans, it is recognized by certain members of the integrin family and serves as a cell-to-substrate adhesion molecule. Also, it binds plasminogen activator inhibitor 1 (PAI-1) and regulates its activity by stabilizing the active PAI-1 conformation as well as potentially regulating PAI-1 clearance.[74] The major physiological role of PAI-1 is to block the conversion of plasminogen to plasmin. And, in fact, PAI-1 deficiency in humans is associated with abnormal bleeding, while, in contrast, elevated plasma PAI-1 levels inhibit endogenous fibrinolysis.[74]

9.4.1.2 Inflammation

9.4.1.2.1 CD5 Antigen-like VLCD increased plasma CD5 antigen-like,[60] while low-calorie diet increased it.[58]

CD5 antigen-like is a secreted protein mainly expressed by macrophages in inflamed tissues It acts as a crucial regulator of lipid synthesis, inhibiting lipid droplet size in adipocytes, and regulates mechanisms in inflammatory responses.[75] CD5 antigen-like increases in blood with the progression of obesity and it is incorporated into adipocytes, inducing lipolysis in adipose tissue by suppressing the activity of fatty acid synthase. CD5L-induced lipolysis is necessary for the recruitment of inflammatory macrophages into adipose tissues, therefore contributing to the initiation of chronic inflammation in obesity.[75]

9.4.1.2.2 Ceruloplasmin Low-calorie diet and VLCD increased plasma ceruloplasmin.[58,60] Ceruloplasmin is a glycoprotein involved in the transport of copper.[76] Also, ceruloplasmin acts as a free-radical scavenger. Ceruloplasmin is an acute-phase protein in the inflammatory response. Conflicting results have been reported on the relation between ceruloplasmin levels and obesity, as increased levels or no differences have been found in obese subjects.[76,77]

9.4.1.2.3 Complement System Proteomics studies have shown that complement system is altered by low-calorie diets and VLCD; however, several components have been reported to be altered in different ways by the different diets (Table 9.2). The complement system is a complex enzymatic cascade consisting of more than 50 circulating and cell surface proteins working in cascades of stepwise protease activation.[78] The complement system is a central component of immune system; it is the first line of defence against pathogens but also it recognizes and opsonizes damaged and apoptotic endogenous cells, and promotes their clearance by macrophages. Complement activation can be triggered through three pathways: classical, alternative and lectin pathways varying in the pattern recognition molecule. The classical pathway is initiated by the activation of the C1 complex by antigen–antibody complexes recognized by C1q.[78] In the lectin pathway, mannose-binding lectin or ficolins bind specifically to pathogen surfaces or to altered self-structures *via* target structures,[79] and lead to autoactivation of mannose-binding lectin-associated serine proteases. After first activation, both the classical and lectin pathways activate C4 and C2, into C4a and C4b and C2a and C2b, respectively. C4b and C2a form the C3 convertase, promoting the cleavage of C3 into C3a and C3b. Together with C4bC2a, C3b then forms the C5 convertase, which cleaves C5 into the anaphylatoxin C5a and fragment C5b. When C5b associates with C6 and C7, the complex becomes inserted into cell membranes and interacts with C8, inducing the binding of several units of C9 to form lytic pores in the plasma membrane of target cells, the terminal membrane attack complex (MAC), which is capable of lysing cells. Several factors modulate the activation of complement system to prevent the negative effects of its overactivation.[79] These factors are named complement regulatory proteins, and include, among others, complement factor I and VN.

Complement plays a major role in metabolic organs and in metabolic diseases; it affects insulin secretion and adipocyte maturation.[80] Previous studies have shown altered complement system in obesity, where excessive activation of the classic pathway of complement commonly occurs.[81]

9.4.1.2.4 Cysteine-rich Secretory Protein-3 Low-calorie diets increased plasma cysteine-rich secretory proteins-3 (CRISP-3).[58,61] CRISP-3 is a member of the CRISP family.[82] CRISP-3 has been shown to have a role in innate immunity, defence response and chronic inflammation. CRISP-3 upregulation is associated with the downregulation of ANXA1, an anti-inflammatory protein, leading to the establishment of an inflammatory state.[83]

9.4.1.2.5 Galectin-3 A low-calorie diet and VLCD decreased plasma galectin-3,[58,60] while a high-calorie diet increased it.[61] Galectin-3 is a β-galactoside-binding protein which interferes with cell adhesion, proliferation, differentiation and angiogenesis.[84] It appears to have both pro- and anti-inflammatory roles depending on the cell type, cellular localization and pathophysiological condition. Extracellular galectin-3 promotes respiratory burst in monocytes, acts itself as a chemoattractant for monocytes and macrophages and mediates adhesion of neutrophils to the ECM protein laminin. Galectin-3 has been implicated in the development of Type 2 diabetes and obesity.[85] Human studies have shown that galectin-3 is increased in obese and diabetic individuals. Moreover, galectin-3 appears to be a modulator of adipogenesis by stimulating differentiation of preadipocytes into mature adipocytes.

9.4.1.2.6 Gelsolin LFHCC diet supplemented with long-chain n-3 PUFAs and low-calorie diet increased cytoplasmic fraction PBMCs and plasma gelsolin,[58,63] while HMUFA decreased nuclear fraction PBMC gelsolin.[63] Gelsolin is a widely distributed actin-modulating protein that exerts a variety of physiological and pathological processes.[86] Intracellular gelsolin participates in the regulation of cellular architecture and motility through its severing, capping and nucleating activities on actin monomers or filaments. Fibroblasts in which gelsolin has been overexpressed display increased motility.[87] The secreted form of gelsolin plays an important role in the inflammation process. It binds actin and a broad spectrum of bioactive mediators that are used in inflammation.[88] Gelsolin has been recognized as a potential biomarker of inflammatory-associated medical conditions, allowing for the prediction of illness severity, efficacy of treatment, recovery and clinical outcome.[88]

9.4.1.2.7 Haptoglobin VLCD decreased plasma haptoglobin (Hp),[60] while low-calorie diet increased it.[61] Hp is an acute-phase protein that is the main carrier of cell-free haemoglobin (Hb), and its circulating levels have been considered a marker of inflammation in clinical practice. Haemolysis often occurs during severe bacterial infection and sepsis, and free Hb increases the effects of infection and amplifies inflammation.[89] Hp levels positively correlate with body mass index and the level of adiposity. In fact, the main contributor to the elevated levels of Hp detected in obesity is white adipose tissue. Such an increase is clearly related to an elevated inflammatory status. Animal studies have also reported that Hp deficiency seems to protect from metabolic complications associated with obesity.[89]

9.4.1.2.8 Heat Shock Protein 70 Kilodalton Heat shock protein 70 kilodalton (HSP70) was decreased by low-calorie diet in plasma[61] and by HSFA in cytoplasmic fraction PBMCs,[64] while it was increased by HMUFA in cytoplasmic fraction PBMCs.[64] HSP70 plays a significant role in protecting against cellular damage and dysfunction. In obesity, it has been reported that the expression levels of HSP70 in skeletal muscle and adipose tissue are reduced.[90]

9.4.1.2.9 Lysozyme VLCD increased plasma lysozyme (LZ),[60] while low-calorie diet decreased it.[61] LZ is a well-characterized, naturally occurring protein that exerts antimicrobial effects through the hydrolysis of bacterial cell wall peptidoglycan.[91] Lysozyme modulates immune responses; although lysozyme is important for driving a proinflammatory response, it also plays a role in limiting inflammation systemically.[92] Severely obese subjects showed decreased protein levels of lysozyme and lysozyme protein levels correlated inversely with body mass index.[93]

9.4.1.2.10 Serum Amyloid A and P VLCD decreased plasma serum amyloid A (SAA) and P (SAP),[60] while low-calorie diet increased them.[61] SAA is a major acute-phase protein in humans, that after secretion into the circulation associates with HDL particles.[94] During acute inflammation, serum SAA levels may rise up to 1000-fold, and SAA displaces ApoA-1 from HDL. SAA induces the synthesis of several cytokines and both stimulate secretion of matrix metalloproteinases from several cell types. Moreover, SAA synergizes with chemokines to increase chemoattraction of monocytes and granulocytes.[94] As a result, these mediators may cooperate to increase the inflammatory response. On the other hand, SAP is an acute-phase protein made by hepatocytes and secreted into the blood.[95] It regulates several aspects of the immune system. SAP interacts with components of the complement pathway to regulate complement activation. Moreover, SAP binds directly to monocytes, neutrophils and macrophages to modify their activation and alter their differentiation to modulate the immune response.[96]

Circulating A-SAA levels have been reported to be elevated in obese compared to lean individuals,[97] and a strong association between body mass index and SAA levels was found in the 11 cross-sectional studies.[98] Moreover weight loss decreases adipose A-SAA expression and secretion as well as circulating A-SAA levels.[98] Elevated SAP levels in obese have also been reported, accompanied by a positive correlation with body mass index.[99]

9.4.1.3 Antioxidants

9.4.1.3.1 Albumin Low-calorie diets and VLCD increased urine and plasma albumin.[60,62] Albumin, the most abundant plasma protein, provides oncotic pressure, transports a large variety of molecules, including fatty acids, drugs and hormones, and acts as an antioxidant agent.[100] The main ligands of serum albumin-implied antioxidant functions of the protein are transition metal ions (copper and iron essentially) that can potentially be extremely pro-oxidant.[101] Moreover, the resulting albumin-bound bilirubin acts as an inhibitor of lipid peroxidation. Two-thirds of serum albumin is reduced working as a free radical scavenger; its Cys34 residue is able to trap multiple reactive species, such as hydrogen peroxide, peroxynitrite, superoxide or hypochlorous acid.[101] Serum albumin concentration was positively associated with the prevalence of metabolic syndrome (MetS), whereas an increase in serum albumin over time might protect against MetS development.[100] Obesity and morbid obesity can be considered as independent predictors of hypoalbuminaemia.[102]

9.4.1.3.2 Selenoprotein P Oller-Moreno *et al.* and Geyer *et al.*[58,60] reported that low-calorie diets and VLCD increased plasma selenoprotein P (SeP). SeP is an abundant extracellular glycoprotein, the circulating form of which is predominantly secreted by the liver, although all tissues express the protein.[103] SeP serves as a major circulating selenium transporter and plays a main role in protecting cells against oxidative stress due to its phospholipid hydroperoxide thiol peroxidase activity. Circulating SeP levels have been reported to be elevated in patients with Type 2 diabetes, overweight and obese.[103]

9.4.1.4 Haemostasis

Haemostatic process is regulated by coagulation, with the formation of a fibrin clot and fibrinolysis leading to the dissolution of fibrin clots. Since clotting is critical for health, imbalances in clot formation, stability or degradation may result in pathology. Excessive clot formation leads to thrombosis and related diseases, while inadequate clot formation causes clinical bleeding. Platelets, produced by megakaryocytes in the bone marrow, play an essential role in haemostasis *via* three major processes: activation, adhesion and aggregation. The clot is a molecular mesh made of polymerized fibrin. The clot is formed *via* a cascade of enzymatic reactions that activates prothrombin to the proteolytic enzyme thrombin, which converts soluble fibrinogen to make insoluble fibrin, the process referred to as blood clotting.[104]

Blood coagulation factors are largely made by the liver and secreted in blood in inactive forms. Several coagulation factors are expressed by extrahepatic tissues, including, among others, coagulation factors 9 and 10. The role of activated factor 9 in the blood coagulation cascade is to activate factor 10 to its active form, while coagulation factor 10 yields the active thrombin, which is responsible for fibrin polymerization to form a thrombus. Proteomics studies have shown that different diets affect coagulation elements (including fibrinogen, coagulation factor 10, platelet glycoprotein Ib alpha) in different ways (Table 9.2).

Previously, alterations in the coagulation process have been reported in obese people, including elevated activity and concentration of the coagulation factors and raised platelet aggregation. There is a strong association between obesity and thrombotic disease.[105] The prothrombotic state of obesity is thought to be driven in large part by chronic inflammation and impaired fibrinolysis. Moreover, there is a relationship between abdominal obesity, platelet blood count and mean platelet volume in patients with MetS.[106]

9.4.1.4.1 Alpha 2-antiplasmin Proteomics studies have shown that plasma alpha 2-antiplasmin (a2-AP) is increased after low-calorie diet and VLCD.[60,61] a2-AP is the most important fibrinolytic regulator in the thrombus formed. Fibrinolysis is initiated by plasminogen activators that cleave the proenzyme plasminogen to plasmin. a2-AP inhibits plasmin in one of the fastest protein–protein reactions described. Previously, it was reported that a2-AP level was increased in patients with morbid obesity.[107]

9.4.1.4.2 Annexin A2 Annexin A2 (ANX2) regulates several functions within diverse cellular compartments and is also secreted.[108] Cell surface ANX2 is implicated in fibrinolysis, coagulation and control of inflammation. It can interact with plasminogen and tissue plasminogen activator, which converts plasminogen to plasmin, thus mediating the lysis of fibrin polymers and ECM degradation, and promotes cell invasion and migration.[109] ANX2-deficient mice showed fibrin accumulation in different tissues and organs.[110] Moreover, secreted ANX-2 can activate macrophage for interleukin (IL)-1, IL-6, and tumour necrosis factor-α secretion.[109]

9.4.1.4.3 Antithrombin III While Oller-Moreno *et al.* and Geyer *et al.*[58,60] reported that low-calorie diets and VLCD increased plasma antithrombin III (AT III), Carayol *et al.*[61] reported that low-calorie diet decreased it. AT III is the most important serine protease inhibitor in plasma that regulates the blood coagulation cascade, as it inhibits the activities of thrombin and factors IXa, Xa, XIa and XIIa.[111] Its deficiency represents a risk factor for thrombotic diseases. AT III is reduced in obesity and normalizes as weight is reduced.[112] Similar to antithrombin, heparin cofactor II inhibits thrombin in a reaction that is accelerated more than 1000-fold by heparin. However, the rate of thrombin inhibition by heparin cofactor II in the absence or presence of heparin or heparin-like molecules is significantly slower than by antithrombin under similar conditions.[113] Proteomics studies have shown that VLCD decreased plasma heparin cofactor II,[60] while low-calorie diet increased it.[61]

9.4.1.4.4 Kininogen Low-calorie diets and VLCD increased plasma kininogen (KNG).[60,61] KNG is the precursor protein of the kallikrein-kinin system.[114] High-molecular-weight (HMW) and low-molecular-weight (LMW) KNGs are formed by alternative splicing of kng1 in mammals. HMW KNG interacts with plasma kallikrein to produce bradykinin and is involved in the contact activation of the blood coagulation process. HMW kininogen plays an important role in blood coagulation by helping to position optimally prekallikrein and factor XI next to factor XII, but also it inhibits the thrombin- and plasmin-induced aggregation of thrombocytes. LMW kininogen inhibits the aggregation of thrombocytes.

9.4.1.5 Glucose Metabolism

9.4.1.5.1 Afamin While Oller-Moreno *et al.* and Geyer *et al.*[58,60] reported that low-calorie diets and VLCD decreased plasma afamin, Carayol *et al.*[61] reported that low-calorie diet increased plasma afamin. Afamin is a glycoprotein, and is a member of the human albumin gene family. Its physiological or pathophysiological functions are still very limited. However, it has been reported that transgenic mice overexpressing the human afamin gene developed increased body weight and elevated blood levels of lipids and glucose.[115]

9.4.1.5.2 Corticosteroid-binding Globulin Low-calorie diets and VLCD increased plasma corticosteroid-binding globulin (CBG).[58,60] CBG is the specific plasma transport glycoprotein for glucocorticoid hormones (GCs).[116] GCs are adrenal cortex steroids that were named for their role in the regulation of glucose metabolism.[117] They also play an intricate role in fatty acid metabolism and storage. Finally, they are important regulators of appetite through complex interactions with orexigenic and anorexigenic neuropeptides and hormones. In humans, plasma CBG was found to be a marker of insulin secretion, being negatively or positively correlated with various markers of obesity depending on the degree of insulin resistance.[118] Indeed, insulin was shown to downregulate CBG levels *in vitro*. Surprisingly, CBG cleavage was found to be reduced in subjects with central obesity and MetS. The mechanism of reduced CBG cleavage in obese patients remains to be described.[119] Furthermore, weight loss in morbidly obese subjects decreased CBG levels.[120]

9.4.1.5.3 Insulin-like Growth Factor-binding Protein-2 Carayol *et al.*[61] and Oller-Moreno *et al.*[58],showed contradictory results, with low-calorie diets decreasing or increasing plasma insulin-like growth factor-binding protein-2 (IGFBP-2). IGFBP-2 is the principal binding protein secreted during adipogenesis of white preadipocytes, but its physiological roles remain poorly understood.[121] IGFBP-2 has been reported to be a key regulator of metabolic diseases, such as diabetes and obesity. It is speculated that IGFBP-1 modulates insulin sensitivity in an insulin-like growth factor-independent manner.[122] Serum IGFBP-2 levels have been reported to be reduced in obese humans.[123] Moreover, IGFBP-2 gene expression was downregulated in visceral white adipose tissue of mice and its circulating levels were reduced in obese *ob/ob*, *db/db* and high-fat-fed mice.[124]

9.4.1.6 Hormones

9.4.1.6.1 Sex Hormone-binding Globulin While Oller-Moreno *et al.* and Geyer *et al.*[58,60] reported that low-calorie diets and VLCD increased plasma sex hormone-binding globulin (SHBG), Carayol *et al.*[61] reported that low-calorie diet decreased plasma SHBG. SHBG is a plasma glycoprotein with high binding affinity for testosterone and dihydrotestosterone and lower affinity for estradiol.[125] SHBG is synthesized in the liver, and its plasma level is important in the regulation of plasma free and albumin-bound androgens and oestrogens. SHBG levels are reduced with obesity and the MetS.[125,126]

9.5 Proteomics Studies on Nutritional Signaling Activities in Diabetes

Actually, just one study[127] has performed proteomics analysis on the effect of a diet in Type 2 diabetic (T2DM) patients. All T2DM patients were also obese and followed VLCD (Modifast ~ 450 kcal day^{-1}) for 16 weeks. Proteomics analysis was carried out in plasma samples. VLCD has been reported to lead to considerable weight loss and subsequently major metabolic improvements in obese T2DM patients.[128] Moreover, several studies have shown that a 16-week VLCD in T2DM patients leads to an increase in quality of life.[129,130] Sleddering *et al.*[127] showed that, after the 16-week VLCD diet, there was a significant decrease in haemoglobin A_{1c} and a reduction of body weight in T2DM obese patients. These results were accompanied by an increased abundance of alpha-1-antitrypsin (ATT). ATT is an acute-phase protein mainly produced by liver, with a broad spectrum of biological functions, including inhibition of proteases, immune modulatory functions and the transport of hormones.[131] Evidence has shown that ATT has broad anti-inflammatory and immunomodulating properties.[132] AAT has the ability to prevent diabetes formation (*in vivo*) and B-cell apoptosis (*in vitro*), meaning that deficiencies in the quantity of AAT may increase the risk of developing diabetes.[133] Furthermore, Type 2 diabetes is associated with decreased levels of plasma AAT, and reduced levels of ATT have been shown to correlate with obesity and increased body mass index.[134–136] This evidence is in line with the results obtained in the proteomics analysis that found increased levels of ATT after VLCD in T2DM obese patients.

Several apolipoproteins were decreased after 16 weeks of VLCD, such as apolipoprotein A-4 (ApoA-4), apolipoprotein B-100 (ApoB-100), apolipoprotein C-3 (ApoC-3) and apolipoprotein E (ApoE). ApoA-4 has antiatherogenic[137] and anti-inflammatory[138] properties. Moreover, it has also been reported to be a satiety factor.[139,140] Interestingly, it has been hypothesized that, as ApoA-4 acts as a satiety factor, lower ApoA-4 levels may be a signal for stimulating feeding behaviour.[141] ApoB-100 is the protein part of low-density lipoprotein (LDL), very-low-density lipoprotein (VLDL) and intermediate-density lipoprotein (IDL).[142] Moreover, increased serum ApoB-100 levels are well established to be solid predictors for ischaemic cardiovascular events[143] and also, several findings observed that T2DM and insulin resistance were associated with elevated ApoB-100 levels.[144,145] ApoC-3 is known as an important element of hypertriglyceridaemia and a risk factor for CVD.[146] Several researchers have reported elevation of plasma ApoC-3 in subjects with Type 2 diabetes.[147] Moreover, ApoE is a major protein of the lipoprotein transport system involved in plasma metabolism and clearance of triglyceride-rich lipoproteins from circulation.[148] Numerous studies showed that ApoE is associated with the development and progression of coronary heart disease in humans,[149–153] and in addition, ApoE has been associated with T2DM.[154]

Moreover, complement C3 (C3) levels were decreased after 16 weeks of VLCD. C3 has a key role in the immune system and levels of C3 have been positively associated with body mass index and linked with development of MetS and diabetes.[155,156] For that reason, the results of Sleddering *et al.*[127] of C3 protein levels after VLCD are in line with other evidence.

9.6 Future Perspectives

Omics platforms are great techniques for the discovery and characterization of nutritional signaling activities defining the nutritional status of healthy, diabetic and obese individuals and identifying the nutritional bioactive compounds responsible for their health outcomes. Systems biology, which aims to integrate several omics disciplines, computational analysis and predictive networks in the interpretation of molecular mechanisms underlying the biological response, is able to provide a more global comprehensive understanding of how diet may influence health and disease.

Among omics approaches, proteomics has the advantage of detecting post-translational modifications and protein interactions that cannot be predicted

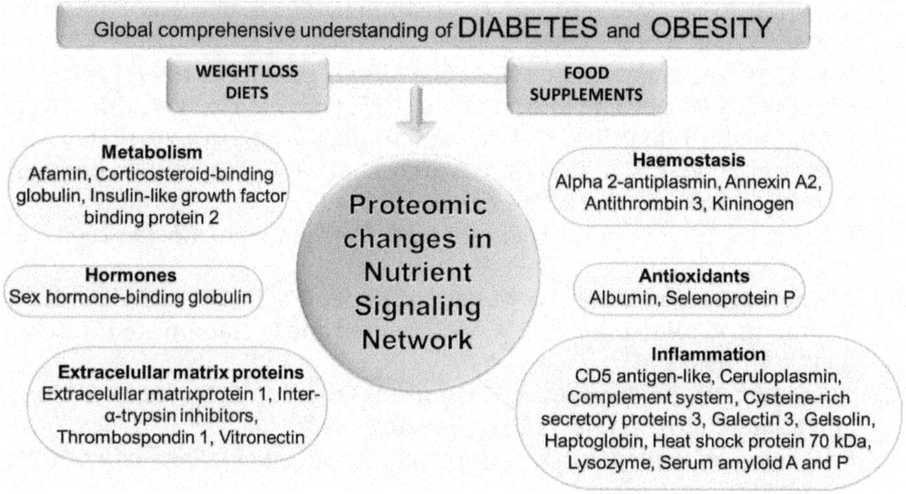

Figure 9.3 Making sense of proteomics data in nutritional research. Proteins associated with obesity and diabetes after a nutritional intervention are involved in several biological processes.

by genome studies and remain outside of the commonly known pathways. Nevertheless, despite the promising potential of this research area, proteomics studies need to overcome the challenge of complex protein analysis and apply more robust and sensitive proteomics strategies.

Few studies have addressed proteomics nutritional signaling in obesity and even less in diabetes. The altered proteins and their biological processes associated with nutritional interventions in obese or diabetic subjects are summarized in Figure 9.3. These studies highlight the complexity of nutritional signaling and the fact that each individual study cannot capture the entire molecular mechanisms and pathways involved in nutritional responses. Further research is needed to better understand the complex interactions between protein networks that cannot be predicted when studying proteins separately. Proteomics may help us to understand how proteomics nutritional signaling activities are becoming important in clinical practice and, in the near future, might allow prevention of comorbidities and may promote health in obese and diabetic patients.

References

1. D. Braconi, G. Bernardini, L. Millucci and A. Santucci, *Expert Rev. Proteomics*, 2018, **15**, 153–164.
2. J. de Toro-Martín, B. J. Arsenault, J.-P. Després and M.-C. Vohl, *Nutrients*, 2017, **9**, 913.
3. W. C. S. Cho, *Genomics. Proteomics Bioinformatics*, 2007, **5**, 77–85.
4. M. Kussmann, A. Panchaud and M. Affolter, *J. Proteome Res.*, 2010, **9**, 4876–4887.

5. M. Unlü, M. E. Morgan and J. S. Minden, *Electrophoresis*, 1997, **18**, 2071–2077.
6. X. Li, W. Wang and J. Chen, *Sci. China: Life Sci.*, 2017, **60**, 1093–1113.
7. J. V. Olsen, S.-E. Ong and M. Mann, *Mol. Cell. Proteomics*, 2004, **3**, 608–614.
8. N. Pappireddi, L. Martin and M. Wühr, *ChemBioChem.*, 2019, **20**, 1210–1224.
9. S. P. Gygi, B. Rist, S. A. Gerber, F. Turecek, M. H. Gelb and R. Aebersold, *Nat. Biotechnol.*, 1999, **17**, 994–999.
10. J.-L. Hsu, S.-Y. Huang, N.-H. Chow and S.-H. Chen, *Anal. Chem.*, 2003, 75, 6843–6852.
11. A. Thompson, J. Schäfer, K. Kuhn, S. Kienle, J. Schwarz, G. Schmidt, T. Neumann, R. Johnstone, A. K. A. Mohammed and C. Hamon, *Anal. Chem.*, 2003, **75**, 1895–1904.
12. C. Lindemann, N. Thomanek, F. Hundt, T. Lerari, H. E. Meyer, D. Wolters and K. Marcus, *Biol. Chem.*, 2017, **398**, 687–699.
13. S. Wiese, K. A. Reidegeld, H. E. Meyer and B. Warscheid, *Proteomics*, 2007, **7**, 340–350.
14. O. T. Schubert, H. L. Röst, B. C. Collins, G. Rosenberger and R. Aebersold, *Nat. Protoc.*, 2017, **12**, 1289–1294.
15. M. P. Washburn, D. Wolters and J. R. Yates, *Nat. Biotechnol.*, 2001, **19**, 242–247.
16. M. C. Wiener, J. R. Sachs, E. G. Deyanova and N. A. Yates, *Anal. Chem.*, 2004, **76**, 6085–6096.
17. D. A. Megger, L. L. Pott, M. Ahrens, J. Padden, T. Bracht, K. Kuhlmann, M. Eisenacher, H. E. Meyer and B. Sitek, *Biochim. Biophys. Acta, Proteins Proteomics*, 2014, **1844**, 967–976.
18. D. A. Megger, T. Bracht, H. E. Meyer and B. Sitek, *Biochim. Biophys. Acta*, 2013, **1834**, 1581–1590.
19. R. B. H. Braakman, M. M. A. Tilanus-Linthorst, N. Q. Liu, C. Stingl, L. J. M. Dekker, T. M. Luider, J. W. M. Martens, J. A. Foekens and A. Umar, *J. Proteomics*, 2012, 75, 2844–2854.
20. E. S. Baker, T. Liu, V. A. Petyuk, K. E. Burnum-Johnson, Y. M. Ibrahim, G. A. Anderson and R. D. Smith, *Genome Med.*, 2012, **4**, 63.
21. E. Niki and M. G. Traber, *Ann. Nutr. Metab.*, 2012, **61**, 207–212.
22. X. Wang and P. J. Quinn, *Prog. Lipid Res.*, 1999, **38**, 309–336.
23. M. G. Traber and J. Atkinson, *Free Radic. Biol. Med.*, 2007, **43**, 4–15.
24. S. Aldred, T. Sozzi, I. Mudway, M. M. Grant, H. Neubert, F. J. Kelly and H. R. Griffiths, *Proteomics*, 2006, **6**, 1695–1703.
25. N. Bekiares, C. G. Krueger, J. J. Meudt, D. Shanmuganayagam and J. D. Reed, *OMICS*, 2018, **22**, 145–153.
26. J. Avorn, M. Monane, J. H. Gurwitz, R. J. Glynn, I. Choodnovskiy and L. A. Lipsitz, *JAMA, J. Am. Med. Assoc.*, 1994, **271**, 751–754.
27. T. Wilson, J. P. Porcari and D. Harbin, *Life Sci.*, 1998, **62**, PL381–PL386.
28. A. P. Singh, T. Wilson, A. J. Kalk, J. Cheong and N. Vorsa, *Food Chem.*, 2009, **116**, 963–968.
29. G. Ruel, S. Pomerleau, P. Couture, S. Lemieux, B. Lamarche and C. Couillard, *Br. J. Nutr.*, 2006, **96**, 357–364.

30. I. T. Lee, Y. C. Chan, C. W. Lin, W. J. Lee and W. H.-H. Sheu, *Diabet. Med.*, 2008, **25**, 1473–1477.
31. M. A. Maher, H. Mataczynski, H. M. Stefaniak and T. Wilson, *J. Med. Food*, 2000, **3**, 141–147.
32. I.-M. Liu, T.-F. Tzeng, S.-S. Liou and T.-W. Lan, *Life Sci.*, 2007, **81**, 1479–1488.
33. A. Kato, N. Nasu, K. Takebayashi, I. Adachi, Y. Minami, F. Sanae, N. Asano, A. A. Watson and R. J. Nash, *J. Agric. Food Chem.*, 2008, **56**, 4469–4473.
34. T. Wilson, J. L. Luebke, E. F. Morcomb, E. J. Carrell, M. C. Leveranz, L. Kobs, T. P. Schmidt, P. J. Limburg, N. Vorsa and A. P. Singh, *J. Food Sci.*, 2010, **75**, H218–H223.
35. S. L. Navarro, E. White, E. D. Kantor, Y. Zhang, J. Rho, X. Song, G. L. Milne, P. D. Lampe and J. W. Lampe, *PLoS One*, 2015, **10**, e0117534.
36. J. Monfort, J.-P. Pelletier, N. Garcia-Giralt and J. Martel-Pelletier, *Ann. Rheum. Dis.*, 2008, **67**, 735–740.
37. Y. Henrotin, A. Mobasheri and M. Marty, *Arthritis Res. Ther.*, 2012, **14**, 201.
38. D. Fuchs, R. Piller, J. Linseisen, H. Daniel and U. Wenzel, *Proteomics*, 2007, **7**, 3278–3288.
39. S. P. B. Caligiuri, H. M. Aukema, A. Ravandi and G. N. Pierce, *Exp. Gerontol.*, 2014, **59**, 51–57.
40. C. M. C. Bassett, R. S. McCullough, A. L. Edel, A. Patenaude, R. K. LaVallee and G. N. Pierce, *Am. J. Physiol.: Circ. Physiol.*, 2011, **301**, H2220–H2226.
41. C. M. C. Dupasquier, E. Dibrov, A. L. Kneesh, P. K. M. Cheung, K. G. Y. Lee, H. K. Alexander, B. K. Yeganeh, M. H. Moghadasian and G. N. Pierce, *Am. J. Physiol.: Circ. Physiol.*, 2007, **293**, H2394–H2402.
42. C. M. C. Dupasquier, A.-M. Weber, B. P. Ander, P. P. Rampersad, S. Steigerwald, J. T. Wigle, R. W. Mitchell, E. A. Kroeger, J. S. C. Gilchrist, M. M. Moghadasian, A. Lukas and G. N. Pierce, *Am. J. Physiol.: Circ. Physiol.*, 2006, **291**, H2987–H2996.
43. E. Burillo, R. Mateo-Gallego, A. Cenarro, S. Fiddyment, A. M. Bea, I. Jorge, J. Vázquez and F. Civeira, *Lipids Health Dis.*, 2012, **11**, 116.
44. L. Calabresi, B. Villa, M. Canavesi, C. R. Sirtori, R. W. James, F. Bernini and G. Franceschini, *Metabolism.*, 2004, **53**, 153–158.
45. R. De Caterina, J. K. Liao and P. Libby, *Am. J. Clin. Nutr.*, 2000, **71**, 213S–223S.
46. W. E. Connor, *Am. J. Clin. Nutr.*, 2000, **71**, 171S–175S.
47. S. Silva, M. R. Bronze, M. E. Figueira, J. Siwy, H. Mischak, E. Combet, W. Mullen and W. Mullen, *Am. J. Clin. Nutr.*, 2015, **101**, 44–54.
48. J. K. Snell-Bergeon, D. M. Maahs, L. G. Ogden, G. L. Kinney, J. E. Hokanson, E. Schiffer, M. Rewers and H. Mischak, *Diabetes Technol. Ther.*, 2009, **11**, 1–9.
49. L. U. Zimmerli, E. Schiffer, P. Zürbig, D. M. Good, M. Kellmann, L. Mouls, A. R. Pitt, J. J. Coon, R. E. Schmieder, K. H. Peter, H. Mischak, W. Kolch, C. Delles and A. F. Dominiczak, *Mol. Cell. Proteomics*, 2008, **7**, 290–298.
50. C. Delles, E. Schiffer, C. von zur Muhlen, K. Peter, P. Rossing, H.-H. Parving, J. A. Dymott, U. Neisius, L. U. Zimmerli, J. K. Snell-Bergeon, D. M.

Maahs, R. E. Schmieder, H. Mischak and A. F. Dominiczak, *J. Hypertens.*, 2010, **28**, 2316–2322.
51. V. Summerhill, V. Karagodin, A. Grechko, V. Myasoedova and A. Orekhov, *Front. Cardiovasc. Med.*, 2018, **5**, 188.
52. D. F. S. Chaves, P. C. Carvalho, E. Brasili, M. M. Rogero, N. A. Hassimotto, J. K. Diedrich, J. J. Moresco, J. R. Yates and F. M. Lajolo, *J. Proteome Res.*, 2017, **16**, 4086–4092.
53. C. E. O'Neil, T. A. Nicklas, G. C. Rampersaud and V. L. Fulgoni, *Nutr. J.*, 2012, **11**, 107.
54. B. de Roos, A. Geelen, K. Ross, G. Rucklidge, M. Reid, G. Duncan, M. Caslake, G. Horgan and I. A. Brouwer, *Proteomics*, 2008, **8**, 1965–1974.
55. R. K. Saini and Y.-S. Keum, *Life Sci.*, 2018, **203**, 255–267.
56. P. N. Duchateau, I. Movsesyan, S. Yamashita, N. Sakai, K. Hirano, S. A. Schoenhaus, P. M. O'Connor-Kearns, S. J. Spencer, R. B. Jaffe, R. F. Redberg, B. Y. Ishida, Y. Matsuzawa, J. P. Kane and M. J. Malloy, *J. Lipid Res.*, 2000, **41**, 1231–1236.
57. T. S. E. Albert, P. N. Duchateau, S. S. Deeb, C. R. Pullinger, M. H. Cho, D. C. Heilbron, M. J. Malloy, J. P. Kane and B. G. Brown, *J. Lipid Res.*, 2005, **46**, 469–474.
58. S. Oller Moreno, O. Cominetti, A. Núñez Galindo, I. Irincheeva, J. Corthésy, A. Astrup, W. H. M. Saris, J. Hager, M. Kussmann and L. Dayon, *Proteomics: Clin. Appl.*, 2018, **12**, 1600150.
59. Ö. Gögebakan, A. Kohl, M. A. Osterhoff, M. A. van Baak, S. A. Jebb, A. Papadaki, J. A. Martinez, T. Handjieva-Darlenska, P. Hlavaty, M. O. Weickert, C. Holst, W. H. M. Saris, A. Astrup and A. F. H. Pfeiffer, *Circulation*, 2011, **124**, 2829–2838.
60. P. E. Geyer, N. J. Wewer Albrechtsen, S. Tyanova, N. Grassl, E. W. Iepsen, J. Lundgren, S. Madsbad, J. J. Holst, S. S. Torekov and M. Mann, *Mol. Syst. Biol.*, 2016, **12**, 901.
61. J. Carayol, C. Chabert, A. Di Cara, C. Armenise, G. Lefebvre, D. Langin, N. Viguerie, S. Metairon, W. H. M. Saris, A. Astrup, P. Descombes, A. Valsesia and J. Hager, *Nat. Commun.*, 2017, **8**, 2084.
62. A. A. Alfadda, A. A. Turjoman, A. S. Moustafa, M. Y. Al-Naami, M. A. Chishti, R. M. Sallam, D. Gibson and M. W. Duncan, *Exp. Biol. Med.*, 2014, **239**, 568–580.
63. O. A. Rangel-Zúñiga, A. Camargo, C. Marin, P. Peña-Orihuela, P. Pérez-Martínez, J. Delgado-Lista, L. González-Guardia, E. M. Yubero-Serrano, F. J. Tinahones, M. M. Malagón, F. Pérez-Jiménez, H. M. Roche and J. López-Miranda, *BMC Genomics*, 2015, **16**, 509.
64. A. Camargo, O. A. Rangel-Zúñiga, P. Peña-Orihuela, C. Marín, P. Pérez-Martínez, J. Delgado-Lista, F. M. Gutierrez-Mariscal, M. M. Malagón, H. M. Roche, F. J. Tinahones, F. Perez-Jimenez and J. Lopez-Miranda, *J. Nutr. Biochem.*, 2013, **24**, 318–324.
65. F. G. Bouwman, P. Wang, M. van Baak, W. H. M. Saris and E. C. M. Mariman, *Obesity*, 2014, **22**, 819–827.
66. C. Gialeli, B. Gungor and A. M. Blom, *Mol. Immunol.*, 2018, **102**, 73–83.

67. S. Sercu, M. Zhang, N. Oyama, U. Hansen, A. E. L. Ghalbzouri, G. Jun, K. Geentjens, L. Zhang and J. H. Merregaert, *J. Invest. Dermatol.*, 2008, **128**, 1397–1408.
68. L. He, W. Gu, M. Wang, X. Chang, X. Sun, Y. Zhang, X. Lin, C. Yan, W. Fan, P. Su, Y. Wang, C. Yi, G. Lin, L. Li, Y. Jiang, J. Lu, C. Dong, H. Wang and B. Sun, *Proc. Natl. Acad. Sci. U. S. A.*, 2018, **115**, 8621–8626.
69. M. Diarra-Mehrpour, N. Sarafan, J. Bourguignon, F. Bonnet, F. Bost and J. P. Martin, *J. Biol. Chem.*, 1998, **273**, 26809–26819.
70. A. Gomez Toledo, J. Nilsson, F. Noborn, C. Sihlbom and G. Larson, *Mol. Cell. Proteomics*, 2015, **14**, 3118–3131.
71. Y. Matsuo, M. Tanaka, H. Yamakage, Y. Sasaki, K. Muranaka, H. Hata, I. Ikai, A. Shimatsu, M. Inoue, T.-H. Chun and N. Satoh-Asahara, *Metabolism*, 2015, **64**, 1490–1499.
72. S. A. Hardy, N. S. Mabotuwana, L. A. Murtha, B. Coulter, S. Sanchez-Bezanilla, M. S. Al-Omary, T. Senanayake, S. Loering, M. Starkey, R. J. Lee, P. P. Rainer, P. M. Hansbro and A. J. Boyle, *PLoS One*, 2019, **14**, e0212230.
73. V. Varma, A. Yao-Borengasser, A. M. Bodles, N. Rasouli, B. Phanavanh, G. T. Nolen, E. M. Kern, R. Nagarajan, H. J. Spencer, M.-J. Lee, S. K. Fried, R. E. McGehee, C. A. Peterson and P. A. Kern, *Diabetes*, 2008, **57**, 432–439.
74. D. T. Eitzman, R. J. Westrick, E. G. Nabel and D. Ginsburg, *Blood*, 2000, **95**, 577–580.
75. J. Kurokawa, H. Nagano, O. Ohara, N. Kubota, T. Kadowaki, S. Arai and T. Miyazaki, *Proc. Natl. Acad. Sci. U. S. A.*, 2011, **108**, 12072–12077.
76. S. M. Safavi, R. Ziaei and M. R. Maracy, *J. Obes.*, 2012, **2012**, 951093.
77. M. Cignarelli, G. DePergola, G. Picca, M. Sciaraffia, N. Pannacciulli, M. Tarallo, E. Laudadio, E. Turrisi and R. Giorgino, *Int. J. Obes. Relat. Metab. Disord.*, 1996, **20**, 809–813.
78. J. M. Moreno-Navarrete and J. M. Fernández-Real, *Semin. Cell Dev. Biol.*, 2019, **85**, 164–172.
79. D. Ricklin, G. Hajishengallis, K. Yang and J. D. Lambris, *Nat. Immunol.*, 2010, **11**, 785–797.
80. J. Phieler, R. Garcia-Martin, J. D. Lambris and T. Chavakis, *Semin. Immunol.*, 2013, **25**, 47–53.
81. J. Zhang, W. Wright, D. A. Bernlohr, S. W. Cushman and X. Chen, *Am. J. Physiol.: Endocrinol. Metab.*, 2007, **292**, E1433–E1440.
82. L. Belardin, M. Camargo, P. Intasqui, M. Antoniassi, R. Fraietta and R. Bertolla, *Andrology*, 2019, **7**, 53–61.
83. B. R. Pathak, A. A. Breed, S. Apte, K. Acharya and S. D. Mahale, *Mol. Cell. Biochem.*, 2016, **411**, 11–21.
84. J. Weigert, M. Neumeier, J. Wanninger, S. Bauer, S. Farkas, M. N. Scherer, A. Schnitzbauer, A. Schäffler, C. Aslanidis, J. Schölmerich and C. Buechler, *J. Clin. Endocrinol. Metab.*, 2010, **95**, 1404–1411.
85. S. Menini, C. Iacobini, C. Blasetti Fantauzzi, C. M. Pesce and G. Pugliese, *Oxid. Med. Cell. Longev.*, 2016, 2016, 9618092.
86. L. D. Burtnick, R. C. Robinson and S. Choe, *Results Probl. Cell Differ.*, 2001, **32**, 201–211.

87. C. C. Cunningham, T. P. Stossel and D. J. Kwiatkowski, *Science*, 1991, **251**, 1233–1236.
88. E. Piktel, I. Levental, B. Durnaś, P. A. Janmey and R. Bucki, *Int. J. Mol. Sci.*, 2018, **19**, 2516.
89. M. Maffei, I. Barone, G. Scabia and F. Santini, *Endocr. Rev.*, 2016, **37**, 403–416.
90. F. C. Di Naso, R. R. Porto, H. S. Fillmann, L. Maggioni, A. V. Padoin, R. J. Ramos, C. C. Mottin, A. Bittencourt, N. A. P. Marroni and P. I. H. de Bittencourt, *Obesity (Silver Spring).*, 2015, **23**, 120–129.
91. F. Zheng, W. Cai, T. Mitsuhashi and H. Vlassara, *Mol. Med.*, 2001, **7**, 737–747.
92. S. A. Ragland and A. K. Criss, *PLoS Pathog.*, 2017, **13**, e1006512.
93. C. M. Hodin, F. J. Verdam, J. Grootjans, S. S. Rensen, F. K. Verheyen, C. H. C. Dejong, W. A. Buurman, J. W. Greve and K. Lenaerts, *J. Pathol.*, 2011, **225**, 276–284.
94. K. K. Eklund, K. Niemi and P. T. Kovanen, *Crit. Rev. Immunol.*, 2012, **32**, 335–348.
95. D. Pilling and R. H. Gomer, *Front. Immunol.*, 2018, **9**, 2328.
96. N. Cox, D. Pilling and R. H. Gomer, *J. Leukoc. Biol.*, 2014, **96**, 739–743.
97. C. Poitou, N. Viguerie, R. Cancello, R. De Matteis, S. Cinti, V. Stich, C. Coussieu, E. Gauthier, M. Courtine, J. D. Zucker, G. S. Barsh, W. Saris, P. Bruneval, A. Basdevant, D. Langin and K. Clément, *Diabetologia*, 2005, **48**, 519–528.
98. Y. Zhao, X. He, X. Shi, C. Huang, J. Liu, S. Zhou and C.-K. Heng, *Inflamm. Res.*, 2010, **59**, 323–334.
99. M. Anwer and M. J. Iqbal, *Indian J. Endocrinol. Metab.*, 2018, **22**, 683–688.
100. J. E. Jun, S.-E. Lee, Y.-B. Lee, J. H. Jee, J. C. Bae, S.-M. Jin, K. Y. Hur, M.-K. Lee and J. H. Kim, *PLoS One*, 2017, **12**, e0176209.
101. M. Taverna, A.-L. Marie, J.-P. Mira and B. Guidet, *Ann. Intensive Care*, 2013, **3**, 4.
102. R. H. Mosli and H. H. Mosli, *Diabetes. Metab. Syndr. Obes.*, 2017, **10**, 467–472.
103. M. Chen, B. Liu, D. Wilkinson, A. T. Hutchison, C. H. Thompson, G. A. Wittert and L. K. Heilbronn, *Obes. Res. Clin. Pract.*, 2017, **11**, 227–232.
104. J. W. Weisel and R. I. Litvinov, *Subcell. Biochem.*, 2017, **82**, 405–456.
105. I. O. Blokhin and S. R. Lentz, *Curr. Opin. Hematol.*, 2013, **20**, 437–444.
106. A. Furman-Niedziejko, P. Rostoff, R. Rychlak, K. Golinska-Grzybala, M. Wilczynska-Golonka, M. Golonka and J. Nessler, *Folia Med. Cracov.*, 2014, **54**, 55–64.
107. I. Iwan-Ziętek, M. Michalska, K. Góralczyk, B. Ruszkowska-Ciastek, S. Dąbrowiecki, Z. Ziętek and D. Rość, *Med. Sci. Technol.*, 2017, **58**, 91–97.
108. A. Salameh, A. C. Daquinag, D. I. Staquicini, Z. An, K. A. Hajjar, R. Pasqualini, W. Arap and M. G. Kolonin, *JCI Insight*, 2016, **1**, e86351.
109. C.-Y. Wang and C.-F. Lin, *Dis. Markers*, 2014, 2014, 308976.

110. Q. Ling, A. T. Jacovina, A. Deora, M. Febbraio, R. Simantov, R. L. Silverstein, B. Hempstead, W. H. Mark and K. A. Hajjar, *J. Clin. Invest.*, 2004, **113**, 38–48.
111. J. J. Feher, *Quantitative Human Physiology: An Introduction*, ed. J. J. Feher, Academic Press, 2012, pp. 419–427.
112. G. Batist, A. Bothe, M. Bern, B. R. Bistrian and G. L. Blackburn, *JPEN. J. Parenter. Enteral Nutr.*, 1983, 7, 447–449.
113. K. Brummel-Ziedins and K. G. Mann, *Hematology*, ed. R. Hoffman, E. J. Benz, Jr, L. E. Silberstein, H. E. Heslop, J. I. Weitz, J. Anastasi, M. E. Salama and S. A. Abutalib, Elsevier, 7th edn, 2018, pp. 1885–1905.
114. M. K. S. Wong, *Handbook of Hormones: Comparative Endocrinology for Basic and Clinical Research*, ed. Y. Takei, H. Ando and K. Tsutsui, Academic Press, 2016, pp. 268–270.
115. F. Kronenberg, B. Kollerits, S. Kiechl, C. Lamina, L. Kedenko, C. Meisinger, J. Willeit, C. Huth, G. Wietzorrek, M. E. Altmann, B. Thorand, A. Melmer, D. Dähnhardt, P. Santer, W. Rathmann, B. Paulweber, W. Koenig, A. Peters, I. M. Adham and H. Dieplinger, *Circ. Cardiovasc. Genet.*, 2014, 7, 822–829.
116. G. E. Mattos, J.-M. Heinzmann, S. Norkowski, J.-C. Helbling, A. M. Minni, M.-P. Moisan and C. Touma, *J. Endocrinol.*, 2013, **219**, 217–229.
117. O. Ousova, V. Guyonnet-Duperat, N. Iannuccelli, J.-P. Bidanel, D. Milan, C. Genêt, B. Llamas, M. Yerle, J. Gellin, P. Chardon, A. Emptoz-Bonneton, M. Pugeat, P. Mormède and M.-P. Moisan, *Mol. Endocrinol.*, 2004, **18**, 1687–1696.
118. M.-P. Moisan and N. Castanon, *Front. Endocrinol. (Lausanne)*, 2016, 7, 160.
119. M. A. Nenke, J. G. Lewis, W. Rankin and D. J. Torpy, *Horm. Metab. Res.*, 2016, **48**, 620.
120. M. Manco, J. M. Fernández-Real, M. E. Valera-Mora, H. Déchaud, G. Nanni, V. Tondolo, M. Calvani, M. Castagneto, M. Pugeat and G. Mingrone, *Diabetes Care*, 2007, **30**, 1494–1500.
121. S. B. Wheatcroft, M. T. Kearney, A. M. Shah, V. A. Ezzat, J. R. Miell, M. Modo, S. C. R. Williams, W. P. Cawthorn, G. Medina-Gomez, A. Vidal-Puig, J. K. Sethi and P. A. Crossey, *Diabetes*, 2007, **56**, 285–294.
122. M. S. Lewitt, M. S. Dent and K. Hall, *J. Clin. Med.*, 2014, **3**, 1561–1574.
123. J. Frystyk, C. Skjaerbaek, E. Vestbo, S. Fisker and H. Orskov, *Diabetes. Metab. Res. Rev.*, **15**, 314–322.
124. Z. Li and F. Picard, *Horm. Metab. Res.*, 2010, **42**, 787–791.
125. A. Hautanen, *Int. J. Obes. Relat. Metab. Disord.*, 2000, **24**, S64–S70.
126. L. A. Cooper, S. T. Page, J. K. Amory, B. D. Anawalt and A. M. Matsumoto, *Clin. Endocrinol. (Oxf).*, 2015, **83**, 828–833.
127. M. A. Sleddering, A. J. Markvoort, H. K. Dharuri, S. Jeyakar, M. Snel, P. Juhasz, M. Lynch, W. Hines, X. Li, I. M. Jazet, A. Adourian, P. A. J. Hilbers, J. W. A. Smit and K. W. Van Dijk, *PLoS One*, 2014, **9**, e112835.
128. I. M. Jazet, A. J. de Craen, E. M. van Schie and A. E. Meinders, *Diabetes Res. Clin. Pract.*, 2007, 77, 70–76.

129. M. Snel, M. A. Sleddering, I. D. vd Peijl, J. A. Romijn, H. Pijl, A. E. Meinders and I. M. Jazet, *Eur. J. Intern. Med.*, 2012, **23**, 143–149.
130. M. Snel, J. T. Jonker, S. Hammer, G. Kerpershoek, H. J. Lamb, A. E. Meinders, H. Pijl, A. de Roos, J. A. Romijn, J. W. A. Smit and I. M. Jazet, *Obesity (Silver Spring).*, 2012, **20**, 1572–1576.
131. N. Kalsheker, *Biology, Diagnosis, Clinical Significance, and Emerging Therapies*, ed. N. Kalsheker and R. A. Stockley, Academic Press, 2017, pp. 25–35.
132. Y. Yuan, B. DiCiaccio, Y. Li, A. S. Elshikha, D. Titov, B. Brenner, L. Seifer, H. Pan, N. Karic, M. A. Akbar, Y. Lu, S. Song and L. Zhou, *Aging Cell*, 2018, **17**, e12694.
133. M. Sandler, B. M. Gemperli, C. Hanekom and S. H. Kühn, *Diabetes Res. Clin. Pract.*, 1988, **5**, 249–255.
134. C. S. Sandström, B. Ohlsson, O. Melander, U. Westin, R. Mahadeva and S. Janciauskiene, *Diabet. Med.*, 2008, **25**, 1370–1373.
135. G. B. Xue, W. L. Zheng, L. H. Wang and L. Y. Lu, *Saudi Med. J.*, 2013, **34**, 34–39.
136. A.-M. Mirea, E. J. M. Toonen, I. van den Munckhof, I. D. Munsterman, E. T. T. L. Tjwa, M. Jaeger, M. Oosting, K. Schraa, J. H. W. Rutten, M. van der Graaf, N. P. Riksen, J. de Graaf, M. G. Netea, C. J. Tack, T. Chavakis and L. A. B. Joosten, *Mol. Med.*, 2019, **25**, 16.
137. F. Kronenberg, M. Stühlinger, E. Trenkwalder, F. S. Geethanjali, O. Pachinger, A. von Eckardstein and H. Dieplinger, *J. Am. Coll. Cardiol.*, 2000, **36**, 751–757.
138. D. Quilliot, E. Walters, B. Guerci, J. C. Fruchart, P. Duriez, P. Drouin and O. Ziegler, *Metabolism*, 2001, **50**, 1019–1024.
139. K. Fujimoto, J. A. Cardelli and P. Tso, *Am. J. Physiol.*, 1992, **262**, G1002-6.
140. P. Tso and M. Liu, *Physiol. Behav.*, 2004, **83**, 631–643.
141. F. Bertile, C. Schaeffer, Y. Le Maho, T. Raclot and A. Van Dorsselaer, *Proteomics*, 2009, **9**, 148–158.
142. V. I. Patel, K. P. Patel, M. G. Makadia, A. D. Shah, K. S. Chaudhari and H. N. Nilayangode, *J. Clin. Diagn. Res.*, 2017, **11**, BC01–BC05.
143. C. Stettler, Y. Suter, S. Allemann, M. Zwahlen, E. R. Christ and P. Diem, *J. Intern. Med.*, 2006, **260**, 272–280.
144. L. J. Hanks, J. H. Pelham, S. Vaid, K. Casazza and A. P. Ashraf, *Diabetes Res. Clin. Pract.*, 2016, **115**, 83–89.
145. J. J. Albers, S. M. Marcovina, G. Imperatore, B. M. Snively, J. Stafford, W. Y. Fujimoto, E. J. Mayer-Davis, D. B. Petitti, C. Pihoker, L. Dolan and D. M. Dabelea, *J. Clin. Endocrinol. Metab.*, 2008, **93**, 735–742.
146. M. Adiels, M. Taskinen, E. Björnson, L. Andersson, N. Matikainen, S. Söderlund, J. Kahri, A. Hakkarainen, N. Lundbom, C. Sihlbom, A. Thorsell, H. Zhou, K. H. Pietiläinen, C. Packard and J. Borén, *Diabetes, Obes. Metab.*, 2019, **21**, 1861–1870.
147. M.-R. Taskinen and J. Borén, *Curr. Atheroscler. Rep.*, 2016, **18**, 59.

148. A. Hatziri, C. Kalogeropoulou, E. Xepapadaki, E. Birli, E. A. Karavia, E. Papakosta, S. Filou, C. Constantinou and K. E. Kypreos, *Biochim. Biophys. Acta, Mol. Basis Dis.*, 2018, 1864, 471–480.
149. V. I. Zannis, P. W. Just and J. L. Breslow, *Am. J. Hum. Genet.*, 1981, **33**, 11–24.
150. G. Utermann, M. Hees and A. Steinmetz, *Nature*, 1977, **269**, 604–607.
151. R. J. Havel and J. P. Kane, *Proc. Natl. Acad. Sci. U. S. A.*, 1973, **70**, 2015–2019.
152. J. E. Eichner, S. T. Dunn, G. Perveen, D. M. Thompson, K. E. Stewart and B. C. Stroehla, *Am. J. Epidemiol.*, 2002, **155**, 487–495.
153. R. W. Mahley, K. H. Weisgraber and Y. Huang, *J. Lipid Res.*, 2009, **50**, S183–S188.
154. D. El-Lebedy, H. M. Raslan and A. M. Mohammed, *Cardiovasc. Diabetol.*, 2016, **15**, 12.
155. T. Nishimura, Y. Itoh, S. Yamashita, K. Koide, N. Harada, Y. Yano, N. Ikeda, K. Azuma and Y. Atsumi, *Diabetes Res. Clin. Pract.*, 2017, **127**, 132–139.
156. Y. Borné, I. F. Muhammad, L. Lorés-Motta, B. Hedblad, P. M. Nilsson, O. Melander, E. K. de Jong, A. M. Blom, A. I. den Hollander and G. Engström, *J. Clin. Endocrinol. Metab.*, 2017, **102**, 4477–4485.

CHAPTER 10

Personalized Nutrition to Treat and Prevent Obesity and Diabetes

ARTURO PANDURO,* SONIA ROMAN, RICARDO G. MILÁN, LUIS ALBERTO TORRES-REYES AND KARINA GONZALEZ-ALDACO

Department of Molecular Biology in Medicine, Civil Hospital of Guadalajara, Fray Antonio Alcalde, and University of Guadalajara, Hospital # 278, Col. El Retiro, Guadalajara, Jalisco 44280, Mexico
*Email: apanduro@prodigy.net.mx

10.1 Obesity and Type 2 Diabetes: An Interaction Between Genes and Environmental Factors

Obesity is a health problem with global expansion. The state of obesity is reached through a gradual weight gain at the expense of the accumulation of body fat. This progression may occur in a period of one or two decades of life initiating with healthy body weight that continues to overweight and obesity, the latter being classified as type 1, 2, or 3 according to the amount of increased body fat.[1]

The WHO estimates that nearly 700 million adults are obese worldwide.[2] Furthermore, because of its prolonged development, the incidence of obesity increases annually due to the high prevalence rates of people who are

Nutritional Signaling Pathway Activities in Obesity and Diabetes
Food Chemistry, Function and Analysis No. 24
Edited by Zhiyong Cheng

Published by the Royal Society of Chemistry, www.rsc.org

overweight.[3] The United States of America, China, India, and Mexico are countries at the top of the global list of populations with obesity and associated comorbidities such as Type 2 diabetes.[4] Currently, more than 400 million people have diabetes, and it is estimated that at least 200 million are prediabetic. This could eventually increase the prevalence of Type 2 diabetes if preventive measures are not implemented.[5]

Regardless of the region, the common denominator of obesity is the triad of raised food portions, being sedentary, and emotional-related disturbances denoted as stress, depression, anxiety, and addictions. Although obesity and Type 2 diabetes have been historically recognized for centuries, in the 1970s, an increment in both conditions began in Mexico due to the migration of people from rural areas to urban cities.[6] This event led to lifestyle changes, mainly in food habits, and a notable decrease in the population's habitual rural-related physical activities. A similar situation prevails in recent times in China, India, and other developing countries.[4] In contrast, in the United States of America, the food portions of sugar-sweetened beverages, high-fat foods, and added sugars increased considerably several decades ago, and a sedentary lifestyle was also accentuated.[7]

Several lifestyle modifications underlay this stage of the nutrition transition.[8,9] However, sparse attention has been given to the changes in the traditional dietary patterns of migrant families and the next generations living in the cities. For example, Mexico, China, and India have their national food culture with interregional differences within each country. In this emerging situation of obesity, the traditional and regional cuisines prepared by rural families are no longer consumed regularly in the great metropolis. Currently, the consumption of fresh and healthy local foods has diminished, whereas the large food companies highly promote the intake of fast foods, sugar-sweetened beverages, processed and ultraprocessed foods at an international level.

In Mexico, the change in the consumption of traditional foods began in the past century, with the stigmatization of several ingredients included in a healthy food pattern. For example, lard, a typical ingredient included in the recipe of several Mexican food dishes, was used initially only for festive occasions; it was not consumed daily. Nonetheless, lard was gradually displaced due to the consumption of more copious amounts of edible oils. Vegetal oil companies' marketing has dramatically influenced consumers on the type, brand, and usage of edible oils by conveying the message that lard is an unhealthy option. Likewise, under the same food company strategies, Mexicans have increased the intake frequency of processed sweetened beverages to the extent that they are the number-one consumer highly addicted to these products in the world.[10]

These and other dietary modifications, such as eating fried meats and dairy products, have led to an increase in two types of dyslipidemias among the Mexican population, predominantly hypertriglyceridemia and hypercholesterolemia.[11] The former predisposes to insulin resistance and the latter to metabolic alterations that eventually generate the main complications associated with Type 2 diabetes, such as cardiovascular disease or diabetic retinopathies.

Dietary food patterns are modifiable environmental factors that play an essential role in the onset, progression, and clinical outcomes of these underlying pathological conditions, as mentioned above. However, genes are also involved. Explicitly, some genes related to metabolism encode SNPs that have arisen through co-evolutionary processes of selection driven by diet exposure over time. These SNPs are now gaining increasing interest to explain the rapid increase in modern-day chronic diseases.[12,13] Notably, the risk of dyslipidemias and insulin resistance will depend upon the predominance of the "non-risk" or "risk" allele and the surrounding nutrients (food habits) because they are distributed in a differential manner among human populations (ethnicity) living in specific geographical regions. Therefore, under this perspective, personalized nutritional strategies should aim to restore a healthy interaction between the individual's genetic profile and the appropriate nutrients.

Several adaptive genes are involved directly or indirectly in carbohydrate and lipid metabolism. Some genes encode proteins that function as sugar and fat taste receptors modulating food preferences, lipid transporters, and starch and milk digestive enzymes, while others are involved in energy intake, energy storage, and food reward systems. The second level of interacting processes is epigenetics that regulates gene expression and gut microbiota-derived mechanisms that are under intensive research.

Several genes associated with obesity and Type 2 diabetes will be reviewed throughout this chapter. An integrated overview of where some nutrients, mainly sugars, and lipids, interact with polymorphic receptors or enzymes throughout different body organs/systems is depicted in Figures 10.1–10.3. Also, a basic recommendation related to nutrients for a personalized diet is highlighted for each genetic polymorphism. In summary, foods (nutrients) that are part of the daily diet are ingested, and complex carbohydrates and lipids are initially cleaved in the oral cavity by salivary AMY1A and lipase, respectively. Sweet TAS1R2 and fat receptor CD36 are activated by their specific chemical ligands. Brain DRD2, modulated by the neurotransmitter dopamine, is activated to integrate the hedonic nature of eating. Carbohydrate and lipid digestion continue in the stomach and intestines. Milk lactose is hydrolyzed by LCT into glucose and galactose; FABP2 transports long-chain fatty acids to assemble chylomicrons. The bidirectional gut microbiota–brain axis will operate to modulate hunger/satiety, and emotions related to food reward with gut microbiota components (Figure 10.1).

Other gene products such as the FTO/alpha-ketoglutarate-dependent dioxygenase are related to the degree of adiposity or amount of fat that can be stored in adipose tissue. MTHFR will exert a nutrigenetic and epigenetic effect on gene expression. Chylomicrons synthesized in the gut will require ApoB48 and ApoE (Figure 10.2) to mobilize triglycerides to the liver or adipocytes, whereas VLDLs contain ApoB100 and ApoE. HDL assembly will require ApoA1 and cholesterol efflux transportation mediated by ABCA1, a key function in macrophages to avoid the formation of foam cells and eventually arteriosclerosis (Figure 10.3). In the context of obesity, hypertriglyceridemia will

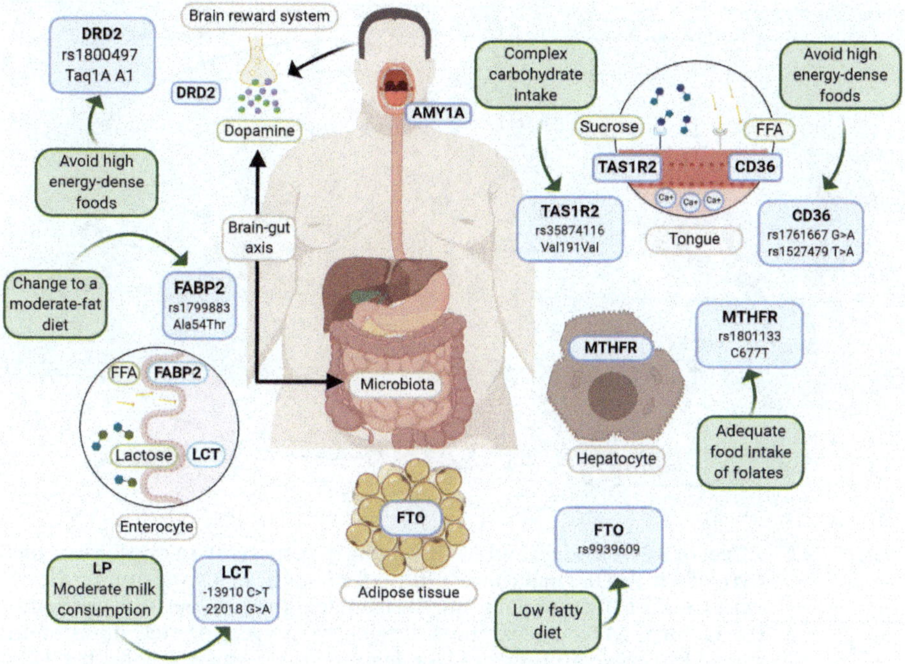

Figure 10.1 Genetic polymorphisms related to the development of hyperglyceridemia, insulin resistance, obesity, and Type 2 diabetes. Blue boxes enclose gene polymorphisms whereas green boxes highlight the nutritional recommendation. *CD36* rs1761667 and *TAS1R2* rs35874116 are associated with a higher consumption of dietary fats and carbohydrates, respectively. *FABP2* rs1799883 is related to hypertriglyceridemia whereas *FTO* rs9939609 increases body fat. High AMY1A copy number protects against obesity whereas lactase persistence increases the consumption of dairy saturated fat. *MTHFR* rs1801133 is associated with hypercysteinemia. *DRD2* rs1800497 may interact with the brain–gut microbiota axis, promoting dysbiosis, obesity, and negative emotions.

arise by the interaction of such polymorphic genes and the components of the diet that eventually could lead to insulin resistance and Type 2 diabetes.

In the following sections, the role of these polymorphic genes in obesity and Type 2 diabetes will be further detailed to explain their effect at the nutrigenetic, nutrigenomic, or epigenetic level.

10.2 Taste Perception, Carbohydrate Digestion, Gut Microbiota, and Food Reward

In this first section, genes and gene products related to sweet and fat taste perception, starch and milk digestion, as well as the relationship between the brain–gut microbiota axis and food reward, will be explained in relationship with dyslipidemia and insulin resistance as illustrated in Figure 10.1.

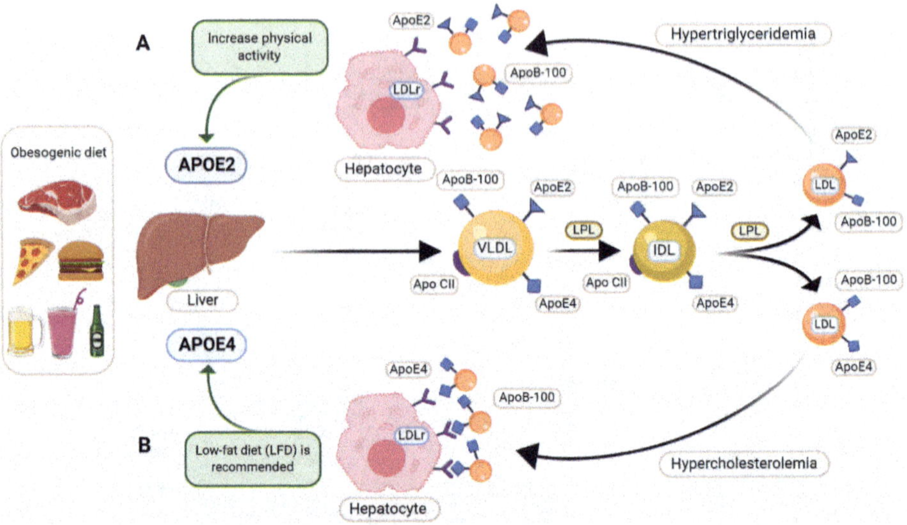

Figure 10.2 Effect of APOE2 allele and APOE4 allele in response to a obesogenic diet. (A) Isoform APOE2 has a lower affinity for hepatic LDLr, promoting a lower capture of VLDLs and IDLs into the liver, leading to hypertriglyceridemia. (B) Isoform APOE4 recaptures efficiently triglyceride-rich lipoproteins due to its higher affinity for LDLr, leading to a negative feedback of LDLr (fewer surface receptors) and hypercholesterolemia.

10.2.1 Taste Perception

Taste receptor genes are chemoreceptors that interact with ligand molecules contained in the diet and are responsible for the gustation of key dietary components.[14] The taste system determines if food is potentially nutritious and safe (non-poisonous) to consume.[15] After the excitation of the taste receptor cells, the gustatory signals are transmitted from the taste buds to the central nervous system through several regulated mechanisms.[16] Taste perception is a genetic trait with high variability throughout populations. Genetic variants in taste receptors are known to modulate interindividual differences in taste perception, which may also affect dietary preferences and chronic disease development.

10.2.1.1 TAS1R2

This protein, expressed in taste bud receptor cells, is encoded by the *TAS1R2* gene (ID : 80834) that senses natural sugars (glucose, fructose, sucrose, maltose), and non-nutritive sweeteners (saccharin, acesulfame-k, aspartate, dulcin, sucralose).[14] Genetic variations of TAS1R2 have been studied in terms of their association with sweet taste perception, sugar/carbohydrate intake, dental caries, and, most recently, with dyslipidemias. In a study carried out in a group of overweight and obese Mexican subjects, the Val191Val genotype

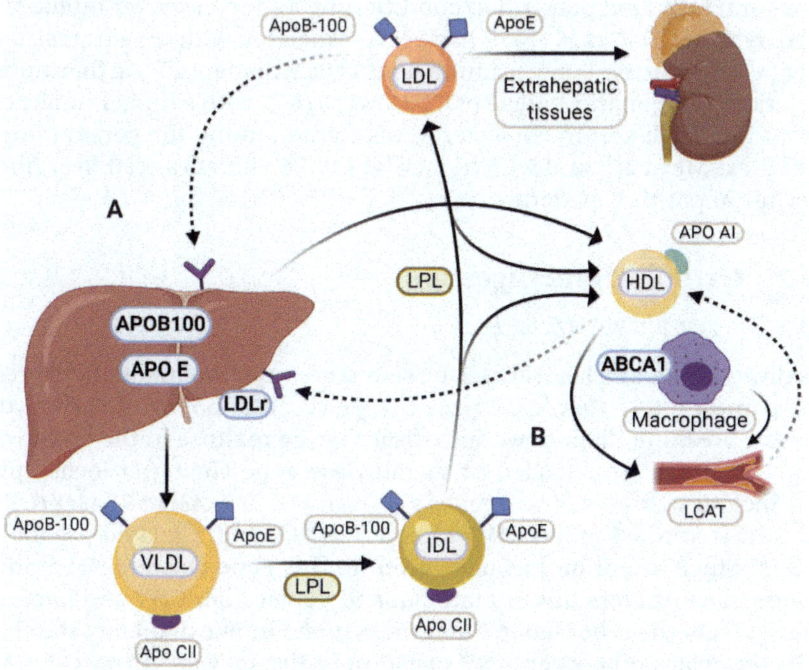

Figure 10.3 Endogenous lipid pathway. (A) Both APOB100 and APOE play an important role in the regulation of plasmatic lipids by binding to liver LDLr. Solid lines represent forward pathways and dotted lines are reverse transportation pathways (B) HDL-c particles are assembled in the liver, and through IDL and LDL. The R230C ABCA1 isoform has been associated with a lower concentration of HDL-c, which diminishes the reverse transportation of cholesterol from extrahepatic tissues to the liver.

(rs35874116) carriers showed a higher intake of total carbohydrates, fiber, cereals, and vegetables compared to the Ile allele carriers, as well as higher risk (OR = 3.26) for hypertriglyceridemia.[10] Diets high in carbohydrates may lead to hyperglyceridemia by activation of *de novo* lipogenesis, reduced fatty acid muscle oxidation and reduced clearance of circulating lipids, a metabolic condition related to obesity, and Type 2 diabetes.[17]

10.2.1.2 CD36

The CD36 protein is a multifunctional receptor in several cell types. In the foliate and circumvallate papillae taste buds, this gustatory receptor is a fatty acid transporter capturing long-chain fatty acids and oxidized lipids.[18] Recently, fat perception has emerged as a great area of interest due to its implication in the consumption of high-energy-dense foods and dietary fat preferences.[19]

Two SNPs, rs1761667 (G > A) and rs1527479 (T > A) in the *CD36* gene (ID : 948) have been associated with a different patron of fatty acid sensitivity that impairs oral fatty acid perception conditioning an increased fat intake.[20] The TT genotype of *CD36* rs1527479 has been associated with insulin resistance, higher fasting glucose concentration, and Type 2 diabetes.[21] Furthermore, an association between the A allele of *CD36* rs1761667 with a higher intake of fat portions and high serum cholesterol was found among the general population of West Mexico,[22] and with higher fat intake and advanced liver fibrosis in chronic hepatitis C patients.[23]

10.2.2 Carbohydrate Digestion

10.2.2.1 Amylase Alpha 1

The salivary alpha-amylase (EC 3.2.1.1) is a secretory protein mainly expressed in the parotid gland that hydrolyzes α-1,4-glycosidic bonds of carbohydrate polymers, breaking them down into disaccharide maltose in the oral cavity.[24] Human amylases are encoded in the amylase gene cluster at locus 1p21.1, which includes three salivary genes (*AMY1A*; ID:275, *AMY1B*, *AMY1C*), two pancreatic α-amylase genes (*AMY2A* and *AMY2B*), and a related pseudogene (*AMYP*).[25] Much attention has been given to this gene cluster. *AMY1* nucleotide sequences are repetitive in tandem and depict high CNV and haplotypic variances.[26] On the other hand, differences in the intake of dietary starch may modify the relation between CNV variation in the salivary amylase gene and BMI.[27] Comparatively, the average copy number of *AMY1* gene is higher in populations that evolved under high-starch diets *versus* low-starch diets, which may reflect the effect of natural selection. Yet, the exact driving force is still under debate.[28] Nonetheless, an all-new field in amylase research (including pancreatic amylase) has emerged, focusing on the association of CNV with salivary α-amylase content, amylase enzymatic activity, BMI (obesity, adiposity), and insulin resistance; thus, reflecting its role in different metabolic states.

Low *AMY1* CNVs correlated with HOMA-IR in asymptomatic Korean men.[29] This observation may be related to the influence that salivary amylase usually has on postprandial cephalic-phase insulin release, which improves glucose tolerance. In contrast, serum amylase may have insulin-sensitizing properties.[26] Furthermore, the relationship between low AMY1 activity and obesity may be due to the colonic microbiota fermentation of oligosaccharides into short-chain fatty acids *via* lactate regulation, which may have a negative impact on energy harvest.[26] In line with these observations, an abundance of *Prevotella* was detected in Mexican children and adults with obesity and low *AMYA1* CN*Vs*.[30] Conversely, a high copy number of the *AMY1* gene may protect against obesity in Mexican children and Portuguese young adults.[31,32]

10.2.2.2 Lactase

Lactase (EC 3.2.1.108), also known as lactase-phlorizin hydrolase (EC 3.2.1.62), is an intestinal brush border enzyme that hydrolyzes the β-1,4-glycosidic linkage of lactose.[33] In mammals, including humans, lactase is particularly abundant in newborns and during infancy. After weaning, a normal physiological phenomenon known as LNP takes place.[33] It is characterized by a downregulation of lactase transcriptional activity and lactase production. Unabsorbed lactose is fermented by bacteria, causing the production of gas and intestinal distress. On the other hand, adult LP has been identified as a recent evolutionary trait among Northern Europeans and Africans with a history of cattle pastoralism.[34] This trait has been associated with several allelic polymorphisms at the promotor region of *LCT*. The most studied SNPs are -13910 C > T (rs4988235) and -22018 G > A (rs182549), located within the non-coding region of the neighboring *MCM6* gene.

The capability to digest lactose has received particular attention regarding its potential effect to restrict or promote the dietary consumption of milk or milk derivatives. Paradoxically, people continue to drink milk regardless of the genotype, although those who are LP tend to drink more servings of milk than LNP.[35] This may suggest that milk consumption may be influenced by food cultural history and marketing of milk across countries. However, it is noteworthy to mention that most of the world's population is LNP, as breast-feeding is the first best option for human babies and later finding alternative food sources of calcium and vitamin D. The history of cow's milk consumption is parallel to the gene–environmental co-evolutionary process that once occurred in LP societies. However, the consumption of cow's milk in adult LNP individuals is debatable and it may be of concern in the era of obesity since the use of processed cow's milk may not be a healthy option for some populations. For example, in a recent study conducted among Mexicans, the LNP trait was highest among Native Amerindian individuals and lowest among genetically admixed subjects with European ancestry, whereas LP was higher in the former. Those who were LP consumed higher servings of milk and higher portions of saturated fat that resulted in a dyslipidemic profile.[36]

10.2.3 Gut Microbiota

Gut microbiota contains nearly 100 trillion microorganisms that share symbiotic properties with the human host.[37] The multiple functions that the gut microbiota carries out are modulation of the host's metabolic and energy homeostasis, intestinal motility, and immune system maturation. It also confers protection against pathogens and toxins, regulates cytokine secretion from adipose tissue, insulin signaling and finally, modulates host emotions and cognition.[38]

The gut microbiota maintains a bi-directional communication with the central nervous system by using hormone, neuropeptide, neurotransmitter, and cytokine signaling pathways that activate the efferent vagus and afferent adrenergic nerve fibers.[38] Since nourishment is a natural and physiological need to obtain energy and seek food from the environment, the gut microbiota–brain axis integrates information from the HPA system, the hunger satiety circuitry and dopaminergic reward systems with the microbiota.

The role of the brain reward system mediated by dopamine receptors will be further explained in the following subsection.

Given the characteristics as mentioned above, the brain–gut axis is influenced by the host's genetic makeup, stress level, and negative emotions related to depression and anxiety, as well as dietary composition.[39] Genetic variations in taste receptors that lead to the consumption of high-sugar and high-fat diets, as well as gene polymorphisms involved in the energy/pleasure circuitry, lead to addiction to several foodstuffs, promote negative emotions and hinder food decision making that may lead to obesity.[37,39] These alterations chronically activate the HPA axis elevating cortisol levels. Cortisol results in dysbiosis, allowing pathogens to permeate the gut barrier and activate inflammation. Unhealthy dietary patterns also lead to dysbiosis, inflammation and negative emotions, anxiety, and low quality of life.[40] Balancing this altered physiological process to regain health may involve personalized medicine and genome-based nutritional strategies.[41]

Metagenomics studies have demonstrated that the gut microbiota of Type 2 diabetes patients is different in comparison with control participants. Generally, patients with diabetes have a decreased diversity and abundance of butyrate-producing bacteria.[42] Additionally, gut microbiome impacts on the insulin resistance-associated metabolic signatures, where branched-chain amino acids were increased in insulin resistance patients.[43] Therefore, characterization of individual gut microbiota signatures in patients with obesity and Type 2 diabetes could predict host and microbiota response to dietary interventions.[44] In this sense, it has been reported that diabetic patients with a higher *Prevotella*/*Bacteroides* ratio showed improved postprandial glucose and insulin resistance after a 3-day intervention with barley kernel bread diet (high in fiber) in comparison with the non-responders, thus, supporting the importance of host microbiota and personalized approaches to improve metabolism.[45]

10.2.4 Food Reward and DRD2

The dopaminergic system plays important roles in neuromodulation of physiological processes that involve motor control, motivation, reward, cognitive function, maternal, and reproductive behaviors.[46] Dopamine is a neurotransmitter involved in the rewarding effects of food synthesized in both the central and peripheral nervous systems.[47] It exerts its actions upon binding to the DRD2, a G-coupled protein with Ser/Thr kinase activity codified by the DRD2/ankyrin repeat domain and content kinase 1 (*DRD2/ANKK1*) gene (ID: 1813).[48] In humans, low DRD2 activity and reduced dopamine activity could contribute to overeating *via* their modulation of striatal prefrontal pathways, which participate in the inhibitory control and salience attribution.[49]

The role of the *DRD2* polymorphisms in addictive behaviors, negative emotions, and cognitive alterations has been well described. One of the most relevant genetic variants is the *DRD2/ANKK1* TaqIA polymorphism (rs1800497). The A1 allele affects DRD2 receptor density and increases L-dopa activity, modifying the rewarding effects of food.[49] However, this association may vary according to the population's genetic composition due to differences in TaqIA A1 allele prevalence. A low prevalence has been reported in Caucasians, whereas Chinese, Indian, and Asian American populations have higher rates. However, the highest TaqIA A1 allele prevalence of 67% in Native Mexicans and 47.3% in Mexican mestizos was reported in a study related to alcoholism.[50] In the context of an obesogenic environment, the Taq1A A1 allele has been associated with higher intake and frequency of carbohydrates, meats, and fried dishes as a lower consumption frequency of legumes than A2 carriers among the Mexican population.[51] As mentioned before, the consumption of this dietary profile could also be favored by a genetic-based choice for sweet (*TAS1R2*) and fat (*CD36*) flavors among the Mexicans,[10,22] which may elicit addictive food behaviors.

Another genetic polymorphism is the synonymous C957T polymorphism (rs6277) located in exon 7 of the *DRD2* gene. Several studies have demonstrated that CC genotype carriers exhibited higher reward sensitivity scores compared to carriers of the T allele and higher consumption of sucrose and carbohydrates.[52,53]

10.3 Lipid Metabolism

In the following section, genes and gene products involved in lipid transportation and fat storage and their relation with dyslipidemia and insulin resistance will be highlighted.

10.3.1 FABP2

FABP2 is a cytosolic protein with a high affinity-binding site for both saturated and unsaturated long-chain fatty acids, indicating that it may have a role in the absorption and intracellular transport of dietary long-chain fatty acids (Figure 10.1). FABP2 is widely expressed in the epithelial cells of the small intestine.[54] Polymorphisms in the *FABP2* promoter region result in transcriptional activation that impacts body composition, insulin sensitivity, and risk of Type 2 diabetes.[55,56]

The Ala54Thr substitution (rs1799883) is the most extensively studied SNP in which the Thr54 carriers express the isoform with twofold higher affinity for long-chain fatty acids compared to the Ala54 isoform. A study carried out in morbidly obese patients showed that Thr54 carriers were more prone to altered insulin resistance, leptin, and adiponectin levels.[57] Moreover, a study performed in Mexican subjects with BMI over 25 kg m^{-2} who underwent a moderate-fat diet intervention with 30% fat (saturated fat <7% of total calories), 15% protein, and 55% carbohydrates for 2 months showed that Thr54 allele carriers were better responders (lower weight, BMI, waist circumference, waist-to-hip ratio, and C-reactive protein levels) than Ala54 allele carriers.[58] However, in contrast, a study performed in Han Chinese with Type 2 diabetes showed that low-glycemic-index diet interventions could decrease plasma triglycerides and cholesterol levels in both Ala54 and Thr54 carriers, in which the Ala54 allele carriers were better diet responders.[59] These discrepancies may be related to ethnicity, study design, risk allele distribution, dietary pattern, and comorbidities among the study group. Such findings underscore the importance of personalized medicine tailored by subpopulations worldwide.

10.3.2 APOB

APOB plays an important role in the catabolism of chylomicrons, VLDL, and LDL by acting as a ligand for LDLR.[60] Thus, high levels of APOB are associated with high levels of lipids, including triglycerides, total cholesterol, and LDL-c. APOB can occur as APOB48 or APOB100 isoforms depending on the molecular size; however, they are encoded by the same gene.[61] *APOB* gene (ID: 338) is highly polymorphic with around over 5 000 polymorphic sites. Recently, a meta-analysis showed that two polymorphisms rs693 (G > A) and rs17240441 (insertion/deletion) were significantly associated with triglycerides, total cholesterol, and high LDL-c levels.[62]

Some dietetic interventions involving different variants in the *APOB* gene have been reported in different populations.[63–65] As with other variants in genes related to metabolism, the results of the dietetic interventions based on the *APOB* gene variants highlight the involvement of gene–nutrient interactions in the success of nutritional interventions (Figures 10.2 and 10.3).

10.3.3 APOE

Transportation and lipid internalization are crucial steps in lipid metabolism. APOE is a key glycoprotein regulator of the catabolism of triglyceride-rich lipoproteins.[66] APOE allows VLDL, LDL-c, and HDL entry into the hepatocyte through the binding to the LDLR expressed on the surface of the hepatocytes.[67] Two polymorphisms in the *APOE* gene (ID: 348), rs7412 and rs429358, change the three-dimensional structure of the protein domain that interacts with the LDLR, affecting the binding capacity of the protein to its receptor. The combination of these two polymorphisms results in three major alleles e2, e3, and e4, and the corresponding isoforms that differ at amino acid positions 112 and 158. e2 contains Cys112/158Cys, while e3 contains Cys112/158Arg, and e4 contains Arg112/158Arg residues. Human *APOE* is primarily expressed in the liver and less extensively in other tissues like the brain, kidney, colon, and white adipose.[66]

It is noteworthy to mention that the e3 allele has a natural affinity for LDLR and is related to normal lipid serum concentration. In contrast, e4 isoform has a higher affinity for LDLR and increases total cholesterol,[67,68] whereas the isoform e2 shows a lower affinity for LDLR and confers genetic susceptibility to hypertriglyceridemia, as shown in Figure 10.2. Given these characteristics, the APOE alleles have been extensively studied to establish an association with insulin resistance and diabetes.[69]

Interestingly, a heterogeneous distribution of *APOE* alleles/genotypes among human populations has been documented,[70] and this variation has been linked to differences in the prevalence of dyslipidemias in combination with other risk factors such as diet and physical activity.[71] These results reinforce the importance of establishing personalized medicine and diet intervention in each specific population based on these characteristics.

10.3.4 ABCA1

ABCA1 plays a key role in reverse cholesterol transport. ABCA1 is implicated in HDL biosynthesis and cellular cholesterol homeostasis. The expression of ABCA1 is ubiquitous in many tissues, but it is mainly expressed in hepatocytes and enterocytes.[72] ABCA1 mediates the release of phospholipids and unesterified cholesterol from the plasma membrane to APOA-1 in the extracellular medium, thus producing nascent HDL particles.[73]

There are over 100 coding variants identified in *ABCA1* (ID: 19); most of them change a single amino acid in the ABCA1 protein.[74] The variant R230C (rs9282541) has been extensively claimed as responsible for the variation of HDL-c concentration, BMI, and obesity-related comorbidities among Mexican descendants.[75] Recently, the R230C variant was associated with low HDL-c levels, obesity, and Type 2 diabetes among Mexican mestizos, suggesting that this variation could be an evolutive adaptive variant from native American populations.[76–78] Nevertheless, recent studies have identified other *ABCA1* variants influencing lipid profiles and adiposity in different populations around the world.[79–81] Regarding dietetic interventions, Mexican patients with metabolic syndrome, carriers of the R230C variant who underwent a diet rich in regional foods such as nopal, chia seed, soy protein, and oat, improved their body weight, triglycerides and adiponectin levels, and glucose tolerance after 2 months of intervention.[82] Similarly, Mexican hyperlipidemic R230C carriers who received a low-saturated-fat diet containing soy protein and 15 g of soluble fiber daily for 2 months were better responders, increasing 21.9% HDL-c in comparison to the R230R genotype subjects.[83] Therefore, these findings suggest that genotype and ethnicity should be considered during the design of a dietetic treatment to reach better therapeutic results.

10.3.5 FTO/alpha-ketoglutarate Dependent Dioxygenase

This enzyme is a nuclear protein of the AlkB-related non-heme iron and 2-oxoglutarate-dependent oxygenase superfamily, and it was identified just recently as a genetic factor for adiposity development (Figure 10.1). In humans, the *FTO* gene (ID: 79068) is ubiquitously expressed in the tissues, but it reaches the highest expression in the hypothalamus site that controls food intake.[84] Early studies suggested that FTO is also associated with dietary fat consumption,[85] increased appetite and reduced satiety,[86] food choices,[87] eating habits, and loss of control over eating.[88]

In 2007, *FTO* was the first gene identified in a GWAS associated with the degree of BMI.[89] This association has been found mainly among subjects with European, Asian, and African ancestries and with fewer data from Native American-derived populations.[89] However, it is known that the associations between *FTO* rs9939609/rs8050136 and Type 2 diabetes are region-related and that they contribute to an increased risk of this disease.[90] A large meta-analysis performed in European individuals ($n = 198\ 502$) examined the effect of the *FTO* rs9939609 SNP on cardiometabolic traits related to increased adiposity. The authors found that this SNP increased the risk for Type 2 diabetes, hypertension, triglycerides, fasting glucose, insulin, 2-hour oral glucose tolerance, liver enzymes, and inflammatory markers.[91] These associations have also been replicated in several individual studies in different populations around the world, including children and adolescent populations.[92,93]

Recently, nutritional interventions have been done in order to avoid the effect of the *FTO* risk alleles. A study found that a higher adherence to the Mediterranean diet is associated with a lower risk of obesity in subject carriers of the genetic *FTO* risk alleles.[94] In addition to diet, lifestyle interventions have had better results among subject carriers of the risk alleles of the *FTO* gene.[95]

10.4 Epigenetics

Epigenetic studies involve understanding the mechanisms of gene expression that do not rely on variations of the coding DNA sequences and how the interaction with environmental factors can turn on and off hundreds of loci under specific physiological conditions.[96] DNA methylation is an epigenetic mechanism that regulates the transcription of many genes by preventing transcription factor binding or promoting the binding of methyl-binding proteins causing inhibition or decreased gene expression.[97] The methyl group bound to a cytosine in CpG dinucleotide islands can be donated by the homocysteine cycle reactions, generated from methionine metabolism. This reaction is carried out by MTHFR (EC 1.5.1.20), an enzyme that catalyzes the irreversible reduction of 5,10-methylenetetrahydrofolate (vitamin B_9) to 5-methyltetrahydrofolate, the most abundant form of plasma folate (Figure 10.1). Hypermethylation of the *MTHFR* gene promoter is associated with the occurrence of diabetic retinopathy and with biochemical, inflammatory, and oxidative stress parameters in the context of chronic complications.[97]

On the other hand, folates play an essential role in one-carbon methyl transfer reactions mediating several biological processes in which DNA methylation is essential, as mentioned before. One of the most common genetic defects of homocysteine metabolism is the C677T substitution (rs 1801133) of the *MTHFR* gene (ID: 4524) that encodes a thermolabile isoenzyme with decreased activity. Studies have shown an association between a high concentration of homocysteine levels and an increased risk factor for neural tube defects, diabetes, and cardiovascular diseases.[98] The C677T polymorphism associated with body fat accumulation could trigger an inflammatory process by elevating homocysteine levels and increasing cytokine production, causing several diseases.[99] Nonetheless, adequate food intake of folates may be sufficient to combat this metabolic condition.

Regarding other dietary factors, high-fat diets change both gene expression and methylation patterns in muscle and adipose tissue.[100] Additionally, the type of dietary fat composition, saturated fatty acids *versus* polyunsaturated fatty acids, modulates CpG site methylation differentially.[101,102] Also, caloric restriction, perhaps the most common approach recommended to treat obesity and Type 2 diabetes, has demonstrated that individuals who are responders have a different epigenetic and transcriptomic signature compared to non-responders.[103] These results suggest that DNA methylation studies could be used to predict which individuals will respond adequately to diet programs.

10.5 What is Next in Personalized Nutrition to Prevent Obesity and Type 2 Diabetes?

Since the inception of the development of a framework for precision medicine in 2008 in the United States, clinical medicine and biomedical research worldwide strive towards the implementation of a new taxonomy of human diseases based on molecular biology.[104] By this new definition, "people are considered to differ in their susceptibility to a particular disease, in the biology and/or prognosis of those diseases they may develop, or in their response to a specific treatment."[104,105] Furthermore, the National Research Council endorses that personalized medicine or precision medicine focuses on the genetic characteristics of subpopulations, which implies that these differences are related to ethnicity. Because several of these diseases involve obesity *per se* and diet-related comorbidities such as Type 2 diabetes, personalized nutrition will play a key role in the prevention and treatment of these abnormal metabolic health conditions.[106] Therefore, the role of nutrition health professionals will be enhanced by an understanding of the growing amount of scientific-based evidence concerning the interaction between genes and nutrients. However, personalized nutrition is still a huge challenge in the clinical setting due to the absence of reproducible results, the high costs of omics technologies, and the scarceness of human resources that can understand and interpret such data.[107] As mentioned in the above sections, several polymorphic genes (SNPs) may influence the onset or progression of obesity and Type 2 diabetes by means of dyslipidemia and insulin resistance pathways at the nutrigenetic, nutrigenomic, and epigenetic level. However, the impact of these SNPs will ultimately depend upon the prevalence of the risk allele among different subpopulations. Furthermore, nutrient sources vary by dietary patterns that include the type and amount of food people eat based on personal and cultural (regional/ethnic) preferences. Therefore, one type of diet will not equally benefit all people.[108]

In this sense, countries which are advocates of the Mediterranean diet have worked extensively to define the role of genes, and dietary behaviors worked best in patients with cardiovascular risk, obesity, and Type 2 diabetes.[93,109,110]

However, as mentioned before, these studies may not be totally extrapolated to regions with different genetic susceptibility and risk components, which is essential to provide personalized therapies. Such is the case of the Latin American region, which in general has an admixed population. An example of the development of a personalized nutrition approach is the GENOMEX diet, which is rooted in the concept of the ancestral relationship between genes, food consumption, and regional food culture[111–113] – a notion that is not explicitly considered in the evolving definition of personalized nutrition.[114] Due to the high rate of overweight and obesity among the Mexican population, they have become a public health priority to be resolved.[11] This situation is driven in part by the dysmetabolic effect of the hepatopathogenic dietary pattern commonly consumed by Mexicans, leading to hyperglycemia, hyperinsulinemia, dyslipidemia, and inflammation, which are known to be associated with obesity and Type 2 diabetes.[6]

In this case, the admixed population of Mexico is comprised of Amerindian, European, and African ancestry with interregional heterogeneity. On the other hand, the food culture based on the Mesoamerican prehispanic diet was enriched by the Old Word products regionally after Spanish colonialization. Therefore, the GENOMEX diet sought to provide nutritional therapy tailored by the population-based genotype profiling and following the history of Mexican regional foods and food culture to prevent obesity-related chronic diseases. The GENOMEX diet was built considering local foods (nutrients) that promote microbiotic diversity, attenuate inflammation, reduce oxidative stress, and contain antifibrogenic and antioxidant properties.[111–113] Genetic profiling of some of the genes mentioned in this chapter revealed the influence of the ancestral Amerindian alleles[112] involved in key metabolic pathways. Subsequently, a study that implemented this diet in a Mexican population with metabolic risk factors for obesity-related chronic diseases showed a significant improvement in metabolic and anthropometric parameters such as high HOMA-IR and BMI, respectively, based on their genetic profile.[113]

Additionally, in alignment with the fact that personalized nutrition seeks to tailor dietary interventions to improve metabolic responses of individuals by subpopulations, some sophisticated algorithms are being implemented to integrate multidimensional data, such as lifestyle, gut microbiome, blood tests, anthropometric and eating behaviors. In this sense, two related studies primarily performed in an Israeli cohort and later replicated with modifications in a Midwestern American population showed that a predictive model of personalized postprandial glycemic response to foods could aid to improve the long-term prevention of cardiovascular disease and Type 2 diabetes.[115,116]

In conclusion, personalized nutrition will be the mainstream that will have an impact on the multiple fields of medical specialties. Omics' sciences big data will transform our knowledge by deciphering the genomic, transcriptomic, and proteomic/metabolic mechanisms underlying diseases. Epigenetics and metagenomic data will scale even higher. Altogether, this trend will lead to the development of novel diagnostic and treatment strategies aimed to prevent the onset and progression of chronic diseases. In conjunction, these innovations will ultimately lead to the updating of conventional medical textbooks, academic curricula, and practical clinical guidelines to keep up with the pace of advancements in omics' sciences applied to personalized therapy, whereas new medical and nutritional specialties will be created using this new framework in the future.[117,118]

Abbreviations

ABCA1	ATP-binding cassette subfamily A member 1
Ala	alanine
AMY1A	amylase alpha 1
AMY1B	amylase alpha 1B
AMY1C	amylase alpha 1C
AMY2A	amylase alpha 2A
AMY2B	amylase alpha 2B
AMYP	amylase pseudogene
ANKK1	ankyrin repeat and kinase domain containing 1
ApoA1	apolipoprotein A1
ApoB100	apolipoprotein B100
ApoB48	apolipoprotein B48
ApoE	apolipoprotein E
Arg	arginine
BMI	body mass index
CD36	cluster of differentiation 36
CNV	copy number variation
CpG	cytosine-phosphate-guanine
Cys	cysteine
DNA	deoxyribonucleic acid
DRD2	dopamine receptor D2
FABP2	fatty acid-binding protein 2
FTO	fat mass and obesity
GENOMEX	genome-based Mexican
GWAS	genome-wide association study
HDL	high-density lipoproteins
HOMA-IR	homeostatic model assessment of insulin resistance
HPA	hypothalamic–pituitary–adrenal
LCAT	lecithin-cholesterol-acyl-transferase
LCT	lactase
LDL	low-density lipoprotein
LDLR	low-density lipoprotein receptor
L-dopa	levodopamine
LNP	lactase non-persistence
LP	lactase persistence
LPL	lipoprotein lipase

MTHFR	methylenetetrahydrofolate reductase
OR	odds ratio
Ser	serine
SNP	single-nucleotide polymorphisms
TAS1R2	taste 1 receptor member 2
Thr	threonine
VLDL	very low-density lipoprotein
WHO	World Health Organization

References

1. http://www.euro.who.int/en/health-topics/disease-prevention/nutrition/a-healthy-lifestyle/body-mass-index-bmi (accessed February 2020).
2. https://www.who.int/news-room/fact-sheets/detail/obesity-and-overweight (accessed February 2020).
3. M. Bluher, *Nat. Rev. Endocrinol.*, 2019, **15**, 288.
4. Y. Zheng, S. H. Ley and F. B. Hu, *Nat. Rev. Endocrinol.*, 2018, **14**, 88.
5. https://www.diabetesatlas.org (accessed February 2020).
6. S. Roman, C. Ojeda-Granados and A. Panduro, *Rev. Endocrinología Nutrición*, 2013, **21**, 42.
7. V. S. Malik, M. B. Schulze and F. B. Hu, *Am. J. Clin. Nutr.*, 2006, **84**, 274.
8. I. Rivera-Iniguez, A. Panduro, O. Ramos-Lopez, S. J. Villasenor-Bayardo and S. Roman, *Eat. Weight Disord.*, 2019, **24**, 835.
9. B. M. Popkin, *Public Health Nutr.*, 2002, **5**, 205.
10. O. Ramos-Lopez, A. Panduro, E. Martinez-Lopez and S. Roman, *Nutrients*, 2016, **8**, 101.
11. C. A. Aguilar-Salinas, F. J. Gomez-Perez, J. Rull, S. Villalpando, S. Barquera and R. Rojas, *Salud Publica Mex.*, 2010, **52**, S44.
12. C. C. Babbitt, L. R. Warner, O. Fedrigo, C. E. Wall and G. A. Wray, *Proc. Biol. Sci.*, 2011, **278**, 961.
13. F. Luca, G. H. Perry and A. Di Rienzo, *Annu. Rev. Nutr.*, 2010, 30, 291.
14. A. A. Baachmanov and G. K. Beauchamp, *Annu. Rev. Nutr.*, 2007, **27**, 389.
15. P. A. Breslin, *Curr. Biol.*, 2013, **23**, R409
16. D. Liu, N. Archer, K. Duesing, G. Hannan and R. Keast, *Prog. Lipid Res.*, 2016, **63**, 41.
17. V. Ormazabal, S. Nair, O. Elfeky, C. Aguayo, C. Salomon and F. A. Zuñiga, *Cardiovasc. Diabetol.*, 2018, **17**, 122.
18. M. Y. Pepino, O. Kuda, D. Samovski and N. A. Abumrad, *Annu. Rev. Nutr.*, 2014, **34**, 281.
19. R. L. Silverstein and M. Febbraio, *Sci. Signal*, 2009, **2**, re3.
20. M. Melis, G. Sollai, P. Muroni, R. Crnjar and I. T. Barbarossa, *Nutrients*, 2015, 7, 2068.
21. E. Corpeleijn, C. J. van der Kallen, M. Kruijshoop, M. G. Magagnin, T. W. de Bruin, E. J. Feskens, W. H. Saris and E. E. Blaak, *Diabetic Med.*, 2006, **23**, 907.

22. O. Ramos-Lopez, A. Panduro, E. Martinez-Lopez, N. A. Fierro, C. Ojeda-Granados, M. Sepulveda-Villegas and S. Roman, *J. Nutr. Food Sci.*, 2015, **5**:2.
23. O. Ramos-Lopez, E. Martinez-Lopez, S. Roman, N. A. Fierro and A. Panduro, *World J. Gastroenterol.*, 2015, **21**, 11552.
24. M. H. Meisler and C.-N. Ting, *Crit. Rev. Oral Biol. Med.*, 1993, **4**, 503
25. D. Carpenter, S. Dhar, L. M. Mitchell, B. Fu and J. Tyson, N. A. A. Shwan, F. Yang, M. G. Thomas and J. A. L. Armour, *Hum. Mol. Genet.*, 2015, **24**, 3472.
26. P. J. D. Elder, D. B. Ramsden, D. Burnett, M. O. Weickert and T. M. Barber, *Exp. Rev. Endocrinol. Metab.*, 2018, **13**, 193.
27. G. Rukh, U. Ericson, J. Andersson-Assarsson, M. Orho-Melander and E. Sonestedt, *Am. J. Clin. Nutr.*, 2017, **106**, 256.
28. C. I. Fernández and A. S. Wiley, *Am. J. Phys. Anthropol.*, 2017, **163**, 645.
29. Y.-J. Choi, Y.-S. Nam, J. M. Yun, J. H. Park, B. L. Cho, H.-Y. Son, J. I. Kim and J. W. Yun, *Diabetic Med.*, 2015 **32**, 1588.
30. P. León-Mimila, H. Villamil-Ramírez, B. E. López-Contreras, S. Morán-Ramos, L. R. Macias-Kauffer, V. Acuña-Alonzo, B. E. del Río-Navarro, J. Salmerón, R. Velazquez-Cruz, T. Villarreal-Molina, C. A. Aguilar-Salinas and S. Canizales-Quinteros, *Nutrients*, 2018, **10**, 1607.
31. M. A. Mejía-Benítez, A. Bonnefond, L. Yengo, M. Huyvaert, A. Dechaume, J. Peralta-Romero, M. Klünder-Klünder, J. García Mena, J. S. El-Sayed Moustafa, M. Falchi, M. Cruz and P. Froguel, *Diabetologia*, 2015, **58**, 290.
32. S. Pinho, C. Padez, and L. Manco, *Ann. Hum. Biol.*, 2018, **45**, 435.
33. R. K. Montgomery, S. D. Krasinski, J. N. Hirschhorn, and R. J. Grand, *J. Pediatr. Gastroenterol. Nutr.*, 2007, **45**, S131.
34. Y. Itan, A. Powell, M. A. Beaumont, J. Burge and M. G. Thomas, *PLoS Comput. Biol.*, 2009, **5**, e1000491.
35. E. Morales, L. Azocar, X. Maul, C. Perez, J. Chianale, J. F. Miquel, *BMJ Open* 2011, **1**:e000125.
36. C. Ojeda-Granados, A. Panduro, J. R. Rebello Pinho, O. Ramos-Lopez, K. Gleyzer, F. M. Malta, K. Gonzalez-Aldaco and S. Roman, *J. Nutrigenet. Nutrigenomics*, 2016, **9**, 83.
37. V. Lazar, L.-M. Ditu, G. G. Pircalabioru, A. Picu, L. Petcu, N. Cucu and M. C. Chifiriuc, *Front. Nutr.*, **6**, 21.
38. M. Carabotti, A. Scirocco, M. A. Maselli and C. Severia, *Ann. Gastroenterol.*, 2015, **28**, 203.
39. A. Panduro, I. Rivera-Iñiguez, M. Sepulveda-Villegas and S. Roman, *World J. Gastroenterol.*, 2017, **23**, 3030.
40. J. D. Galley, M. C. Nelson, Z. Yu, S. E. Dowd, J. Waltere, P. S. Kumay, M. Lyte and M. T., Bailey, *BMC Microbiol.*, 2014, **14**, 189.
41. I. Rivera-Iñiguez, S. Roman, C. Ojeda-Granados, A. Panduro in *Microbiome and Metabolome in Diagnosis, Therapy and Other Strategic Applications*, ed. J. Faintuch and S. Faintuch, Elsevier, Academic Press, 2019, p. 379.
42. N. Larsen, F. K. Vogensen, F. W. van den Berg, D. S. Nielsen, A. S. Andreasen, B. K. Pedersen, W. A. Al-Soud, S. J. Sorensen, L. H. Hansen and M. Jakobsen, *PLoS One*, 2010, **5**, e9085.

43. H. K. Pedersen, V. Gudmundsdottir, H. B. Nielsen, T. Hyotylainen, T. Nielsen, B. A. Jensen, K. Forslund, F. Hildebrand, E. Prifti, G. Falony, E. Le Chatelier, F. Levenez, J. Dore, I. Mattila, D. R. Plichta, P. Poho, L. I. Hellgren, M. Arumugam, S. Sunagawa, S. Vieira-Silva, T. Jorgensen, J. B. Holm, K. Trost, H. I. T. C. Meta, K. Kristiansen, S. Brix, J. Raes, J. Wang, T. Hansen, P. Bork, S. Brunak, M. Oresic, S. D. Ehrlich and O. Pedersen, *Nature*, 2016, **535**, 376.
44. K. Korpela, H. J. Flint, A. M. Johnstone, J. Lappi, K. Poutanen, E. Dewulf, N. Delzenne, W. M. de Vos and A. Salonen, *PLoS One*, 2014, **9**, e90702.
45. P. Kovatcheva-Datchary, A. Nilsson, R. Akrami, Y. S. Lee, F. De Vadder, T. Arora, A. Hallen, E. Martens, I. Bjorck and F. Backhed, *Cell Metab.*, 2015, **22**, 971.
46. M. O. Klein, D. S. Battagello, A. R. Cardoso, D. N. Hauser, J. C. Bittencourt, and R. G. Correa, *Cell Mol. Neurobiol.*, 2019, **39**, 31.
47. R. A. Wise, *Philos. Trans. R. Soc., B*, 2006, **361**,1149
48. C. Missale, S. R. Nash, S. W. Robinson, M. Jaber and M. G. Caron, *Physiol. Rev.*, 1998, **78**, 189.
49. J.-H. Baik, *BMB Rep.*, 2013, **46**, 519.
50. A. Panduro, O. Ramos-Lopez, O. Campollo, E. A. Zepeda-Carrillo, K. Gonzalez-Aldaco, R. Torres-Valadez and S. Roman, *Drug Alcohol Depend.*, 2017, **172**, 66.
51. I. Rivera-Iñiguez, A. Panduro, O. Ramos-Lopez, S. J. Villaseñor-Bayardo and S. Roman, *Eat. Weight Disord.*, 2019, **24**, 835.
52. O. Ramos-Lopez, A. Panduro, I. Rivera-Iniguez and S. Roman, *Physiol. Behav.*, 2018, **194**, 532.
53. K. M. Eny, P. N. Corey and A. El-Sohemy, *Nutr. Nutrigen*, 2009, 2, 235.
54. Y. Li, E. Fisher, M. Klapper, H. Boeing, A. Pfeiffer, J. Hampe, S. Schreiber, B. Burwinkel, J. Schrezenmeir and F. Doring, *Horm. Metab. Res.*, 2006, **38**, 300.
55. L. Bu, L. M. Salto, K. J. De Leon and M. De Leon, *Diabetes Res. Clin. Pract.*, 2011, **92**, 82.
56. E. P. Weiss, M. D. Brown, A. R. Shuldiner and J. M. Hagberg, *Physiol. Genomics*, 2002, **10**, 145.
57. D. A. de Luis, M. G. Sagrado, R. Aller, O. Izaola, R. Conde and E. Romero, *Obes. Surg.*, 2009, **19**, 1691.
58. E. Martinez-Lopez, M. R. Garcia-Garcia, J. M. Gonzalez-Avalos, M. Maldonado-Gonzalez, B. Ruiz-Madrigal, B. Vizmanos, Z. Hernandez-Nazara, S. Roman and A. Panduro, *Nutrition*, 2013, **29**, 46.
59. P. J. Liu, Y. P. Liu, H. K. Qin, T. Xing, S. S. Li and Y. Y. Bao, *Diabetes Metab. Syndr. Obes.*, 2019, **12**, 413.
60. S. Y. Morita, *Biol. Pharm. Bull.*, 2016, **39**, 1.
61. A. J. Whitfield, P. H. Barrett, F. M. van Bockxmeer and J. R. Burnett, *Clin. Chem.*, 2004, **50**, 1725.
62. C. Niu, Z. Luo, L. Yu, Y. Yang, Y. Chen, X. Luo, F. Lai and Y. Song, *Lipids Health Dis.*, 2017, **16**, 166.

63. M. Ilmonen, T. Helio, R. Butler, A. Palotie, P. Pietinen, J. K. Huttunen and M. J. Tikkanen, *Arterioscler. Thromb. Vasc. Biol.*, 1995, **15**, 1287.
64. C. M. Phillips, L. Goumidi, S. Bertrais, M. R. Field, R. McManus, S. Hercberg, D. Lairon, R. Planells and H. M. Roche, *Atherosclerosis,* 2011, **214**, 408.
65. M. Rantala, T. T. Rantala, M. J. Savolainen, Y. Friedlander and Y. A. Kesaniemi, *Am. J. Clin. Nutr.*, 2000, **71**, 713.
66. M. C. Phillips, *IUBMB Life*, 2014, **66**, 616.
67. A. D. Marais, *Pathology*, 2019, **51**, 165.
68. I. A. Larson, J. M. Ordovas, C. DeLuca, J. R. Barnard, G. Feussner and E. J. Schaefer, *Atherosclerosis*, 2000, **148**, 327.
69. P. G. Anthopoulos, S. J. Hamodrakas and P. G. Bagos, *Mol. Genet. Metab.*, 2010, **100**, 283.
70. D. Aceves, B. Ruiz, P. Nuño, S. Roman, E. Zepeda, A. Panduro, *Hum. Biol.*, 2006, **78**, 65.
71. P. Abondio, M. Sazzini, P. Garagnani, A. Boattini, D. Monti, C. Franceschi, D. Luiselli and C. Giuliani, *Genes (Basel)*, 2019, **10**, 222.
72. L. R. Brunham, J. K. Kruit, J. Iqbal, C. Fievet, J. M. Timmins, T. D. Pape, B. A. Coburn, N. Bissada, B. Staels, A. K. Groen, M. M. Hussain, J. S. Parks, F. Kuipers and M. R. Hayden, *J. Clin. Invest.*, 2006, **116**, 1052.
73. S. Wang and J. D. Smith, *Biofactors*, 2014, **40**, 547.
74. L. R. Brunham, R. R. Singaraja and M. R. Hayden, *Annu. Rev. Nutr.*, 2006, **26**, 105.
75. M. T. Villarreal-Molina, C. A. Aguilar-Salinas, M. Rodriguez-Cruz, D. Riano, M. Villalobos-Comparan, R. Coral-Vazquez, M. Menjivar, P. Yescas-Gomez, M. Konigsoerg-Fainstein, S. Romero-Hidalgo, M. T. Tusie-Luna, S. Canizales-Quinteros and Metabolic Study Group, *Diabetes*, 2007, **56**, 1881.
76. M. T. Villarreal-Molina, M. T. Flores-Dorantes, O. Arellano-Campos, M. Villalobos-Comparan, M. Rodriguez-Cruz, A. Miliar-Garcia, A. Huertas-Vazquez, M. Menjivar, S. Romero-Hidalgo, N. H. Wacher, M. T. Tusie-Luna, M. Cruz, C. A. Aguilar-Salinas, S. Canizales-Quinteros and Metabolic Study Group, *Diabetes*, 2008, 57, 509.
77. V. Acuna-Alonzo, T. Flores-Dorantes, J. K. Kruit, T. Villarreal-Molina, O. Arellano-Campos, T. Hunemeier, A. Moreno-Estrada, M. G. Ortiz-Lopez, H. Villamil-Ramirez, P. Leon-Mimila, M. Villalobos-Comparan, L. Jacobo-Albavera, S. Ramirez-Jimenez, M. Sikora, L. H. Zhang, T. D. Pape, A. Granados-Silvestre Mde, I. Montufar-Robles, A. M. Tito-Alvarez, C. Zurita-Salinas, J. Bustos-Arriaga, L. Cedillo-Barron, C. Gomez-Trejo, R. Barquera-Lozano, J. P. Vieira-Filho, J. Granados, S. Romero-Hidalgo, A. Huertas-Vazquez, A. Gonzalez-Martin, A. Gorostiza, S. L. Bonatto, M. Rodriguez-Cruz, L. Wang, T. Tusie-Luna, C. A. Aguilar-Salinas, R. Lisker, R. S. Moises, M. Menjivar, F. M. Salzano, W. C. Knowler, M. C. Bortolini, M. R. Hayden, L. J. Baier and S. Canizales-Quinteros, *Hum. Mol. Genet.*, 2010, **19**, 2877.

78. M. H. Yao, J. He, R. L. Ma, Y. S. Ding, H. Guo, Y. Z. Yan, J. Y. Zhang, J. M. Liu, M. Zhang, D. S. Rui, Q. Niu and S. X. Guo, *Int. J. Environ. Res. Public Health*, 2016, **13**, 220.
79. V. Kolovou, A. Marvaki, M. Boutsikou, G. Vasilopoulos, D. Degiannis, C. Marvaki and G. Kolovou, *Open Cardiovasc. Med. J.*, 2016, **10**, 233.
80. D. Jung, S. Cao, M. Liu and S. Park, *Horm. Metab. Res.*, 2018, **50**, 308.
81. M. Guevara-Cruz, A. R. Tovar, C. A. Aguilar-Salinas, I. Medina-Vera, L. Gil-Zenteno, I. Hernandez-Viveros, P. Lopez-Romero, G. Ordaz-Nava, S. Canizales-Quinteros, L. E. Guillen Pineda and N. Torres, *J. Nutr.*, 2012, **142**, 64.
82. M. Guevara-Cruz, A. R. Tovar, E. Larrieta, S. Canizales-Quinteros and N. Torres, *Mol. Genet. Metab.*, 2010, **101**, 268.
83. J. S. McTaggart, S. Lee, M. Iberl, C. Church, R. D. Cox and F. M. Ashcroft, *PLoS One*, 2011, **6**, e27968.
84. E. Sonestedt, C. Roos, B. Gullberg, U. Ericson, E. Wirfalt and M. Orho-Melander, *Am. J. Clin. Nutr.*, 2009, **90**, 1418.
85. J. Wardle, S. Carnell, C. M. Haworth, I. S. Farooqi, S. O'Rahilly and R. Plomin, *J. Clin. Endocrinol. Metab.*, 2008, **93**, 3640.
86. L. Brunkwall, U. Ericson, S. Hellstrand, B. Gullberg, M. Orho-Melander and E. Sonestedt, *Food Nutr. Res.*, 2013, **57**, 20028.
87. M. Tanofsky-Kraff, J. C. Han, K. Anandalingam, L. B. Shomaker, K. M. Columbo, L. E. Wolkoff, M. Kozlosky, C. Elliott, L. M. Ranzenhofer, C. A. Roza, S. Z. Yanovski and J. A. Yanovski, *Am. J. Clin. Nutr.*, 2009, **90**, 1483.
88. A. Scuteri, S. Sanna, W. M. Chen, M. Uda, G. Albai, J. Strait, S. Najjar, R. Nagaraja, M. Orru, G. Usala, M. Dei, S. Lai, A. Maschio, F. Busonero, A. Mulas, G. B. Ehret, A. A. Fink, A. B. Weder, R. S. Cooper, P. Galan, A. Chakravarti, D. Schlessinger, A. Cao, E. Lakatta and G. R. Abecasis, *PLoS Genet.*, 2007, **3**, e11.
89. T. M. Frayling, N. J. Timpson, M. N. Weedon, E. Zeggini, R. M. Freathy, C. M. Lindgren, J. R. Perry, K. S. Elliott, H. Lango, N. W. Rayner, B. Shields, L. W. Harries, J. C. Barrett, S. Ellard, C. J. Groves, B. Knight, A. M. Patch, A. R. Ness, S. Ebrahim, D. A. Lawlor, S. M. Ring, Y. Ben-Shlomo, M. R. Jarvelin, U. Sovio, A. J. Bennett, D. Melzer, L. Ferrucci, R. J. Loos, I. Barroso, N. J. Wareham, F. Karpe, K. R. Owen, L. R. Cardon, M. Walker, G. A. Hitman, C. N. Palmer, A. S. Doney, A. D. Morris, G. D. Smith, A. T. Hattersley and M. I. McCarthy, *Science*, 2007, **316**, 889.
90. Y. Yang, L. Boyang, W. Xia, J. Yan, H.-Y. Liu, L. Hu and S.-M. Liu, *Genes*, 2017, **8**, 70.
91. T. Fall, S. Hagg, R. Magi, A. Ploner, K. Fischer, M. Horikoshi, A. P. Sarin, G. Thorleifsson, C. Ladenvall, M. Kals, M. Kuningas, H. H. Draisma, J. S. Ried, N. R. van Zuydam, V. Huikari, M. Mangino, E. Sonestedt, B. Benyamin, C. P. Nelson, N. V. Rivera, K. Kristiansson, H. Y. Shen, A. S. Havulinna, A. Dehghan, L. A. Donnelly, M. Kaakinen, M. L. Nuotio, N. Robertson, R. F. de Bruijn, M. A. Ikram, N. Amin, A. J. Balmforth, P. S. Braund, A. S. Doney, A. Doring, P. Elliott, T. Esko, O. H. Franco, S.

Gretarsdottir, A. L. Hartikainen, K. Heikkila, K. H. Herzig, H. Holm, J. J. Hottenga, E. Hypponen, T. Illig, A. Isaacs, B. Isomaa, L. C. Karssen, J. Kettunen, W. Koenig, K. Kuulasmaa, T. Laatikainen, J. Laitinen, C. Lindgren, V. Lyssenko, E. Laara, N. W. Rayner, S. Mannisto, A. Pouta, W. Rathmann, F. Rivadeneira, A. Ruokonen, M. J. Savolainen, E. J. Sijbrands, K. S. Small, J. H. Smit, V. Steinthorsdottir, A. C. Syvanen, A. Taanila, M. D. Tobin, A. G. Uitterlinden, S. M. Willems, G. Willemsen, J. Witteman, M. Perola, A. Evans, J. Ferrieres, J. Virtamo, F. Kee, D. A. Tregouet, D. Arveiler, P. Amouyel, M. M. Ferrario, P. Brambilla, A. S. Hall, A. C. Heath, P. A. Madden, N. G. Martin, G. W. Montgomery, J. B. Whitfield, A. Jula, P. Knekt, B. Oostra, C. M. van Duijn, B. W. Penninx, G. D. Smith, J. Kaprio, N. J. Samani, C. Gieger, A. Peters, H. E. Wichmann, D. I. Boomsma, E. J. de Geus, T. Tuomi, C. Power, C. J. Hammond, T. D. Spector, L. Lind, M. Orho-Melander, C. N. Palmer, A. D. Morris, L. Groop, M. R. Jarvelin, V. Salomaa, E. Vartiainen, A. Hofman, S. Ripatti, A. Metspalu, U. Thorsteinsdottir, K. Stefansson, N. L. Pedersen, M. I. McCarthy, E. Ingelsson, I. Prokopenko, European Network for Genetic and Genomic Epidemiology, *PLoS Med.*, 2013, **10**, e1001474.

92. F. Lauria, A. Siani, K. Bammann, R. Foraita, I. Huybrechts, L. Iacoviello, A. C. Koni, Y. Kourides, S. Marild, D. Molnar, L. A. Moreno, I. Pigeot, Y. P. Pitsiladis, T. Veidebaum, P. Russo and IEFICS Consortium, *PLoS One*, 2012, 7, e48876.
93. P. Ferreira Todendi, A. R. de Moura Valim, E. Klinger, C. P. Reuter, S. Molina, J. A. Martinez and M. Fiegenbaum, *Obes. Res. Clin. Pract.*, 2019, **13**, 137.
94. F. Hosseini-Esfahani, G. Koochakpoor, M. S. Daneshpour, B. Sedaghati-Khayat, P. Mirmiran and F. Azizi, *Nutrients*, 2017, **9**, 1064.
95. L. Xiang, H. Wu, A. Pan, B. Patel, G. Xiang, L. Qi, R. C. Kaplan, F. Hu, J. Wylie-Rosett and Q. Qi, *Am. J. Clin. Nutr.*, 2016, **103**, 1162.
96. M. Loh, L. Zhou, H. K. Ng and J. C. Chambers, *Mol. Metab.*, 2019, **27S**, S33.
97. M. K. dos Santos Nunes, A. S. Silva, I. W. de Queiroga Evangelista, J. M. Filho, C. N. Alves Pegado Gomes, R. Anderson Ferreira do Nascimento, R. C. Pordeus Luna, M. J. de Carvalho Costa, N. F. Paulo de Oliveira and D. Camati Persuhn, *Diabetol. Metab. Syndr.*, 2017, **9**, 84.
98. Y. Meng, X. Liu, K. Ma, L. Zhang, M. Lu, M. Zhao, M.-X. Guan and G. Qin, *Mol. Genet. Genomic Med.*, 2019, 7, e1020.
99. J. V. de Carvalho Lisboa, M. Ribeiro, R. C. Pordeus Luna, R. P. Ataíde Lima, R. Anderson Ferreira do Nascimento, M. Gomes Cavalcante Alves Monteiro, K. Querino de Farias Lima, C. P. Novaes dos Santos Fechine, N. F. Paulo de Oliveira, D. Damati Persuhn, R. Cavalcante Veras, M. C. Rodrigues Gonçalves, F. E. Leite de Lima Ferreira, R. Teixeira Lima, A. da Silva, A. da Silva Diniz, A. T. Cavalcanti de Almeida, R. Marcos de Moraes, E. Verly Junior and M. J. de Carvalho Costa, *Nutrients*, 2020, **12**, 361.
100. S. C. Jacobsen, C. Brons, J. Bork-Jensen, R. Ribel-Madsen, B. Yang, E. Lara, E. Hall, V. Calvanese, E. Nilsson, S. W. Jorgensen, S. Mandrup, C.

Ling, A. F. Fernandez, M. F. Fraga, P. Poulsen and A. Vaag, *Diabetologia*, 2012, **55**, 3341.
101. F. Rosqvist, D. Iggman, J. Kullberg, J. Cedernaes, H. E. Johansson, A. Larsson, L. Johansson, H. Ahlstrom, P. Arner, I. Dahlman and U. Riserus, *Diabetes*, 2014, **63**, 2356.
102. A. Perfilyev, I. Dahlman, L. Gillberg, F. Rosqvist, D. Iggman, P. Volkov, E. Nilsson, U. Riserus and C. Ling, *Am. J. Clin. Nutr.*, 2017, **105**, 991.
103. L. Bouchard, R. Rabasa-Lhoret, M. Faraj, M. E. Lavoie, J. Mill, L. Perusse and M. C. Vohl, *Am. J. Clin. Nutr.*, 2010, **91**, 309.
104. https://bigdatawg.nist.gov/PCAST_Personalized_Medicine_Priorities.pdf (accessed March 2020).
105. https://www.ncbi.nlm.nih.gov/books/NBK92140 (accessed March 2020).
106. D. D. Wang and F. B. Hu, *Lancet Diabetes Endocrinol.*, 2018, **6**, 416.
107. J. M. Ordovas, L. R. Ferguson, E. S. Tai and J. C. Mathers, *BMJ*, 2018, **361**:k2173.
108. M. Verma, R. Hontecillas, N. Tubau-Juni, V. Abedi and J. Bassaganya-Riera, *Front. Nutr.*, 2018, **5**, 117.
109. F. J. Basterra-Gortari, M. Ruiz-Canela, M. A. Martínez-González, N. Babio, J. V. Sorlí, M, Fito, E. Ros, E. Gómez-Gracia, M. Fiol, J. Lapetra, R, Estruch, L. Serra-Majem, X. Pinto, J. I. González, M. Bulló, O. Castañer, Á. Alonso-Gómez, L. Forga, and F. Arós, for the PREDIMED Study Investigators, *Diabetes Care*, 2019, **42**, 1390.
110. J. Dıez-Espino, P. Buil-Cosiales, M. Serrano-Martinez, E. Toledo. J. Salas-Salvado and M. A. Martınez-Gonzalez, *Ann. Nutr. Metab.*, 2011, **58**, 74.
111. S. Roman, C. Ojeda-Granados, O. Ramos-Lopez and A. Panduro, *World J. Gastroenterol.*, 2015, **21**, 3449.
112. C. Ojeda-Granados, A. Panduro, K. Gonzalez-Aldaco, M. Sepulveda-Villegas, I. Rivera-Iñiguez, and S. Roman, *J. Pers. Med.*, 2017, 7, 16.
113. C. Ojeda-Granados, A. Panduro, I. Rivera-Iniguez, M. Sepulveda-Villegas and S. Roman, *Nutrients*, 2020, **12**, 645.
114. C. L. Bush, J. B. Blumberg, A. El-Sohemy, D. M. Minich, J. M. Ordovás, D. G. Reed and V. A. Yunez Behm, *J. Am. Coll. Nutr.*, 2020, **39**, 5.
115. D. Zeevi, T. Korem, N. Zmora, D. Israeli, D. Rothschild, A. Weinberger, O. Ben-Yacov, D. Lador, T. Avnit-Sagi, M. Lotan-Pompan, J. Suez, J. A. Mahdi, E. Matot, G. Malka, N. Kosower, M. Rein, G. Zilberman-Schapira, L. Dohnalova, M. Pevsner-Fischer, R. Bikovsky, Z. Halpern, E. Elinav and E. Segal, *Cell*, 2015, **163**, 1079.
116. H. Mendes-Soares, T. Raveh-Sadka, S. Azulay, Y. Ben-Shlomo, Y. Cohen, T. Ofek, J. Stevens, D. Bachrach, P. Kashyap, L. Segal and H. Nelson, *Am. J. Clin. Nutr.*, 2019, **110**, 63.
117. A. Panduro and S. Roman, *Ann. Hepatol.*, 2020, **19**, 1.
118. S. Roman and A. Panduro, *World J. Gastroenterol.*, 2015, **21**, 8227.

CHAPTER 11

Allometric Scaling of Dietary Bioactives in Metabolic Research: The Present and Future

JAY WHELAN*

Department of Nutrition, 1215 West Cumberland Avenue, 229 Jessie Harris Building, and the Tennessee Agricultural Experiment Station, University of Tennessee, Knoxville TN 37996, USA
*Email: jwhelan@utk.edu

11.1 Introduction

The use of preclinical experimental animal models in biomedical research is an integral part of the scientific process. The use of these models provides the necessary preliminary data that is needed prior to moving to clinical trials. These models are critical in testing safety and efficacy. Yet, while preclinical biomedical research is a necessary preliminary step to clinical trials, it is not without criticism. These include the use of poorly characterized cell lines when using xenograph models, lack of comparable physiological environments between humans and rodents, poor applications of pharmacodynamics/pharmacokinetics, inappropriate biomarkers, insufficient negative and positive controls, homogeneity of animal populations, inappropriate/

Nutritional Signaling Pathway Activities in Obesity and Diabetes
Food Chemistry, Function and Analysis No. 24
Edited by Zhiyong Cheng

Published by the Royal Society of Chemistry, www.rsc.org

inadequate randomization and statistical analyses, insufficient sample sizes, and so on.[1–6] All of these can lead to the lack of reproducibility of results and the inability to translate these results to humans and potentially the clinical setting. In fact, a checklist has been recommended to increase the quality of reporting of key methodological information.[4]

The role of diet in health promotion and disease prevention has been at the forefront of biomedical research since Hippocrates (400 BC) said, "Let thy food be thy medicine and medicine be thy food." Since then, we have discovered essential nutrients that are required to maintain health and investigated many other non-essential dietary bioactives that promote health and combat disease. At the forefront of these investigations is the use of preclinical experimental animal models. However, in the pursuit of "raising standards for preclinical research,"[1] the diet of these animals is never mentioned despite the fact that the investigation of dietary bioactives on health promotion and disease prevention is a robust area of preclinical/clinical exploration.

A recent review investigating publication rates in animal research reports that as much as half of all animal research may not be published, resulting in the repetition of animal experiments.[7] Furthermore, publication bias contributes to the concept that publication results are biased towards a previously reported outcome.[7] As such, researchers design experiments that give them the greatest chance of generating positive results, an experimental design bias not considering translational fidelity, all of which can contribute to the failure rates of subsequent clinical trials that are based on these preclinical experiments. Some of the key challenges of insuring the translational fidelity of dietary results generated in preclinical experimental models is understanding the metabolic differences between these models and humans, the factors that influence the generation of "false-positive" results, and how to properly translate dietary designs between species so that results have a greater probability of being reproduced in humans. Therefore, this chapter will review the theoretical bases for various allometric scaling models for dietary nutrients between species, in particular, experimental rodent models and humans, in an effort to minimize "false-positive" results and insure translational fidelity.

11.2 Contributing Factors Using Rodents that Impede Reproducibility of Results in Humans

1. Rodents typically are derived from the same strain where the genetic background is known and is similar or identical from one mouse to another, while humans are genetically diverse with unknown genetic profiles. Thus, rodents are much more genetically homogenous.
2. Typically, rodents are purchased from the same company and the same breeding colony, while humans are randomly selected or self-selected.
3. It is critically important that the rodents are fed the same food and consume the same amounts of food, while day-to-day human diets vary in amount and content.

4. Rodents are housed in a very controlled environment, while living conditions for humans vary considerably.
5. Although there are exceptions, the ages of the rodents are routinely the same for dietary studies, while clinical trials select ranges, such as adults between 18 and 65, and so forth.
6. For a given experiment, rodents will have very similar body weights and body compositions, while the diversity of the physical anatomy of human subjects can be vast.
7. Rodents are housed under strict, programmed light: dark cycles, while humans have diverse awake: sleep cycles (circadian rhythms).
8. The exercise patterns of rodents are typically limited, while humans have diverse and individualized exercise patterns.
9. Rodents are typically provided a pharmacological dose of a dietary bioactive, many times based on dose–response curves (or because the dose generates positive results), while humans are limited to "human equivalent doses" (HEDs).

Thus, this chapter attempts to address the issue of assessing how to determine an HED for nutrients in rodent diets.

11.3 Allometric Scaling in Biology

Allometric scaling is a term that refers to interspecies dose extrapolations. It is most commonly applied to pharmacology and drug treatments. Since it is unethical to experiment on humans as to the appropriate level or dose of a drug, we use experimental models to test safety and efficacy and apply those results based on mathematical conversions to other species. The conversion factors associated with the extrapolation can be influenced by absorption and transport processes, metabolic rates, mass, and volume of the species. As an illustration, rodents and humans have similar metabolic pathways for the conversation of dietary polyunsaturated fatty acids (PUFAs) to their more highly unsaturated derivatives. For example, the conversion of alpha-linolenic acid (ALA, the most common omega-3 PUFA in the Western diet) to its highly unsaturated product, docosahexaenoic acid (DHA), utilizes the same enzymatic machinery in both humans and rodents. However, as it turns out, tissue requirements for DHA are dependent upon metabolic activity, meaning, animals with higher metabolic rates have higher requirements for DHA in their membranes.[8] In fact, the levels of DHA in tissues of various species can be used as a biomarker for metabolic activity.[8] Therefore, if you are conducting a feeding trial investigating health benefits of dietary ALA or DHA, the rodent tissues will need to be sufficiently replete in DHA at the start of the study in order to accurately assess the biological impact of these fatty acids (for translational fidelity) because human tissues are sufficiently replete due to our Western diet. To illuminate this point, it is not uncommon for background diets of rodents to be based on the AIN76A semi-purified diet, whose sole lipid source is derived from corn oil. Corn oil is, for all practicality, devoid of omega-3 fatty acids.[9] Therefore, the tissues of rodents will be omega-3-deficient when fed this diet, exaggerating the results when ALA or DHA is supplemented in the experimental group.

11.3.1 Allometric Scaling Models

11.3.1.1 Body Weight

The use of differences in body weight (typically in kg) is a common mathematical model when extrapolating oral and non-oral doses of bioactives. Part of this is because it is a very simple mathematical process with the assumption that the linear relationship with body weight (*viz.* mass) also applies to nutrient needs. However, this model doesn't take into account the multidimensional aspects of surface area and volume (both contributing to metabolic activity). As such, when extrapolating nutrients from a rodent to a human using body weight as the conversion factor, it overestimates the HED. For example, a 25-g mouse, consuming 4.18 g day^{-1} of AIN93M diet (diet for mature animals), will consume an equivalent of 0.09 g day^{-1} of linoleic acid (LA, the major PUFA in the Western diet). Extrapolating this to a 70-kg man, this would be an HED of 252 g day^{-1} (2800×0.09). This is ~17 times higher when compared to the mean intake of LA in the US diet of 12–17 g day^{-1}. Similarly, when extrapolating an HED down to a mouse, this model underestimates the amounts needed in the mouse diet. For example, using the scenario outlined above except in reverse, if we wanted to calculate an HED for a mouse diet using the mean intake of LA of 12–17 g day^{-1}, this would translate to a daily amount of 0.004–0.006 g day^{-1} for the mouse (12–17 g/2800), far below what is typically provided to the mouse for this essential nutrient in this diet (0.09 g day^{-1}).

11.3.1.2 Body Surface Area and Metabolic Rate

In the early part of the 20th century, a group of pioneers recognized that the relationship between body weight and body surface area was non-linear and could be explained by the equation $Y = a(BW_{kg})^b$, where Y is the surface area, a is a constant, and b is an exponent.[10,11] Through oxygen consumption and calorimetric studies, they also determined that body surface area was correlated with metabolic activity (heat production through unit time) using a similar equation. It appears that the relationship between body weight and metabolic rate was best defined with an exponent between 0.67 and 0.75, with 0.75 being the more accepted exponent.[11–14]

11.3.1.3 Body Surface Area and Toxicology

In 1966, Freireich *et al.* published a paper that developed a mathematical relationship for the extrapolation of chemotherapeutic drugs between rodents and humans based on surface area (mg m^{-2} of surface area) using the following formula (dose in mg $m^{-2} = K_m \times$ dose in mg kg^{-1}), where K_m was a constant for each species for converting mg kg^{-1} to mg m^{-2}.[15] The K_m factor for a human, rat, and mouse was determined to be 37, 6, and 3, respectively. Therefore, the ratios of the K_m factors for human: rat and human: mouse are 6.2 and 12.3, respectively, and these ratios are commonly used for the allometric scaling of drugs.[16] Based on this earlier paper, the formula was refined in a follow-up manuscript to better translate doses between animals and humans, hoping to clarify any misunderstandings in those extrapolations.[16] This was again based on body surface area and the results published by Freireich *et al.* with his K_m factors for humans, rats, and mice of 37, 6, and 3 respectively.[15] Therefore, their formula for dose translation based on body surface area was: HED (mg kg^{-1}) = animal dose (mg kg^{-1}) × $Km_{animal}\ Km_{human}^{-1}$.[17]

Unfortunately, this formula and these K_m ratios are commonly used for the allometric scaling of dietary nutrients.[16] When using these formulas for the extrapolation of doses of nutrients between species, it could be assumed that the methodologies and data have been appropriately validated and applied to nutrients in the same way they were for chemotherapeutic drugs. That is not the case. The K_m factors generated by Freireich *et al.* were never intended to be applied to diet and dietary constituents:

1. These were for the extrapolation of chemotherapeutic drugs between rodents and humans, not dietary nutrients.
2. The formula was designed to predict toxicity of chemotherapeutic drugs, not the health benefits of nutrients.
3. Freireich *et al.* used different endpoints in rodents and humans. With rodents, they determined the LD_{10} (lethal dose for 10% of the animals) and compared this with the maximum tolerable dose in humans. For riboflavin, how does one determine the LD_{10} in rodents and the maximum tolerable dose in humans? How about fiber?
4. They used different routes of administration between the species. For dietary studies, the route of administration should be oral for both rodents and humans. For studies involving Freireich's formula, the major route of administration for rodents was intraperitoneally, and intravenously for humans.
5. When using the the K_m converstion factors, while better than extrapolation with body weight, you overestimate the amounts in rodent diets and underestimate the amounts that humans consume. For example, for mice consuming 4.18 g day^{-1} of AIN93M diet (containing 0.09 g day^{-1} of LA), the HED for LA when extrapolated to humans based on the K_m ratios would be 1.1 g day^{-1} (well under the US daily reference intake (DRI) of 12–17 g day^{-1}). If the calculation was performed in the other direction for a human consuming 12–17 g day^{-1}, the mouse daily equivalence would be 0.98–1.4 g day^{-1} (10 times higher than their typical intake of 0.09 g day^{-1}). Therefore, this would provide an inappropriate justification to treat animals with a dose that was 10 times more than reasonable, potentially generating false-positive results. Using this formula provides an acceptable justification for scientists to use unreasonably high doses of dietary

bioactives in rodents (*i.e.*, resveratrol) because you underestimate the actual HED. This can lead to results that cannot be reproducible in clinical trials.

11.3.1.4 Caloric Consumption, Differences in Energy Needs

This allometric scaling model takes into account body surface area, metabolic rates, size of the animal, and food intake. It is based on the premise of ad-libitum food intake needed to maintain body weight or for the appropriate growth of an animal. It uses the assumption of a standard human whose body weight is ~70 kg and who is consuming 2 000 kcal day^{-1}. There is evidence that it can be applied to both non-energy-producing and energy-producing nutrients. For example, it is estimated that a 25-g mouse will consume an average of 15 kcal day^{-1}. A larger mouse or rat will consume more food. The difference between 2000 kcal (human) and 15 kcal (mouse) is 133.3-fold. A rat weighing 250 g will consume on average 99.3 kcal day^{-1} (a difference of 20.2-fold *versus* the human). A larger animal consumes more, but the conversion factor with humans is proportionally less. This concept was born out of a paper published by Rucker and Storms (2002), where they stated that: "We suggest that when making interspecies comparisons from a nutrition perspective, the strongest case is made when a measure of metabolic body size or *food intake*, rather than body weight, is used to extrapolate dosages required for a given response."[18] They validated this concept by allometric scaling the levels of a variety of vitamins and minerals between species using caloric consumption.[18,19] We validated this method using a variety of dietary PUFAs and herbal extracts.[9,20–22]

As it turns out, differences in caloric intake is the most conservative of all models. It is 10 times more conservative than the ratios of the K_m factors that are based on Freireich's formulas (as recommended elsewhere).[17] As articulated earlier, every phase of an experiment using preclinical models (rodents) is so controlled that it is designed to reduce variability. As such, an experimental design for rodent models should not overexaggerate the levels of nutrients being tested. If rodents, under exceedingly controlled conditions (as described above), can't robustly respond to a "human equivalent dose," how are we to expect humans to respond to an HED? Therefore, how do these different models compare when extrapolating nutrients between species?

11.3.2 Comparison of the Different Allometric Scaling Models

11.3.2.1 Standard Research Diets

The standard maintenance diet that rodents are placed on is "chow." Chow is a variable formulated diet that is grain and animal byproduct-based composed of ground wheat/corn/oats, fish/soybean/corn gluten/alfalfa meals, with a variety of lipid sources.[23] Due to the lack of standardization of its components, it was replaced by the American Institute of Nutrition (AIN) series of diets. There are three major AIN semi-purified rodent diets that provide the framework for the design of other diets, AIN76A, AIN93G, and AIN93M. These were formulated to maintain health and proper growth, and support reproduction and lactation in rodents. Their ingredients are highly controlled, with little to no contaminants, making them ideal as standard diets that can be manipulated for targeted experiments. Because their nutrients are based on physiological need, there is little incentive to overly enrich the diets with an excess of the nutrients. The AIN76A diet was the original diet formulated for nutritional and toxicological research; however, some adverse effects were reported.[24–26] At the time, it was not well understood that ALA was an essential omega-3 PUFA that was required in the diet. To address these issues, the AIN76A diet was reformulated, by modifying the fiber content, and corn oil was replaced by soybean oil at two doses, one for the growing animal (AIN93G, 7% soybean oil), and one for the mature animal (AIN93M, 4% soybean oil).[25] The vitamin and mineral contents, while not identical, are very similar in the diets, with the biggest differences between the diets being the type and level of fat used (corn oil *versus* soybean oil at two doses).

11.3.2.2 Allometric Scaling of Nutrients from the Standard Research Diets to Humans

In an effort to compare the various allometric scaling models on their ability to predict human equivalent levels, the quantities of nutrients in the AIN76A, AIN93G, and AIN93M diets were extrapolated to humans based on the different mathematical formulas and then compared to the recommended levels of these nutrients for humans using the US RDI values (Table 11.1, as presented elsewhere with modifications).[9] The US DRIs are nutrient-based reference values established by the Institute of Medicine that provide guidelines for daily intakes of nutrients to reduce the risk of deficiencies in a large majority of the population (replacing the previously used recommended dietary allowances) and as estimates for intakes in the healthy population for those nutrients where an estimated adequate requirement can't be determined.[27] We compared differences in energy, body weight (BW), the formula "HED (mg kg^{-1}) = animal dose (mg kg^{-1}) × Km_{animal} Km_{human}^{-1}," and metabolic rate ($BW^{0.75}$).[9] For example, if we were to extrapolate a nutrient by body weight, the daily intake (by a mouse) would be multiplied by 2800 (the difference between the body weight of a 25-g mouse and a 70-kg man). That would give us a theoretical HED. The same process would be repeated for each nutrient in the mouse diet with each formula (as described above) and compared to recommended (US RDIs) or actual intakes of nutrients by humans. In this way, we can get insight as to which of the allometric scaling formulas may better predict interspecies conversions of nutrients.

When looking at the data presented in Table 11.1, the first column lists each of the nutrients in the commercial rodent diets. The second column lists the US RDIs for each nutrient (for humans) as a basis for comparison, or in the case of LA and ALA, the median intake in the US population.[28] The third, eighth, and 13th columns represent the daily intake of each nutrient for a mouse when

Table 11.1 Nutrients associated with the commercial rodent diets AIN76A, AIN93G, and AIN93M (at a level of 15 kcal day^{-1}) that are converted to human equivalent doses (HED) based on the various allometric scaling models.[a] Adapted from reference 9, https://doi.org/10.31989/ffhd.v7i5.338, under the terms of the CC BY 2.0 licence, https://creativecommons.org/licenses/by/2.0/.

Nutrient		AIN76A diet					AIN93G diet					AIN93M diet				
	US DRI	Rodent diet g/4.6 g diet	Energy needs (HED)	BW(kg) (HED)	Formula (HED)	Metabolic rate (HED)	Rodent diet g/4 g diet	Energy needs (HED)	BW(kg) (HED)	Formula (HED)	Metabolic rate (HED)	Rodent diet g/4.18 g diet	Energy needs (HED)	BW(kg) (HED)	Formula (HED)	Metabolic rate (HED)
Linoleic acid	12–17 g	0.14	18	386	31	53	0.16	21	442	36	61	0.09	12	253	20	35
α-Linolenic acid	1.1–1.6 g	0	0.0	0	0	0	0.02	2.7	57	4.6	7.9	0.01	1.6	33	2.7	4.5
		mg or μg 4.6 g^{-1} diet					mg or μg 4 g^{-1} diet					mg or μg 4.18 g^{-1} diet				
Thiamin	1.2 mg	0.025	3.3	69	6	9	0.02	2.7	56	5	8	0.021	2.8	59	5	8
Riboflavin	1.3 mg	0.028	3.7	77	6	11	0.024	3.2	67	5	9	0.025	3.3	70	6	10
Vitamin B_6	1.3 mg	0.026	3.5	74	6	10	0.024	3.2	67	5	9	0.025	3.3	70	6	10
Niacin	16 mg	0.138	18	386	31	53	0.12	16	336	27	46	0.125	17	351	28	48
Folate	400 μg	9.2	1227	25760	2089	3542	8	1067	22400	1816	3080	8.36	1115	23408	1898	3219
Biotin	30 μg	0.92	123	2576	209	354	0.8	107	2240	182	308	0.836	111	2341	190	322
Vitamin E	15 mg	0.161	21	451	37	62	0.2	27	560	45	77	0.209	28	585	47	80
Vitamin D	5 mg	0.115	15	322	26	44	0.1	13	280	23	39	0.105	14	293	24	40
Vitamin A	900 μg	5.52	736	15456	1253	2125	4.8	640	13440	1090	1848	5.016	669	14045	1139	1931
Vitamin C	90 mg	0	0	0	0	0	0	0	0	0	0	0	0	0	0	0
Vitamin K	120 μg	0.23	31	644	52	89	3.6	480	10080	817	1386	3.6	479	10065	816	1384
Vitamin B_{12}	2.4 μg	0.046	6	129	10	18	0.1	13	280	23	39	0.105	14	293	24	40
Pantothenate	5 mg	0.068	9	189	15	26	0.06	8	168	14	23	0.063	8	176	14	24
Ca	1000 mg	21	3189	66976	5430	9209	20	2667	56000	4541	7700	21	2787	58520	4745	8047
P	700 mg	13	2453	51520	4177	7084	12	1600	33600	2724	4620	13	1672	35112	2847	4828
K	4700 mg	15	2208	46368	3760	6376	14	1920	40320	3269	5544	15	2006	42134	3416	5793
Na	1500 mg	4	626	13138	1065	1806	4.1	551	11570	938	1591	4	576	12090	980	1662
Cl	2300 mg	6.7	957	20093	1629	2763	6.5	870	18267	1481	2512	6.7	899	18879	1531	2596
Mg	400 mg	2.1	307	6440	522	886	2.1	274	5746	466	790	2.1	285	5981	485	822

S	NA	1.3	613	12880	1044	1771	1.2	160	3360	272	462	1.3	167	3511	285	483
Cr	35 µg	4.2	1227	25760	2089	3542	4.0	533	11200	908	1540	4.2	557	11704	949	1609
Cu	900 µg	25	3680	77280	6266	10626	24	3200	67200	5449	9240	25	3344	70224	5694	9656
Fl	4 mg	0.004	0	0	0	0	0.004	1	11	1	2	0.004	1	12	1	2
I	150 µg	0.8	123	2576	209	354	0.8	107	2240	182	308	0.8	111	2341	190	322
Fe	8 mg	0.2	21	451	37	62.0	0.2	24	504	41	69	0.2	25	527	43	72
Mn	2.3 mg	0.04	33	696	56	96	0.04	5	112	9	15	0.04	6	117	9	16
Mo	45 µg	0.6	0	0	0	0	0.6	80	1680	136	231	0.6	84	1756	142	241
Se	55 µg	0.7	61	1288	104	177	0.7	96	2016	163	277	0.7	95	1990	161	274
Zn	11 mg	0.1	18	386	31	53	0.2	20	426	35	59	0.1	20	410	33	56

[a]BW, body weight; US DRI, US daily reference intake.

consuming 15 kcal day^{-1} of each commercial diet. The subsequent columns represent the HEDs of those nutrients. For example, the US RDI for LA is 12–17 g day^{-1} (row 3, column 2). If a mouse consumed 15 kcal of AIN76A, that would be equivalent to 4.6 g of diet day^{-1}, with a daily intake of 0.14 g LA day^{-1} (row 3, column 3). Extrapolating 0.14 g day^{-1} based on each of the formulas yields HED for LA (row 3, columns 4–7). Thus, the HED for LA for a mouse consuming 15 kcal day^{-1} of AIN76A would be 18 g day^{-1} (row 3, column 4) when extrapolated based on differences in energy and this compares to the US RDI of 12–17 g day^{-1} (row 3, column 2). The same process applies for each of the mathematical models and using the other commercial diets, AIN93G and AIN93M.

What is clearly evident is that in almost every case for every nutrient (with the exception of vitamin K, potassium, sodium, and fluorine), the allometric scaling model that comes the closest to the US RDIs is "energy needs"; that is, extrapolations based on differences in caloric consumption (columns 4, 9, and 14). As it turns out, with dietary lipids the use of percent energy (% energy) aligns with differences in energy needs.[9,29] For example, dietary LA comprises ~6% energy in the US diet. This also would be equivalent to 6% energy in the rodent diet. This model was validated using various levels of PUFAs in rodents and comparing those results to humans.[9,29] The model least appropriate for extrapolations of nutrients is body weight, as this model grossly overestimates human levels. While the other two models are much better than body weight, they still overestimate theoretical intakes for a human.

11.3.2.2 Comparing Energy as the Allometric Scaling Model between Mice and Humans

In an effort to examine the use of energy needs as a mathematical model for extrapolation of nutrients between preclinical experimental rodent models and humans, we investigated the impact of supplementing diets with various PUFAs on changes in phospholipid fatty acid composition and compared these effects between rodents (mice) and humans. The hypothesis was: "If you design a rodent diet to mimic a human diet based on % energy of the nutrients, you will get a human equivalent response." Rodent diets were designed to mimic the % distribution of macronutrients, based on energy, found in the US diet (protein: 16%en; carbohydrates: 50%en; fat: 34%en) (Table 11.2).[30]

The distribution within the lipid class of nutrients was also comparable with regard to saturated fatty acids, monounsaturated fatty acids, and PUFAs. Furthermore, the PUFAs were further delineated with regard to the omega-6 PUFAs (LA, 18:2 n-6; arachidonic acid, 20:4 n-6) and omega-3 PUFAs (ALA, 18:3 n-3; eicosapentaenoic acid, 20:5 n-3; DHA, 22:6 n-3) at levels typically consumed in the US diet (based on energy). To this background diet, mice were then supplemented with increasing levels of various PUFAs at HEDs, based on % energy (*i.e.*, 0.3–1.45%).

Changes in tissue phospholipid fatty acid composition were determined and compared to comparable data generated from human clinical trials.[9,29–31] Figure 11.1 represents a sample of that data. The *x* axis represents increasing

Table 11.2 Dietary designs of the background/control diets as compared to the US diet[1] (approximate macronutrient distribution, % energy)

Nutrient	US diet	Rodent diet
Protein	16	16
Carbohydrate	50	50
Fat	34	34
SFA	13	13
MUFA	14	14
PUFA	7	7
PUFA (% energy)		
18:2 n-6	6–7	6.9
20:4 n-6	0.08	0.08
18:3 n-3	0.63	0.6
20:5 n-3	0.034	0.034
22:6 n-3	0.054	0.054

[a]Human equivalent dose (HED) based on energy (en) %, equivalent in a 2000-kcal day^{-1} human diet. (See reference 30.)
MUFA, monounsaturated fatty acid; PUFA, polyunsaturated fatty acid; SFA, saturated fatty acid.

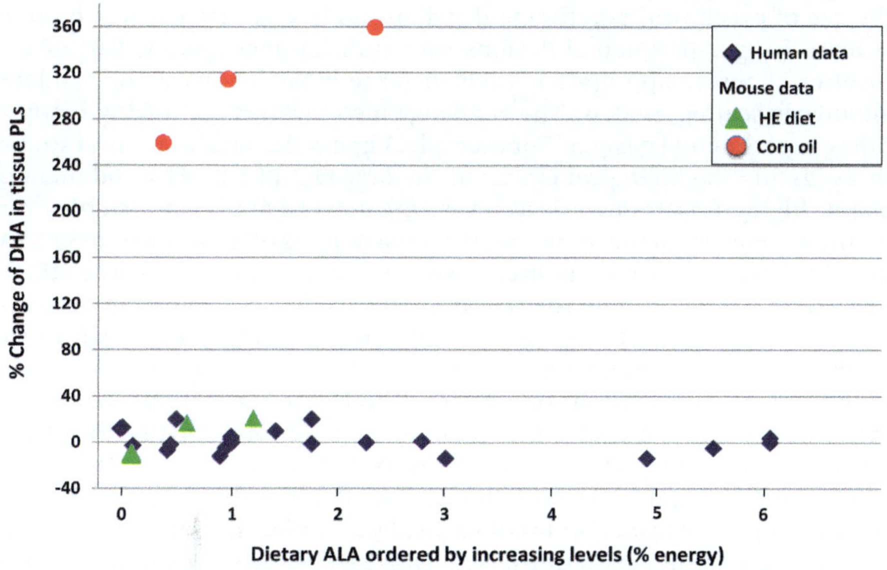

Figure 11.1 The effects of increasing dietary alpha-linolenic acid (ALA) on changes in tissue docosahexaenoic acid (DHA) levels. Diamonds represent data generated from human clinical trials. Triangles respresent data from mice where the composition of the background diet mimicked that of the US diet based on % energy. Circles represent data generated from mice consuming a background diet devoid of omega-3 fatty acids, whose sole lipid source was corn oil (AIN76A) HE, human equivalent.

supplemental levels of dietary ALA (based on % energy), and the *y* axis represents the % change in tissue DHA levels. The diamonds are data from human clinical trials.[29] The triangles represent data from mice fed a background diet that mimicked the US diet, *viz.*, the human equivalent diet (Table 11.2), following supplementation with the three levels of ALA. The circles represent data from mice fed AIN76A as the background diet (corn oil-based devoid of omega-3 fats), also supplemented with three levels of ALA.[32] The human data demonstrates that, when the levels of ALA are increased in the diet, there are no changes in DHA levels, regardless of the amounts supplemented.[33,34] These results are recapitulated in the mouse model when they are fed diets that mimic the nutrient distribution of humans based on energy (triangles). However, when the background diet is devoid of omega-3 fats (circles), there is an exaggerated response in changes in tissue DHA levels as represented by the circles even when HEDs were used. Comparable results between mice and humans were generated with other dietary PUFAs examining the same endpoints,[9,29–31] and validating, in part, the results presented by Rucker and Storms with micronurients.[18]

11.4 Summary

A major challenge for biomedical research involving dietary bioactives and the use of preclinical experimental rodent models as surrogates to humans is identifying experimental designs with translational validity. Part of that involves choosing appropriate levels of these bioactives based on validated allometric scaling models. The challenge also involves identifying inappropriate experimental designs. For example, Figure 11.2 lists a variety of studies investigating the biological effects of the omega-3 PUFAs, ALA and its long-chain, highly unsaturated derivatives (LC n-3 PUFAs).[9] Using studies randomly selected from the literature, the dietary levels of ALA and LC n-3 PUFAs used in these studies were plotted based on % energy in the diet. The HED of ALA (1.1–1.6 g day^{-1} being the typical intake in the US population based on the US DRI) is signified with a horizontal line at the bottom of the left side of the figure. Similarly, the HED of LC n-3 PUFAs (500 mg day^{-1} or 0.23%en, as based on recommendations by the American Heart Association) is identified with a horizontal line at the bottom of the right side of the figure. This figure illustrates the supra-pharmacological levels that are commonly used in diets when investigating the effects of these PUFAs. Because there are no guidelines framing appropriate levels, authors typically take liberties with regards to levels used, potentially generating false-positive results and questioning translational validity to humans for those results.

In summary, this chapter describes weaknesses and strengths of the major allometric scaling models for dietary bioactives when trying to translate doses between species. It emphasizes that, because the use of rodent models is so highly controlled as compared to the human circumstances in clinical trials, if rodents don't respond to an HED, what is the likelihood that that humans will respond to an HED? To address this issue, nutrients in the standard AIN diets were extrapolated to HEDs using the various mathematical models. The

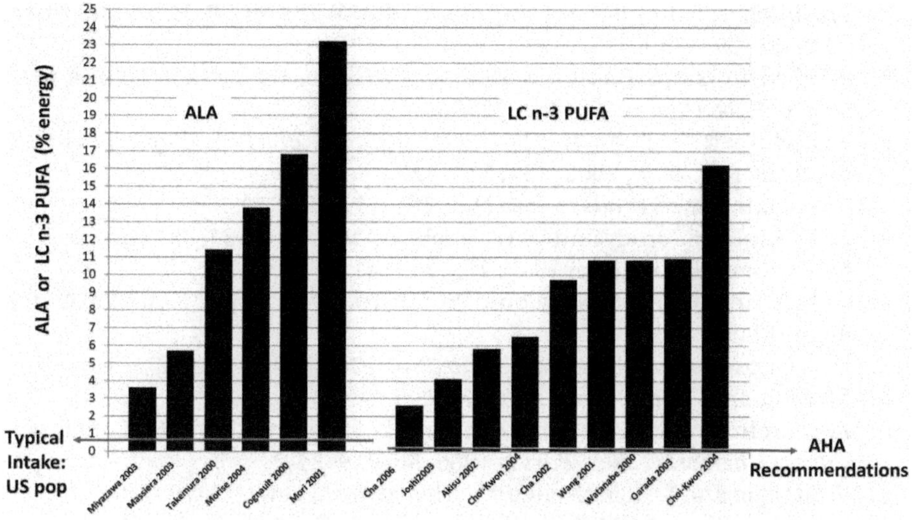

Figure 11.2 Levels of alpha-linolenic acid (ALA) or long-chain n-3 polyunsaturated fatty acids (LC n-3 PUFA) supplemented to corn oil-based rodent background diets (randomly selected studies).[9] The line at the bottom left represents the typical intake of ALA in the US diet (~0.6%en). The line at the bottom right represents the recommended intakes of LC n-3 PUFAs by the American Heart Association (AHA) (500 mg day^{-1}).[35] The US population consumes ~200 mg day^{-1} of LC n-3 PUFAs. Adapted from reference 9, https://doi.org/10.31989/ffhd.v7i5.338, under the terms of the CC BY 2.0 licence, https://creativecommons.org/licenses/by/2.0/.

results demonstrated that extrapolations based on differences in caloric consumption most closely translated to human levels, and thus, should be the preferred method for this process. This model was then tested using the same biological endpoint, changes in tissue fatty acid composition following supplementation with various PUFAs, and the results in mice were comparable to data generated in human clinical trials.

References

1. C. G. Begley and L. M. Ellis, *Nature*, 2012, **483**, 531–533.
2. G. Llovera and A. Liesz, *J. Neurochem.*, 2016, **139**, 271–279.
3. F. D. Ramirez, P. Motazedian, R. G. Jung, P. Di Santo, Z. D. MacDonald, R. Moreland, T. Simard, A. A. Clancy, J. J. Russo, V. A. Welch, G. A. Wells and B. Hibbert, *Circ. Res.*, 2017, **120**, 1916–1926.
4. S. Han, T. F. Olonisakin, J. P. Pribis, J. Zupetic, J. H. Yoon, K. M. Holleran, K. Jeong, N. Shaikh, D. M. Rubio and J. S. Lee, *PLoS One*, 2017, **12**, e0183591.
5. J. B. Schulz, M. R. Cookson and L. Hausmann, *J. Neurochem.*, 2016, **139**, 253–270.
6. F. D. Ramirez, R. G. Jung, P. Motazedian, D. Perry-Nguyen, P. Di Santo, Z. MacDonald, A. A. Clancy, A. Labinaz, S. Promislow, T. Simard, S.

Provencher, S. Bonnet, I. D. Graham, G. A. Wells and B. Hibbert, *Stroke*, 2020, **51**, 291–299.
7. S. Wieschowski, S. Biernot, S. Deutsch, S. Glage, A. Bleich, R. Tolba and D. Strech, *PLoS One*, 2019, **14**, e0223758.
8. A. J. Hulbert, T. Rana and P. Couture, *Comp. Biochem. Physiol., Part B: Biochem. Mol. Biol.*, 2002, **132**, 515–527.
9. J. Whelan, *Funct. Foods Health Dis.*, 2017, 7, 314–328.
10. H. H. Kibler, S. Brody and D. Worstell, *J. Nutr.*, 1947, **33**, 331–338.
11. M. Kleiber, *Physiol. Rev.*, 1947, **27**, 511–541.
12. C. R. White and R. S. Seymour, *Proc. Natl. Acad. Sci. U. S. A.*, 2003, **100**, 4046–4049.
13. D. S. Glazier, *Biol. Rev. Camb. Philos. Soc.*, 2005, **80**, 611–662.
14. M. Kleiber, *J. Theor. Biol.*, 1975, **53**, 199–204.
15. E. J. Freireich, E. A. Gehan, D. P. Rall, L. H. Schmidt and H. E. Skipper, *Cancer Chemother. Rep. Part 1*, 1966, **50**, 219–244.
16. V. Sharma and J. H. McNeill, *Br. J. Pharmacol.*, 2009, **157**, 907–921.
17. S. Reagan-Shaw, M. Nihal and N. Ahmad, *FASEB J.*, 2008, **22**, 659–661.
18. R. Rucker and D. Storms, *J. Nutr.*, 2002, **132**, 2999–3000.
19. R. B. Rucker and F. M. Steinberg, *Biochem. Mol. Biol. Educ.*, 2002, **30**, 86–89.
20. E. C. Huang, M. F. McEntee and J. Whelan, *Nutr. Cancer*, 2012, **64**, 749–760.
21. E. D. Tague, A. K. Bourdon, A. MacDonald, M. S. Lookadoo, E. D. Kim, W. M. White, P. D. Terry, S. R. Campagna, B. H. Voy and J. Whelan, *J. Med. Food*, 2017.
22. M. A. Bilen, S. H. Lin, D. G. Tang, K. Parikh, M. H. Lee, S. C. Yeung and S. M. Tu, *Case Rep. Oncol. Med.*, 2015, **2015**, 471861.
23. D. E. Barnard, S. M. Lewis, B. B. Teter and J. E. Thigpen, *J. Am. Assoc. Lab. Anim. Sci.*, 2009, **48**, 709–713.
24. J. G. Bieri, G. S. Stoewsand, G. M. Briggs, R. W. Phillips, J. C. Woodard and J. J. Knapka, *J. Nutr.*, 1977, **107**, 1340–1348.
25. P. G. Reeves, *J. Nutr.*, 1997, **127**, 838S–841S.
26. M. A. Medinsky, J. A. Popp, T. E. Hamm and J. G. Dent, *Toxicol. Appl. Pharmacol.*, 1982, **62**, 111–120.
27. Institute of Medicine, *The Development of DRIs 1994–2004: Lessons Learned and New Challenges – Workshop Summary*, The National Academies Press, Washington, DC, 2008.
28. Institute of Medicine, *Dietary Reference Intakes for Energy, Carbohydrate, Fat, Fatty Acids, Cholesterol, Protein, and Amino Acids*, The National Academies Press, Washington, DC, 2005.
29. J. Whelan, R. Hardy, R. S. Wilkes and H. E. Valentin, in *Convergence of Food Security, Energy Security and Sustainable Agriculture*, ed. D. D. Songstad, Springer-Verlag, Berlin, 2014, vol. Biotechnology in Agriculture and Forestry, pp. 129–168.
30. K. A. Weldon and J. Whelan, *Nutr. Metab (Lond.)*, 2011, **8**, 43.
31. B. S. Rett and J. Whelan, *Nutr. Metab. (Lond.)*, 2011, **8**, 36.

32. J. Whelan, K. S. Broughton and J. E. Kinsella, *Lipids*, 1991, **26**, 119–126.
33. L. M. Arterburn, E. B. Hall and H. Oken, *Am. J. Clin. Nutr.*, 2006, **83**, 1467S–1476S.
34. J. T. Brenna, N. Salem, Jr., A. J. Sinclair and S. C. Cunnane, *Prostaglandins Leukotrienes Essent. Fatty Acids*, 2009, **80**, 85–91.
35. E. B. Rimm, L. J. Appel, S. E. Chiuve, L. Djousse, M. B. Engler, P. M. Kris-Etherton, D. Mozaffarian, D. S. Siscovick and A. H. Lichtenstein, *Circulation*, 2018, **138**, e35–e47.

Subject Index

Nutritional Signaling Pathway Activities in Obesity and Diabetes
Food Chemistry, Function and Analysis No. 24
Edited by Zhiyong Cheng

Published by the Royal Society of Chemistry, www.rsc.org